TP 1069

Buildings Sealants: Materials, Properties, and Performance

Thomas F. O'Connor, editor

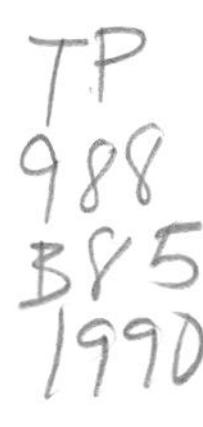

ASTM
1916 Race Street
Philadelphia, PA 19103

Library of Congress Cataloging-in-Publication Data

Building sealants: materials, properties, and performance
Thomas F. O'Connor, editor.
(ASTM special technical publication: 1069)
Papers from a symposium held 31 Jan.-Feb., 1990 in Fort Lauderdale, Fla., sponsored by ASTM Committee C-24 on Building Seals and Sealants.
"ASTM publication code number (PCN) 04-010690-10."
Includes bibliographical references and index.
ISBN 0-8031-1282-3
1. Sealing compounds—Congresses. I. O'Connor, Thomas F., 1941- . II. ASTM Committee C-24 on Building Seals and Sealants.
TP988.B85 1990
691'.99—dc20 90-9
C

NOTE

The Society is not responsible, as a body, for the statements and opinions advanced in this publication.

Peer Review Policy

Each paper published in this volume was evaluated by three peer reviewers. The auth addressed all of the reviewers' comments to the satisfaction of both the technical editor and the ASTM Committee on Publications.

The quality of the papers in this publication reflects not only the obvious efforts of t authors and the technical editor(s), but also the work of these peer reviewers. The AST Committee on Publications acknowledges with appreciation their dedication and contributi of time and effort on behalf of ASTM.

Printed in Baltimore, MD
August 1990

Foreword

The symposium on Building Sealants: Materials, Properties, and Performance was held in Fort Lauderdale, Florida, on 31 Jan.–1 Feb. 1990. ASTM Committee C-24 on Building Seals and Sealants sponsored the symposium. Thomas F. O'Connor, Smith, Hinchman & Grylls Associates, Inc., served as chairman of the symposium and editor of the resulting publication.

Contents

Overview

Elastomeric sealants have a relatively short history. The early 1950s saw the introduction of the first elastomeric sealants based on a polysulfide polymer. Since then acrylic, urethane, and silicone polymer based sealants, among others, have been developed. Presently, the range and types of sealant products available is extensive, with many excellently performing products on the market to fulfill the most demanding joint sealing requirements.

Shortly after the introduction of modern sealants, ASTM Committee C-24 on Building Seals and Sealants was organized in 1959 to develop standards for sealants. Since then, C-24 has developed over seventy standards, most of which relate to sealant testing, usage, and performance. Although a considerable amount of work has been performed in developing standards for sealants, there are additional standards that need to be developed. Presently C-24 has about 25 task groups working on new standards that are needed in the sealant industry.

In conjunction with its standards making activities, C-24 initiated a series of symposia in 1986 based on the C-24 scope of interest; the sealant, waterproofing, and pipe gasket and coupling industries. The first symposium held in 1988 resulted in Special Technical Publication (STP) 1054, "Science and Technology of Glazing Systems", the 1989 symposium in STP 1084, "Building Deck Waterproofing" and the 1990 symposium in publication of this book. At least two more symposia are planned, one each in 1991 and 1992.

The purpose of the current series of C-24 symposia is the presentation of the latest developments in the sealant, waterproofing and pipe gasket and coupling industries. Additionally, the identification of areas of concern as evidenced by laboratory testing, field observation, and theoretical study and the presentation of this information in the hopes of stimulating additional research as well as providing a basis for new standard development.

When this book and the symposium upon which it was based was organized, a concerted effort was made to obtain authors from around the world. The intent was to produce a symposium with a range of viewpoints and international concerns. That effort resulted in eight papers being submitted from Australia, Belgium, England, and Sweden, and sixteen from the United States. As evidenced by the submitted papers, the use of sealants for structural sealant glazing, the identification and quantification of the effects of movement on sealants, continuing laboratory investigation of sealant performance under various environmental factors, and the in-service performance of sealants appear to be major concerns of the sealant industry. The following lists the topics of the submitted papers, as well as identifies areas where additional research and standard development could be performed.

STRUCTURAL SEALANT GLAZING

The continuing and timely development of standards for structural sealant glazing must continue if this segment of the sealant industry is to grow while maintaining adequate quality control over the end product. The papers present new information on structural sealants relative to insulating glass edge seal performance, the use of structural glazing techniques to attach stone panels, the effects of long term creep rupture and fatigue loading on structural seals, stresses that occur in structural glazing, testing and qualification of structural sealants, and structural glazing system design and quality control.

There are many other areas of structural sealant glazing that need to be addressed in the future, including the effects of combinations of secondary forces in addition to the primary force of wind loading, such as, thermal movement, dead loading, and seismic forces on the structural seals. This could lead to development of a practical theoretical model for the design professional that will allow calculation of structural glazing seal performance when secondary forces are included.

Presently, quality control procedures are the result of an agreement between the parties involved on a particular project and, therefore, can vary from project-to-project. Uniform-industry wide recommendations for quality control of structural sealant glazing needs to be developed. This would result in a minimum level of inspection and quality control for the structural glazing industry.

MOVEMENT CAPACITY OF SEALANTS

Papers on the ability of sealants to absorb movement, the identification and quantification of sealant stresses during movement, establishing the movement capability of sealants, and designing sealant joints for movement are included.

Additional work needs to be performed in defining the factors that effect the movement capability of a sealant joint. This effort should be multi-disciplinary since it will involve establishment of properties for materials other than sealants, such as, thermal and moisture movement effects of sealed materials. Also, realistic construction tolerances need to be established for materials and systems that form the sides of a sealant joint and that indirectly effect a joint's ability to perform and remain durable. Combined movements of a sealant joint, such as, compression and extension in conjunction with longitudinal shear should be investigated to establish movement capability for use in sealant joint design.

LABORATORY INVESTIGATION

Laboratory testing includes papers on temperature and moisture effects on sealant performance, adhesion properties of sealants, heat aging effects, the water resistance of sealants, compatibility issues with laminated glass, new materials for sealant formulation and usage, and sealant usage with membrane waterproofing systems.

Other areas that could be investigated are the effects of movement that occur during sealant cure, the development of laboratory test methods that will be predictive of actual in-service performance, better test methods to predict sealant compatibility with other materials, and the staining or dirtying potential of sealants with other materials.

SEALANT JOINT PERFORMANCE

In-service performance is represented by papers on sealant usage with exterior insulation systems, correlation of in-service performance with laboratory testing, and the identification of sealant in-service failure mechanisms, among others. This area should receive increased attention in the future so that better correlation between laboratory testing and prediction and actual in-service performance can be obtained. Periodic assessment of in-service performance will also assist in identifying potential failure mechanisms or workmanship deficiencies so that corrective procedures can be implemented and information provided to the entire sealant industry.

In-service performance is also subject to both design and workmanship deficiencies. Sadly, at least in the United States, there is no recognized nor uniform national training or certification program for the designers of sealant joints or the workmen who install the sealants and related accessories. Sealant joints, which are a small part of a building's total cost, when improperly designed or installed can, among other effects, be the source of leakage and deterioration of materials that can result in very costly correctional action. The development of standards and certification programs that will educate the design professionals and workmen is overdue and would help immensely in decreasing the incidence of premature sealant joint failure.

PRESENT AND FUTURE TRENDS

Lastly, two papers complete the book. The first discusses the present status of the sealant industry and the second is an overview of challenges to the sealing and other related building envelope industries.

The quality and quantity of world-wide research work being performed in the sealant industry today is impressive and this book is evidence of a continuing quest for knowledge of sealant performance. It is hoped that these papers will stimulate further research in the areas described, as well as others, and contribute to the development of new standards that are needed in the industry.

The symposium chairman would like to thank the authors and peer reviewers, who by their timely and well considered contributions, made the symposium a success, and importantly, resulted in timely publication of this book.

Thomas F. O'Connor, AIA, FASTM
Smith, Hinchman & Grylls Associates, Inc.
Detroit, MI 48226
Symposium Chairman and Editor

Structural Sealant Glazing

L. Bogue Sandberg and Amy E. Rintala

RESISTANCE OF STRUCTURAL SILICONES TO CREEP RUPTURE AND FATIGUE

REFERENCE: Sandberg, L. B. and Rintala, A. E., **"Resistance of Structural Silicones to Creep Rupture and Fatigue,"** Building Sealants: Materials, Properties, and Performance, ASTM STP 1069, Thomas F. O'Connor, editor, American Society for Testing and Materials, Philadephia, 1990.

ABSTRACT: Structural silicone sealants are subjected to repeated wind and thermal load cycles and, in certain applications, to sustained loads. Data from two types of creep tests are presented, one using constant loading and the other ramp loading. Both were done on an acetoxy silicone. Fatigue data are reported for tension on a neutral cure sealant, tension on a two-part silicone in insulating glass joints, and shear on an acetoxy silicone. All fatigue testing used full reversal, constant load amplitude cycling. Regression analyses on failure stress versus time or cycles to failure and on stress versus the logarithm of time or cycles to failure are discussed in relation to current design practice. Some of the problems associated with creep and fatigue testing of silicone sealants are explained.

KEYWORDS: silicone, sealant, creep rupture, fatigue, structural glazing

Structural silicone sealants function as adhesives, bonding glass and other panel materials to the exteriors of buildings. The sealant joints are subjected to repeated wind loadings and to stresses caused by daily and seasonal thermal movements. In some designs the sealant may be subject to sustained dead loads which are present for the entire life of the structure. Less frequently, snow or other transient loads may last for several days to months. Given these possibilities, it is important for a designer of a structurally glazed system to have some knowledge of the behavior of silicone

Dr. Sandberg is a Professor of Civil and Environmental Engineering at Michigan Technological University, Houghton, MI 49931; Ms. Rintala is an Assistant Project Engineer with Rowe Engineering, Flushing, MI 48433.

sealants under sustained and cyclic loads.

For situations involving possible creep rupture, loads of various durations must be considered. Design snow loads are often assumed to have cumulative durations of three months while dead loads are treated as being constant for thirty years [1].

For many engineering problems, fatigue implies millions or even billions of cycles. In structural glazing design the situation is somewhat different. Thermal stresses stem from the temperature changes brought on by the seasons, the daily cycle, and short term weather and cloud cover changes. But, even if changing cloud cover caused an average of five major cycles every day for thirty years, the total would only be about 55,000 cycles. For wind loading, suppose an average of four gusts every hour for thirty years. About one million cycles would result. Wind induced vibration might greatly increase this number, but in most cases the number of cycles over a thirty year life is comparatively small.

Balanced against the need for long term creep data and high cycle fatigue data is the need for timely, economical testing. Even two years of creep testing is a very long time for a sealant manufacturer or user anxious to put a new product into service. When fatigue testing, we must take care to avoid affecting results by testing at too high a frequency. Silicones are strain rate dependent to some extent, and testing at 5 Hz has been shown to cause a noticable rise in specimen temperature [2]. Thus, testing for reasonable times and at reasonable frequencies will always leave us in the position of having to make extrapolations and apply judgment in developing design guidelines.

An interesting problem with silicone creep and fatigue testing is the fact that the sealant is likely to continue to cure during the test. Cure is asymptotic in nature, and may take a year or more to go to something approximating complete cure. How this slow residual gain in strength and stiffness interacts with stress is not known. One thing is clear. If all specimens are not put under load at the same time after sample preparation, differences in the degree of cure may produce misleading results.

The design of a creep rupture or fatigue testing program involves a number of choices. Do we take the traditional approach of subjecting specimens to constant load or cycling conditions? Or, do we use a linear increase with time, guaranteeing a failure of every specimen? Will the results be comparable?

In fatigue testing we can choose constant strain or constant load amplitude cycling and we must decide on a zero or nonzero mean strain or load for the cycle. For many materials, a strain cycle is preferrable [3]. With silicone joints, experience [2] has shown that failure propagation may be arrested if a strain controlled cycle is used. Also, wind induced stresses are load, rather than strain, limited. These factors point to a constant load amplitude approach for silicone joints. Current thinking for sealant joints, embodied in ASTM Test for Adhesion and Cohesion of Elastomeric Joint Sealants

Under Cyclic Movement (Hockman Cycle) (C 719-86), seems to point to a zero-mean stress (full reversal) cycle as the worst case condition.

There are choices as to how we plot and analyze data. The most common approach is probably a plot of failure stress versus the log of time or cycles to failure. The usual equation fit to the data is of the form

$$\sigma_f = \sigma_0 + A \log (t \text{ or } N) \tag{1}$$

where
σ_f = failure stress
t or N = time or number of cycles to failure
A and σ_0 = constants from regression analysis.

This implies a finite life at zero stress. Unless the material is aging by some non-stress process, this lack of any endurance limit (stress below which the specimen will never fail) does not make sense. However, Eq. 1 often gives a reasonable approximation when a finite life requirement can be defined. Another approach is to plot failure stress versus time and fit an equation of the type

$$\sigma_f = \sigma_e + K t^{-\eta} \tag{2}$$

where
σ_f, t, and N areas defined for Eq. 1
σ_e = endurance limit predicted by regression
K and η = material constants, predicted by regression or assumed.

If η = 1 we have a simple hyperbolic fit, commonly known as the Wöhler curve [4]. This presumes the existence of an endurance limit. While this makes intuitive sense, it is difficult to accurately predict the endurance limit given the usual data scatter and time or cycle range.

TEST METHODS AND SPECIMENS

The five types of data presented in this paper are summarized in Table 1. Fig. 1 shows the specimens used, while Fig. 2 illustrates the constant, quasi-ramp, and full reversal loadings employed. All testing was done under typical room temperature and humidity conditions.

The basic tension specimen of Fig. 1a is a variation of the H-block specimen used in ASTM C 719 and other sealant testing. The two insulating glass (IG) joint specimens, Fig. 1b, are identical to ones which were a part of a previous study of static strength effects of IG spacer geometry and setback [5] The lap shear specimens of Fig. 1c were devised to give a small specimen having a typical sealant bead cross section.

The loading for the constant load creep tests (Fig. 2a) was done using gravity loads amplified by a lever arm. The quasi-ramp loading of Fig. 2b also used lever amplified gravity loading. The approximation of a ramp load (increasing linearly with time) was done by adding 25.4mm (1 inch) diameter, steel balls to a bucket at regular time intervals. The balls were fed by computer control for the higher stress rates and by hand for the slower rates.

TABLE 1 -- Creep rupture and fatigue tests.

Type of Test	Sealant	Specimen	Loading	Data
Tensile creep, constant load	Acetoxy	Fig. 1a	Fig. 2a	Fig. 3
Tensile creep, ramp load	Acetoxy	Fig. 1a	Fig. 2b	Fig. 4
Tensile fatigue	Neutral	Fig. 1a	Fig. 2c	Fig. 5
Tens. fatigue, IG	Two-part	Fig. 1b&c	Fig. 2c	Fig. 6
Shear fatigue	Acetoxy	Fig. 1d	Fig. 2c	Fig. 7

All of the fatigue testing was done on a rotating device which employed gravity to produce a full-reversal, sinusoidal cycle with constant load amplitude. The details of the device are given elsewhere [6].

RESULTS AND ANALYSIS

In Fig. 3 through Fig. 7, we have shown results for each of the five tests in two forms, failure stress versus the log of time (or cycles) to failure and failure stress versus time (or cycles) to failure. Equations in the form of Eq. 1 were fit to the semi-log plots and in the form of Eq. 2 to the second type of plot. The semi-log equations and key statistics from the regressions are shown in Table 2. The hyperbolic equations (Eq. 1) are summarized similarly in Table 3. In both tables, where the number of observations is less than indicated on the corresponding plot, it simply reflects the fact that very high stressed, short lived data was omitted from the regression analysis. Justification for this lies in the fact that the failure mechanism is often different for the high stress, short life data. In all cases, failure stress was the dependent variable in the analysis. The log of time or cycles and the reciprocal of time or cycles were the independent variables for Eqs. 1 and 2, respectively. A linear regression was performed on the resulting linearized forms.

The statistical values in the tables are given to allow the reader the opportunity to assess the goodness of fit and to make estimates of confidence limits [7]. The equations have been shown on the data plots. The cutoffs at the left of these curves

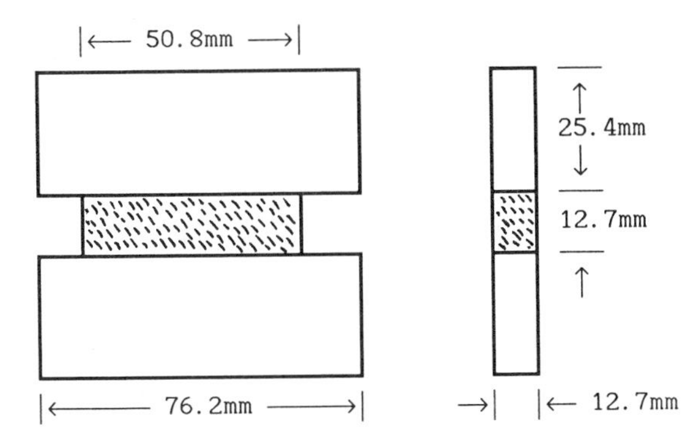

(a) Tensile specimen for creep and fatigue.

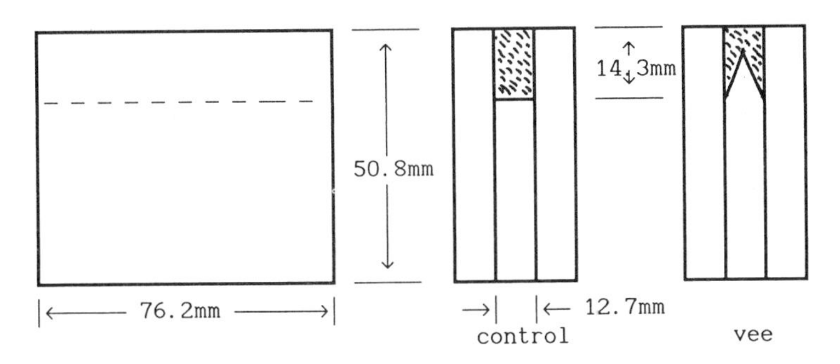

(b) Insulating glass tensile fatigue specimen.

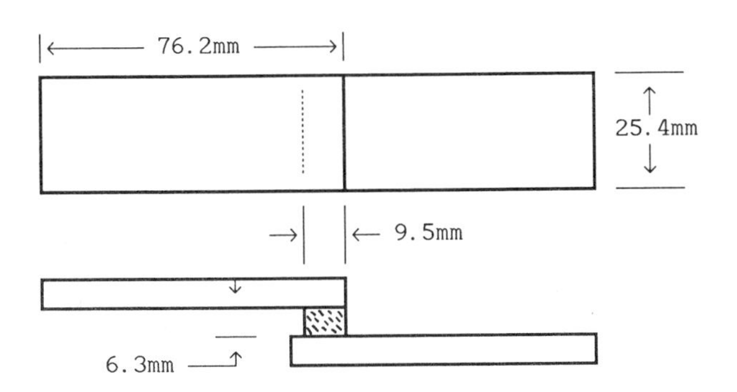

(c) Shear fatigue specimen.

NOTE: All substrates were aluminum.

FIG. 1 -- Test specimens used for this study.

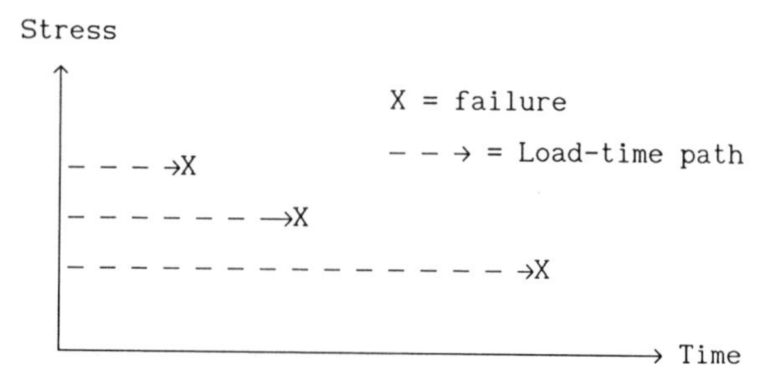

(a) Constant stress creep rupture test.

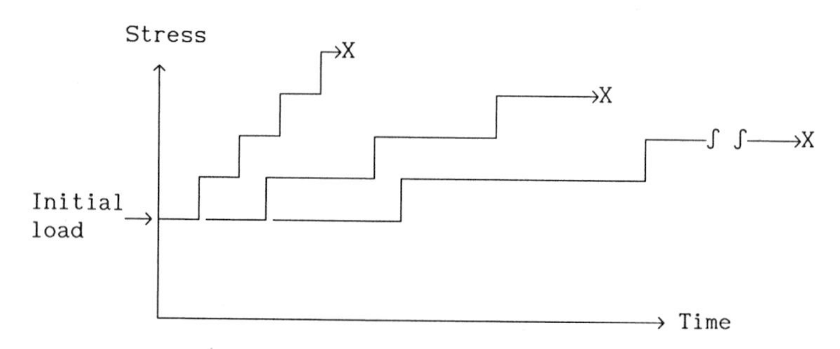

(b) Quasi-ramp or step load history.

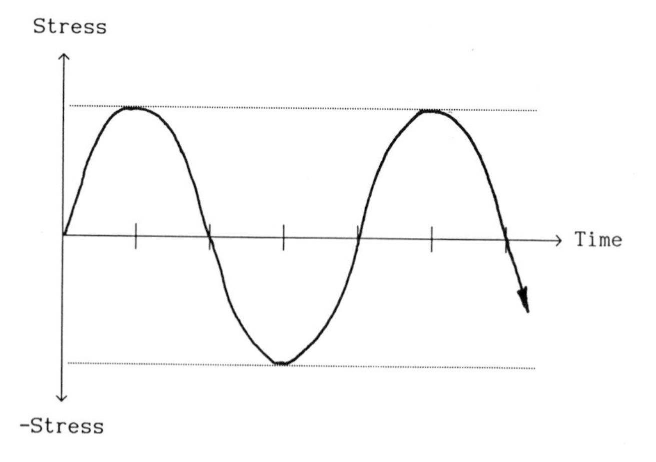

(c) Full reversal fatigue cycle.

FIG. 2 -- Load histories used for this study.

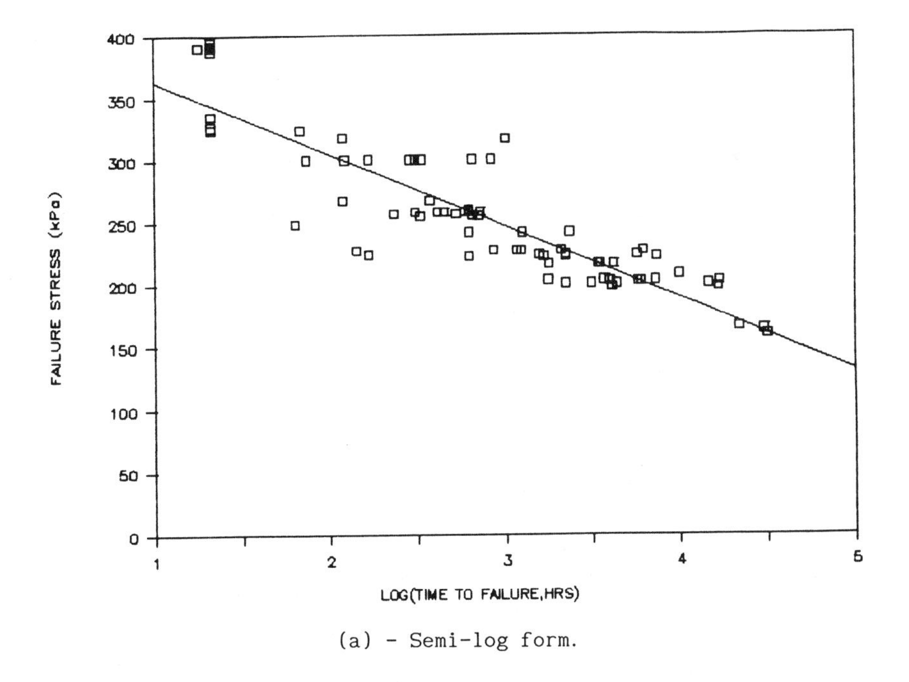

(a) - Semi-log form.

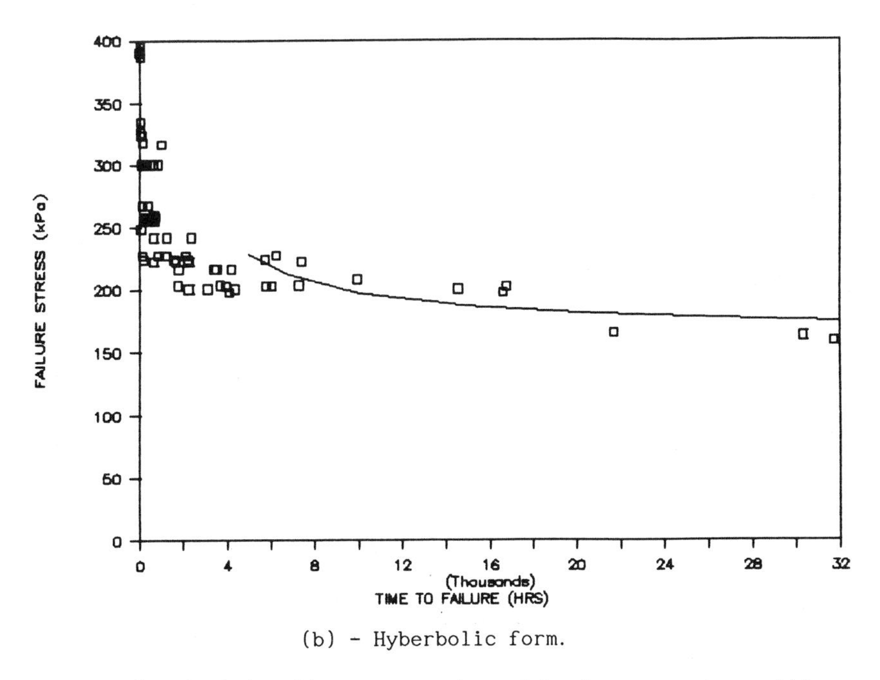

(b) - Hyberbolic form.

FIG. 3 -- Constant tensile creep rupture data for an acetoxy silicone.

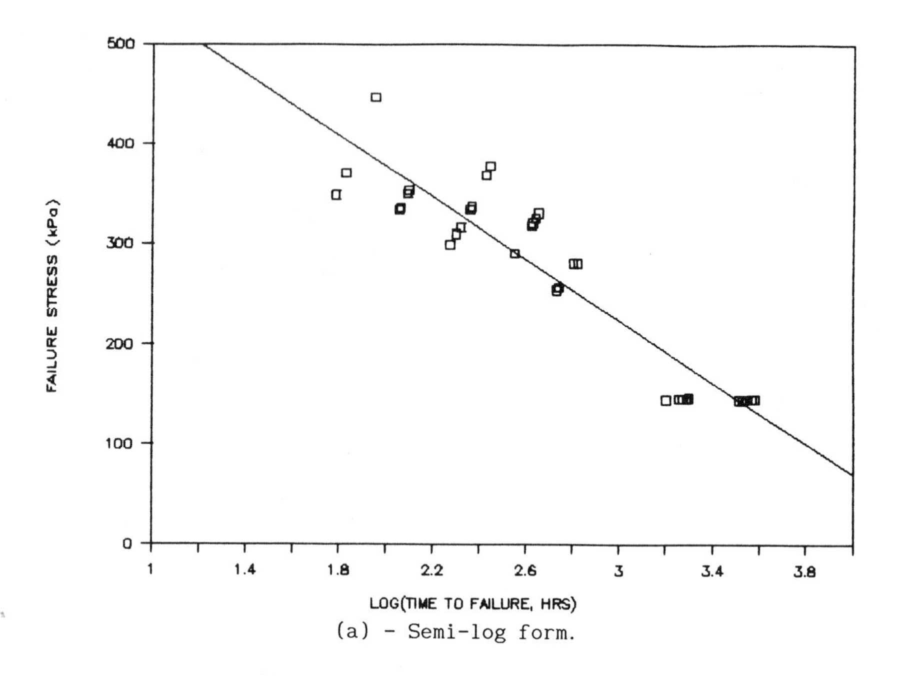

(a) - Semi-log form.

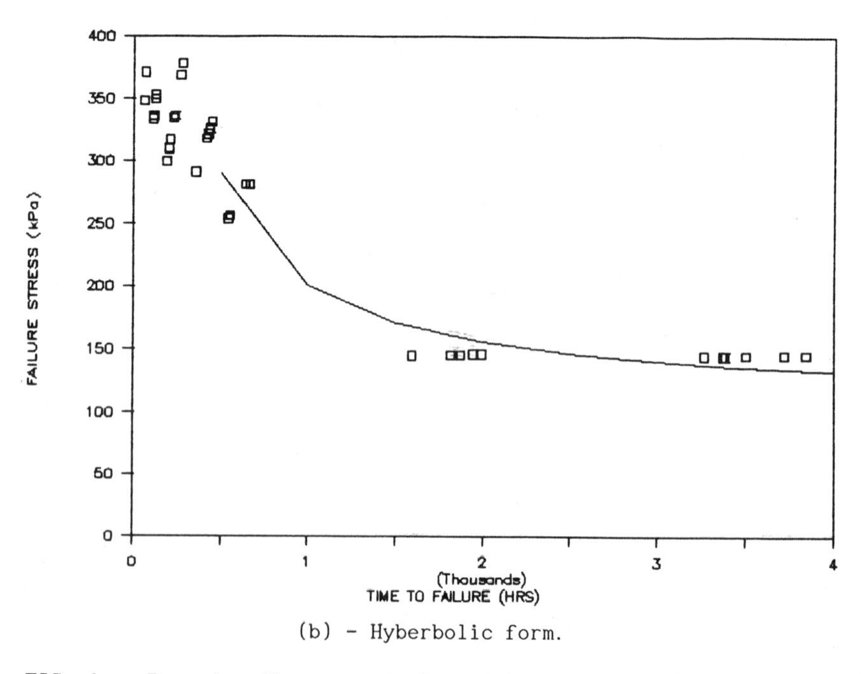

(b) - Hyberbolic form.

FIG. 4 -- Ramp tensile creep rupture data for an acetoxy silicone.

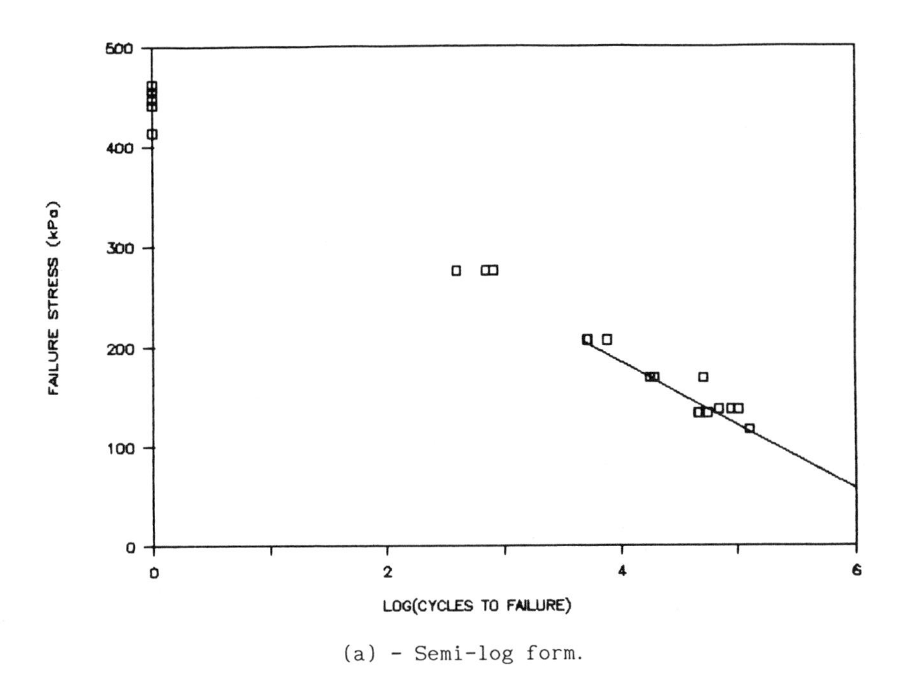

(a) - Semi-log form.

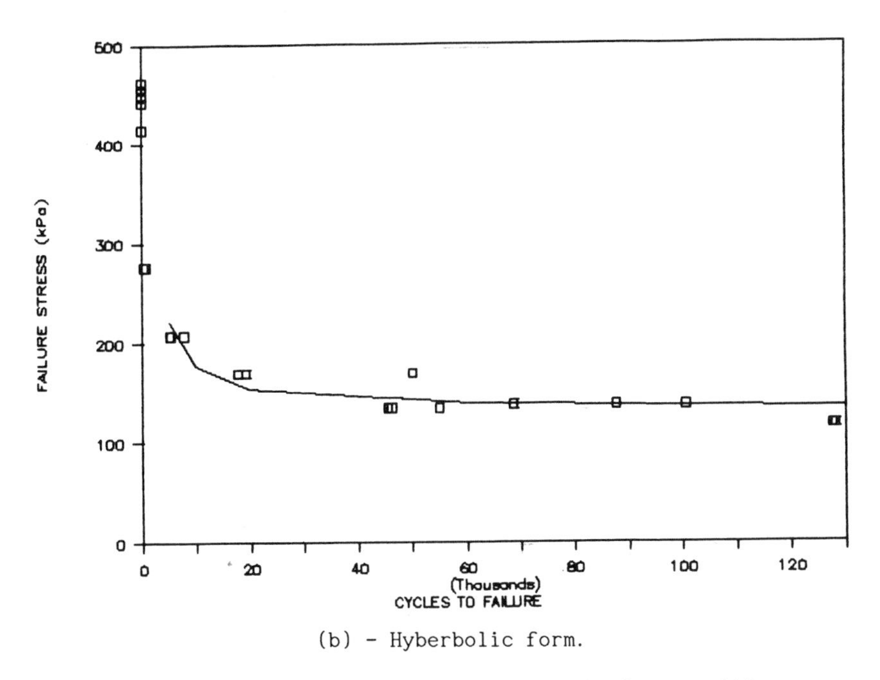

(b) - Hyberbolic form.

FIG. 5 -- Tensile fatigue data for a neutral cure silicone.

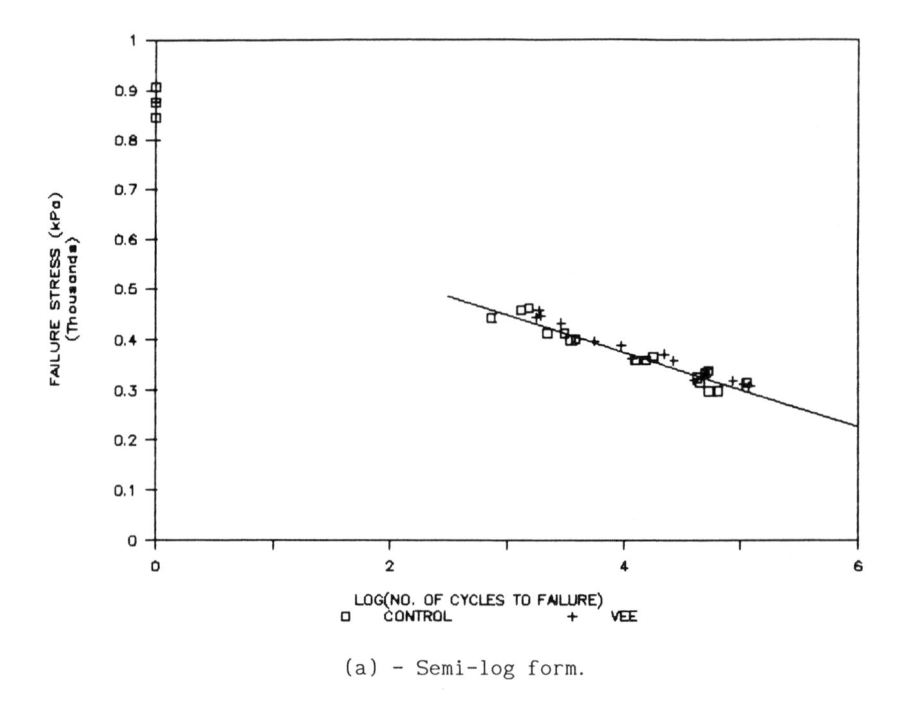

(a) - Semi-log form.

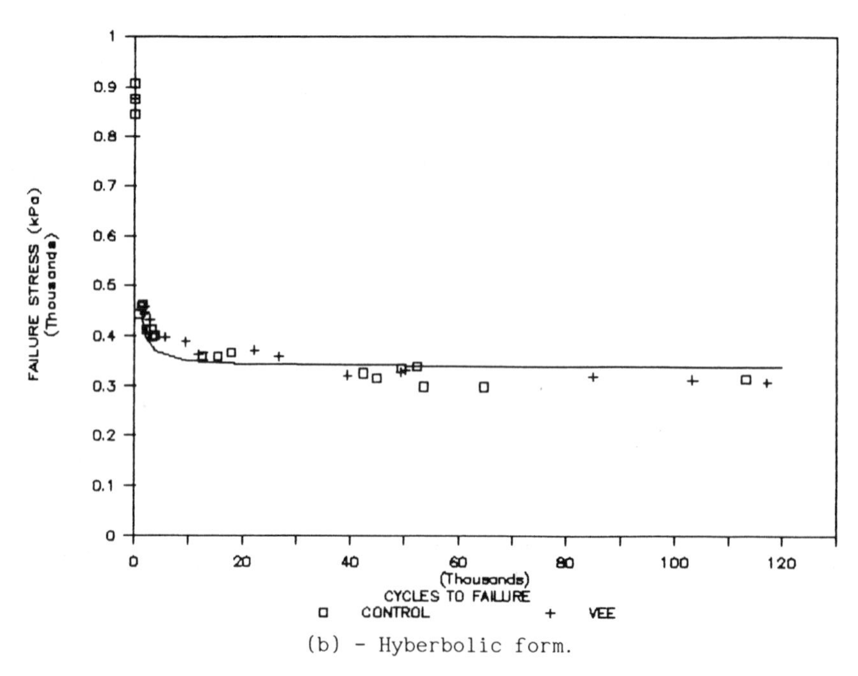

(b) - Hyberbolic form.

FIG. 6 -- Tensile fatigue data for a two-part silicone in IG joints.

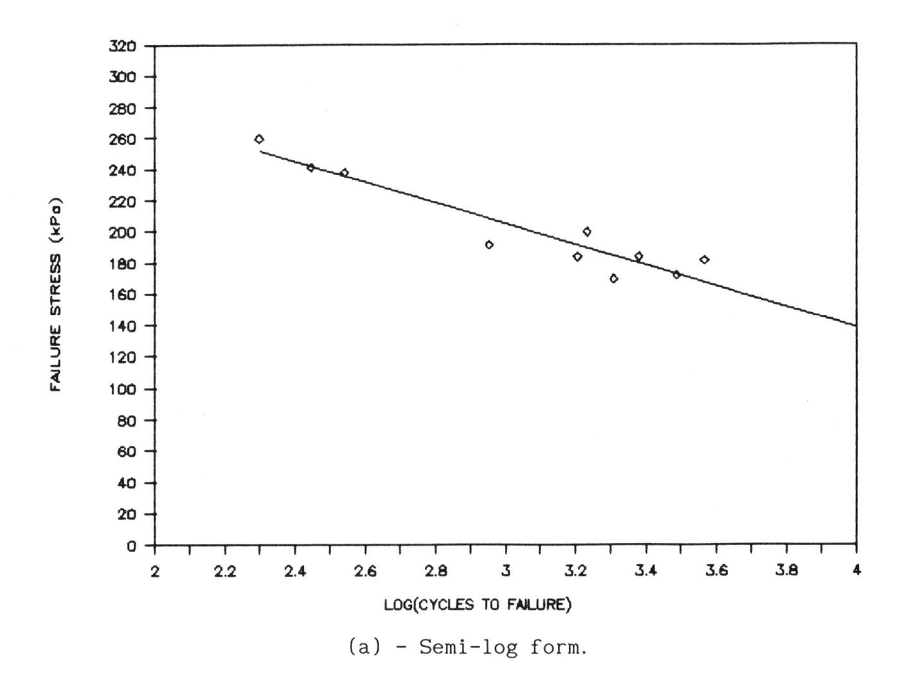

(a) - Semi-log form.

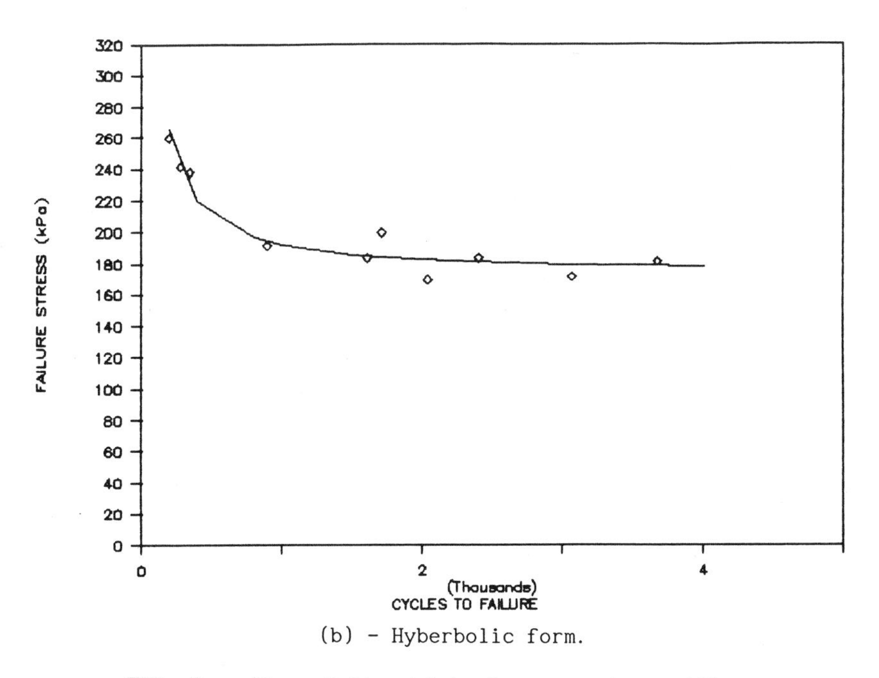

(b) - Hyberbolic form.

FIG. 7 -- Shear fatigue data for an acetoxy silicone.

TABLE 2 -- Regressions in semi-log form, Eq. 1

Test	Equation, kPa	Obs.	R^2	s_e	s_c	$\bar{x}$
Cons. load creep rupt.	$\sigma_f = 421-57.9 \cdot \log(t)$	74	0.78	26.1	3.6	2.88
Ramp load creep rupt.	$\sigma_f = 686-154 \cdot \log(t)$	36	0.87	34.5	10.4	2.71
T. fatigue ntrl cure	$\sigma_f = 441-62.9 \cdot \log(N)$	14	0.90	10.8	6.1	4.54
IG fatigue crtl, 2 pt.	$\sigma_f = 672-74.4 \cdot \log(N)$	17	0.93	15.2	5.4	4.06
IG fatigue vee, 2 pt.	$\sigma_f = 701-78.6 \cdot \log(N)$	15	0.97	9.9	4.0	4.19
Shr. fat. acetoxy	$\sigma_f = 405-66.6 \cdot \log(N)$	10	0.90	10.9	8.0	3.04

t = time, hrs
N = no. of cycles
R^2 = goodness of fit parameter
Obs. = number of observations used in regression
s_e = standard error of estimate
s_c = standard error of coefficient
$\bar{x}$ = mean of t or N values for observations
1 kPa = 0.145 psi.

TABLE 3 -- Regressions in hyperbolic form, Eq. 2

Test	Equation, kPa	Obs.	R^2	s_e	s_c	$\bar{x}$
Cons. load creep rupt.	$\sigma_f = 166 + \frac{314000}{t}$	13	0.62	14.5	75000	2950
Ramp load creep rupt.	$\sigma_f = 111 + \frac{89500}{t}$	16	0.90	18.2	8020	265
T. fatigue ntrl cure	$\sigma_f = 132 + \frac{442000}{t}$	14	0.82	14.5	60400	18500
IG fatigue ctrl, 2 pt.	$\sigma_f = 338 + \frac{120000}{t}$	17	0.67	32.5	21900	3840
IG fatigue vee, 2 pt.	$\sigma_f = 333 + \frac{233000}{N}$	15	0.87	19.9	25400	6010
Shr. fat. acetoxy	$\sigma_f = 174 + \frac{18400}{N}$	10	0.93	9.2	1820	655

See note on TABLE 2

give an indication of the range of data used in the regression analyses.

If we take thirty years as the design life of the structural glazing joint, along with the corresponding numbers of cycles posed earlier, we can look at predictions for survivable stress levels. For use with the semi-log equation, we chose to use the lower 5% confidence level. This was computed using very rudimentary methods for making extrapolated limits of prediction, as outlined in Miller and Fruend, pp. 235-236 [7]. The 5% level should be quite conservative, since the equation assumes a finite life at zero stress. For the hyperbolic equation, the more severe lower 0.5% confidence level was used. The hyperbolic equation does predict an endurance limit and, given our limited data, the error in the extrapolated least squares fit could be large. The results of these calculations are summarized in Table 4.

In examining the values in Table 4, keep in mind the typical design values for structural glazing practice. Tensile stress for full design wind load is usually 138 kPa (20 psi). Dead load stresses are often limited to about 7 kPa (1 psi).

TABLE 4 -- Lower bound failure stress predictions

Type of Test	Exposure	Equation	Failure Stress kPa
Creep rupture constant load	30 yrs.	semi-log	71.6
		hyperbolic	87.7
Creep rupture ramp load	30 yrs.	semi-log	**
		hyperbolic	18.3
Tension fatigue neutral cure	1 million cycles	semi-log	44.7
		hyperbolic	91.1
IG fatigue controls, 2-part	1 million cycles	semi-log	200.2
		hyperbolic	251.0
IG fatigue vees, 2-part	1 million cycles	semi-log	212.1
		hyperbolic	277.5
Shear fatigue acetoxy	60,000 cycles	semi-log	61.9
		hyperbolic	146.6

** Zero strength predicted at 5 months.
1 kPa = 0.145 psi.

The constant load data from both equations is indicating that the 7 kPa (1 psi) dead load stress limit is quite conservative. The ramp load data must be regarded as much less reliable. The authors have spent considerable effort over several years trying to develop and correlate an accelerated creep rupture test based on ramp loading. That effort has not succeeded. Part of the difficulty may rest with the residual curing problem discussed earlier.

Static tests of the neutral cure sealant gave strengths about 15 percent lower than would be expected from past experience with that sealant. Fatigue values may also be somewhat low. Even so, the semi-log value implies survival of one million cycles at a loading of 0.32 times the maximum design load, which is usually based on a 100 yr recurrence. The hyperbolic value predicts one million cycle survival at 0.66 times maximum design load.

Table 4 predictions for the IG joints are very interesting. First, there is little difference between the semi-log and hyperbolic predictions or between the control and vee geometries. Second, all predictions are well above the 138 kPa (20 psi) value for maximum design load. It should be emphasized here that these specimens contained no PIB seals, so the silicone took the full load.

The shear data, though limited, says that if a joint were designed so that thermal movements were limited to between 20 and 40 percent of the joint width (gap distance between substrates), the joint should survive for thirty years. This can be deduced by calculating the nominal shear strain, by dividing the stresses in Table 4 by the approximate shear modulus of the sealant in the joint, about 340 kPa (49 psi).

CONCLUSIONS

The data presented are limited and our analyses and interpretations are by no means infallible. Overall, however, the results are reassuring in light of current design practice. In any design involving long term loading, the final answers must always come from experience in service. Here too, the indications are that properly designed and installed structually glazed systems have stood the test of time.

As the use of structural silicones evolves toward new applications and products, attention should be given to the creep rupture and fatigue behavior of the joints. Future research should address interactions between stress and environmental factors, such as temperature and moisture.

The need remains for an reliable method of obtaining creep rupture and fatigue data on a timely basis. Ramp loading approaches probably offer the best hope, but residual cure and cycle induced temperature effects must be carefully considered.

ACKNOWLEDGEMENT

This work was funded by unrestricted grant support from the Dow Corning Corporation, Midland Michigan. The authors deeply appreciate this assistance.

REFERENCES

[1] _____, National Design Specification for Wood Construction, National Forest Products Association, Washington, D.C., 1986.

[2] Sandberg, L. B. and Albers, M. P., Durability of Structural Adhesives for Use in the Manufacture of Mobile Homes - Task VIII Sealant Test Report, report for U.S. Dept. of HUD under contract no. H-2817, 1980.

[3] McClintock, F. A. and Argon, A. S., Mechanical Behavior of Materials, Addison-Wesley Publ. Co., Reading, MA, 1966.

[4] Loveless, H. S., Deeley. C. W., and Swanson, D. L., "Prediction of Long-Term Strength of Reinforced Plastics," Society of Plastics Engineers Transactions, Vol. 2, No. 2, April 1962, pp.126-134.

[5] Sandberg, L. B., and Carbary, T. M., "Spacer Geometry Effects on Strength of Insulating Glass Joints for Structural Glazing Applications," Science and Technology of Glazing Systems, ASTM STP 1054, C. Parise, editor, (in press).

[6] Sandberg, L. B. "A Fatigue-Testing Device for Elastomeric Adhesives and Sealants," Experimental Techniques, Vol. 9, No. 10, October 1985, pp. 28-29.

[7] Miller, I., and Freund, J. E., Probability and Statistics for Engineers, Prentice-Hall, Englewood Cliffs, NJ, 1965.

C.V. Girija Vallabhan and Gee David Chou

SEALANT STRESSES IN STRUCTURAL GLAZING: A MATHEMATICAL MODEL

REFERENCE: Vallabhan, C. V. G., and Chou, G. D., "Sealant Stresses in Structural Glazing: A Mathematical Model," Building Sealants: Material, Properties, and Performance, ASTM STP 1069, Thomas F. O'Connor, editor, American Society for Testing and Materials, Philadelphia, 1990.

ABSTRACT: Silicone sealants are used to structurally connect the glass plates of an insulating glass (IG) unit and they are used between the IG unit and the supporting mullion. Thus, the sealants become an integral part of the overall glazing system. Design engineers have to compute stresses and deformations of the glass plates as well as the forces in the sealants when the overall system is subjected to lateral pressures such as wind. By use of von Karman's nonlinear theory of plates and Boyle's law of ideal gas, the complex interaction between the plates, the air trapped inside the IG unit, and the elastic behavior of the sealants is modelled mathematically. Some numerical examples are shown as illustrations.

KEYWORDS: Sealants, Insulating Glass Units, Structural Glazing, Wind Pressure, Nonlinear Von Karman Plate Theory.

Dr. Vallabhan is a Professor of Civil Engineering at Texas Tech University, P.O. Box 4089, Lubbock, TX 79409; Dr. Chou is a Research Engineer at Computerized Structural Analysis Research Corporation, CSA-NASTRAN, Los Angeles, CA 91335.

The architectural glazing industry introduced a new dimension in the glazing technique by providing a flushed glazed appearance; i.e., no structural component extends beyond the outer glass line as illustrated in Fig. 1 [1]. Promotion of this conspicuous architectural feature was materialized by the development of silicone sealants, the properties of which are the main theme of this symposium. A new term, "structural glazing," emerged as the silicone sealants became an integral structural part of the glazing system. They are used to structurally connect the two glass plates of an insulating glass (IG) unit and the IG unit and the supporting mullion (structural seal). The overall structural glazing system now comes under the scrutiny of structural design engineers, as these systems are subjected to wind pressures; the engineers, besides computing the maximum stresses and displacements of the glass plates, have to consider the forces in the silicone sealants also.

An IG unit which is connected to a supporting mullion on four sides is a complex structural-mechanical system. The two glass plates which comprise the IG unit enclose a sealed airspace, thus producing a sharing of load between glass plates as pressures are placed across the unit. Further, the structural seals (IG unit seal and structural seal) are elastic, thus providing deformable, elastic supports for the thin glass plates.

The complex behavior of the IG unit has been studied by Solvason [2] and by Vallabhan and Chou [3]. Chou, Vallabhan, and Minor [4] used these works to develop a mathematical model of an IG unit which is supported on an elastic structural seal. This model describes the interactive behavior of the two glass plates in the IG unit, the IG unit seal and the structural seal. Here, the model is used to define stresses in glass plates, and forces which occur in the IG unit seal and the structural seal as a lateral pressure is placed across the structural glazing system.

THE STRUCTURAL GLAZING SYSTEM MODEL

In this model, it is assumed that the structural seal is deformable; hence, the IG unit interacts with the seal like a plate on an elastic foundation. It is also assumed that the building frame that supports the IG unit through the seal is rigid. Sealants which support the glass plates are assumed to have linear load-deflection characteristics; they are modelled as a series of continuous springs along the boundaries of the plates. In other words, these springs have constant spring stiffness, i.e., k values. Two sets of springs,

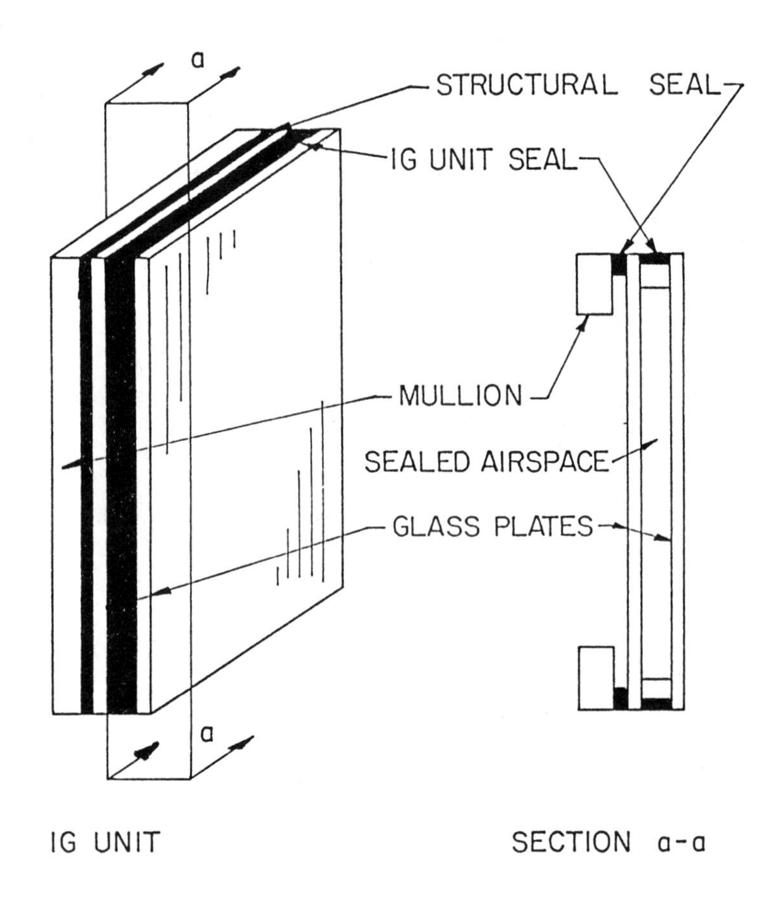

FIGURE 1 -- Structural Glazing System

k_1 and k_2, are used in the model. Spring k_1 represents the combined stiffness of the IG unit seal and the spacer in between the glass plates of the IG unit. Spring k_2 represents the stiffness of the structural seal.

The mathematical model is constructed using the associated equations of its components and solving them in a compatible manner. The components are:

(1) the two thin glass plates,

(2) the sealed airspace, and

(3) the two linear springs, k_1 and k_2, representing the stiffness of the IG unit seal and spacer, and the structural seal.

These components are illustrated in Fig. 2. The size of the glass plates is 2a x 2b, with thicknesses t_1 and t_2.

Equations for the Plates

The two glass plates are modelled by the von Karman nonlinear plates equations, as they have been proven by researchers [4,5,6] to be suitable for stress analysis of thin glass plates. The bending phenomenon of plates is described by the von Karman equations as follows [7,8]:

$$D_1 \nabla^4 w_1 = \Delta p_1 + t_1 L(w_1, \phi_1)$$

$$D_2 \nabla^4 w_2 = \Delta p_2 + t_2 L(w_2, \phi_2), \text{ in } -a \leq x \leq a \text{ and } -b \leq y \leq b \tag{1}$$

where w = the lateral displacement of the plate,

ϕ = the Airy stress function,

D = the flexural rigidity of the plate

p = the lateral pressure on the unit,

Δp_1, Δp_2 = the corresponding lateral pressures on the the individual plates.

The subscripts 1 and 2 represent outer (loaded) and inward plates respectively. The operator L is defined as:

$$L(w, \phi) = w_{,xx}\, \phi_{,yy} - 2w_{,xy}\, \phi_{,xy} + w_{,yy}\, \phi_{,xx}$$

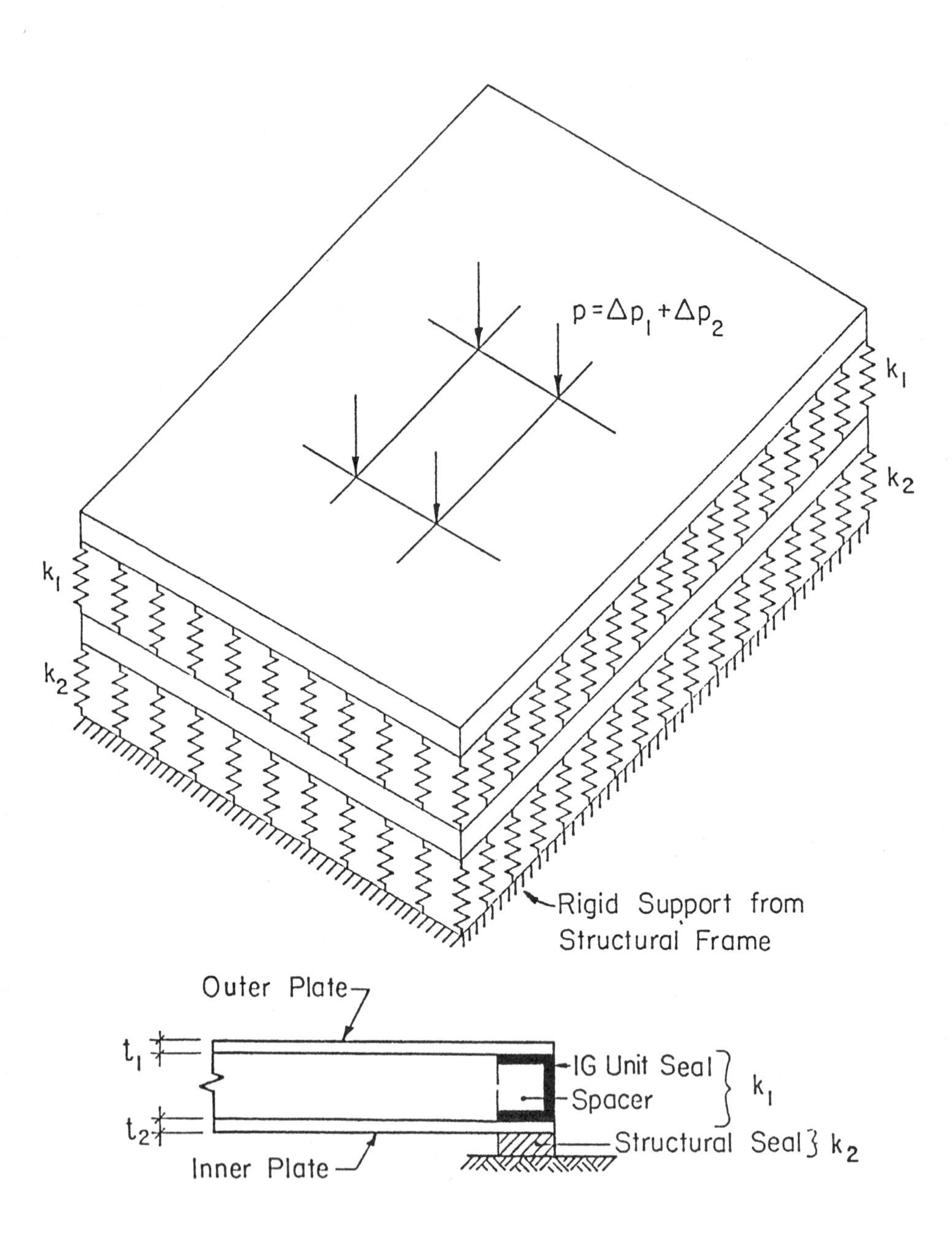

FIGURE 2 -- Model for Structural Glazing System

where the subscripts after the comma represent differentiation with respect to the variable following the comma. The Airy stress functions ϕ_1 and ϕ_2 yield the membrane stresses in the plates, such that the respective membrane stresses are given by:

$$\sigma_x^m = \phi_{,yy}$$
$$\sigma_y^m = \phi_{,xx} \quad (2)$$
$$\tau_{xy}^m = -\phi_{,xy}$$

The membrane phenomena in the plates are described by two compatibility equations,

$$\nabla^4 \phi_i = -\frac{E}{2} L(w_i, w_i) \text{ for } i = 1, 2 \quad (3)$$

where E is the Young's modulus of elasticity of the plate.

Boundary Conditions for the Plates

It is further assumed that the IG unit spacer-sealer assembly along the edges of the glass plates does not produce resistance to bending and membrane forces; i.e., the bending moment perpendicular to the edges is zero for both plates, and the membrane stresses (normal and shear) along the edges are also zero. The interaction between the plates and the springs k_1 and k_2 is represented by the classical Kirchoff shear forces along the edges. A free body diagram, illustrated in Fig. 3, is helpful in writing the shear force (normal to the plane of the glass) as:

For outer plate (t_1),

$$V_{xz} = \pm [k_1 w_1 - k_1 w_2], \text{ @ } x = \pm a, \text{ and}$$

$$V_{yz} = \pm [k_1 w_1 - k_1 w_2], \text{ @ } y = \pm b \quad (4)$$

For inner plate (t_2),

$$V_{xz} = \pm [-k_1 w_1 + (k_1 + k_2) w_2], \text{ @ } x = \pm a, \text{ and}$$

$$V_{yz} = \pm [-k_1 w_1 + (k_1 + k_2) w_2], \text{ @ } y = \pm b \quad (5)$$

In Eqs 4 and 5, a coupling of the behavior of the two plates caused by the boundary deflections of each plate can be observed.

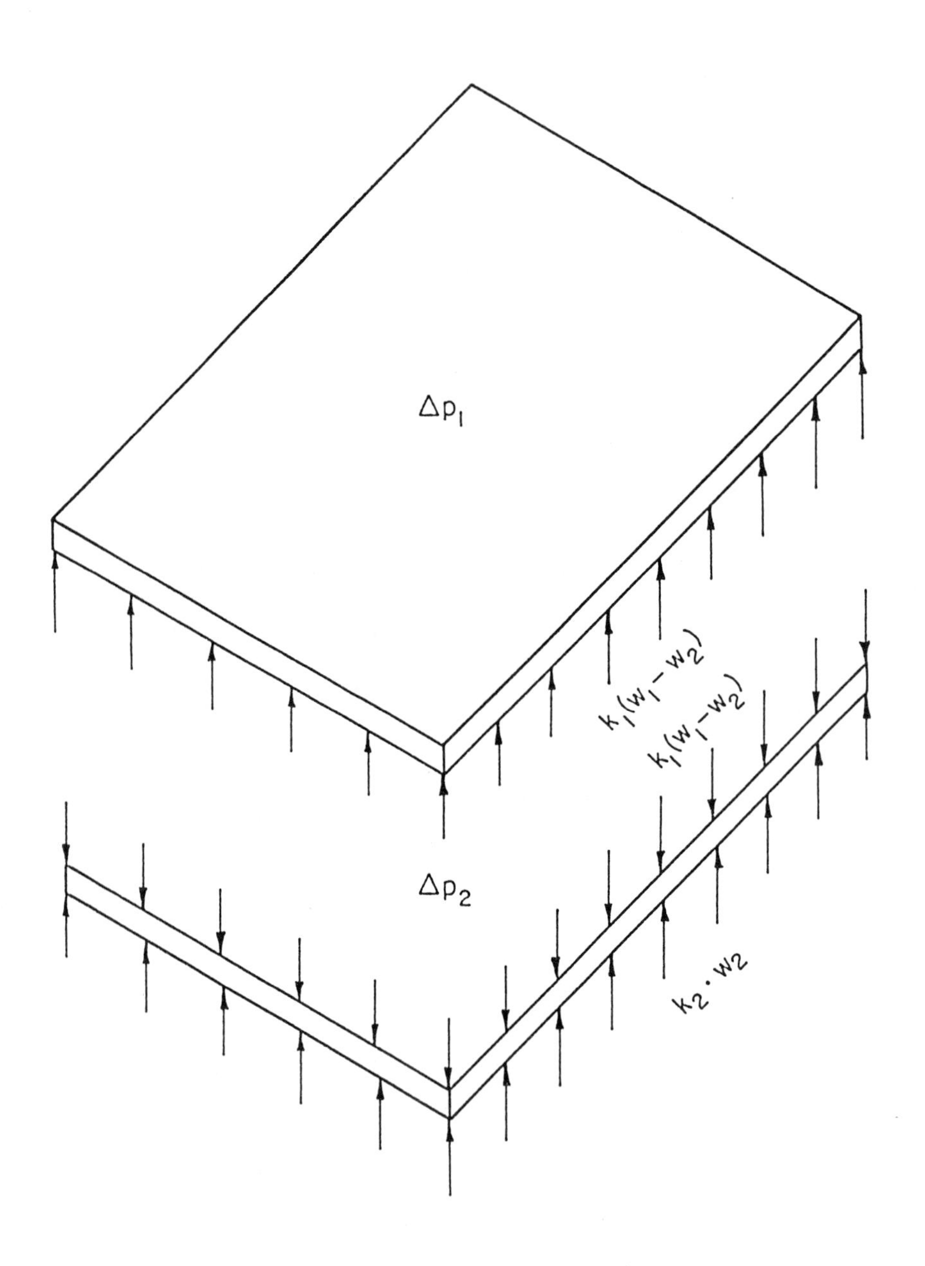

FIGURE 3 -- Free Body Diagram of Structural Glazing System Model

Vallabhan [5] has shown that the von Karman plate equations can be solved numerically using the finite difference technique combined with an iterative procedure. Using his procedure, and defining the discrete values of w and ϕ in the finite difference model by the vectors {W} and {F} with proper subscripts for the two plates, the finite difference equivalents of Eqs 1 and 2 are represented by two equations as shown in Fig. 4. The matrix $[A_1]$ corresponds to the differential operator $D_1\nabla^4$ which is modified for the effects of the spring constant k_1 on the boundary of the outer plate. The matrix $[A_1]$ contains the stiffness coefficient k_1 on the diagonals corresponding to the boundary nodes. The coefficient matrix $[A_2]$ similarly represents the differential operator $D_2\nabla^4$ modified for the effects of spring constant $k_1 + k_2$ on the diagonal corresponding to the nodes representing the boundary. The matrices $[K_1]$ are diagonal matrices with $-k_1$ on the diagonal corresponding to the nodes on the boundary and zero on other diagonal elements. Matrices $[B_1]$ and $[B_2]$ are identical, since they both represent ∇^4 operator with the same prescribed boundary conditions for membrane actions. In the computer program developed for the model, only $[B_1]$ is kept in high-speed memory in a half-banded form; and $[B_1]$ is used to solve both $[F_1]$ and $[F_2]$. Since the iterative technique used to solve these nonlinear equations has been published, it is not repeated here. Details of this technique are described by Chou [9].

Interaction Between the Plates and the Sealed Air

The interaction between the plates and the airspace for a given set of IG unit dimensions subjected to a series of specific lateral pressures is modelled in the following manner. When the outer glass plate is subjected to a uniform positive pressure (inward acting), it deflects and compresses the air inside the sealed airspace. In turn, the compressed air applies pressure on both plates. Thus, the sealed airspace serves as a load transferring medium between the plates. The change in volume of the air within the sealed airspace due to the induced inside pressure is calculated using Boyle's ideal gas law, i.e.,

$$\frac{P_o V_o}{T_o} = \frac{P_a V_a}{T_a} \tag{6}$$

where P, V, T are the pressure, volume, and temperature

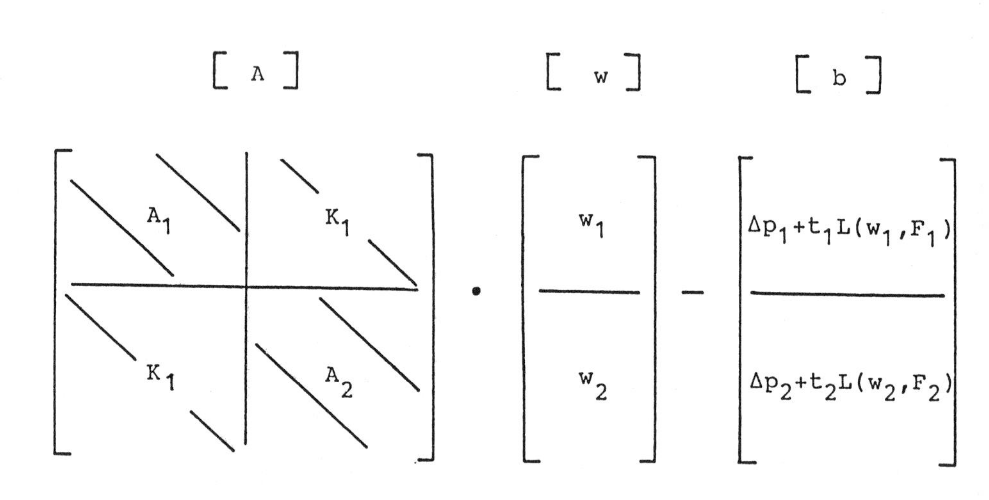

(a) Bending Equation

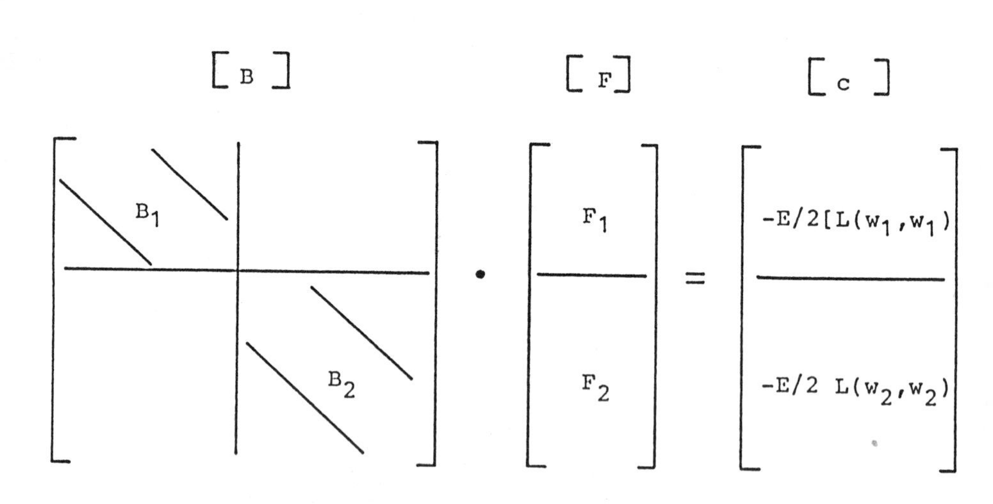

(b) Membrane Equation

FIGURE 4 -- System Equations for Model of Structural Glazing System

of the airspace. The subscripts o and a denote initial and new values of the parameters. Compatibility of displacements of the entire IG unit is achieved when the above volume change inside the IG unit is made exactly equal to the differences in volumes between the deformed and undeformed surfaces of each plate. If v_1 and v_2 represent the volumes of the undeformed and deformed middle surfaces of the two plates, respectively, then

$$V_o = V_a - v_1 + v_2 \quad (7)$$

The finite difference plate analysis program is used to compute the volume between the deformed and undeformed plate surfaces. Since the behavior of the plates and the change in volume of the air inside are both nonlinear functions of pressure, an iterative scheme has to be employed to achieve compatibility and equilibrate the system. The interaction or sharing of loads between the plates depends on plate dimensions, respective plate thicknesses, the thickness of the spacer, and variations in temperature and barometric pressures on the two faces of the unit.

Numerical Examples

The model of the structural glazing system described above is used here to analyze two typical IG unit examples. The analyses include the computation of boundary reactions in the IG unit seal and the structural seal, as well as the magnitude and locations of maximum principal tensile stresses in the plates. The examples included herein address two sizes of IG units: a large unit 60 x 96 in. (1.524 x 2.438 m), with equal thickness 1/4 in. (6.35 mm) plates and a smaller unit 30 x 48 in. (0.762 x 1.219 m), with equal and unequal thickness plates. Pressures applied to these units result in various degrees of nonlinearities in the plates. It is interesting to note that the distribution of the boundary reactions depends very much on the degree of nonlinearity developed in the plate by the applied pressure.

Example 1: This example involves an IG unit 60 x 96 in. (1.524 x 2.438 m) in size, with equal thickness plates. The thicknesses of the plates and the spacer are 0.225 in. (5.72 mm) and 0.5 in. (12.7 mm), respectively. To simplify the analysis, the airspace pressure and barometric pressures both indoors and outdoors are assumed to be equal to 14.7 psi (101.4 kPa) initially. The lateral pressures used are 10 psf (0.48 kPa) and 100 psf (4.79 kPa) applied as a positive pressure to the windward plate.

Spring constants k_1 representing the IG unit sealant spacer system are 10^5 and 10^4 psi (689.5 and 68.95 MPa). A spring constant of 10^3 psi (6.895 MPa) is assigned to k_2 which represents the structural seal.

The spring constant, k_1, and lateral pressure, p, assigned to each case are shown in Table 1 along with the results of the analysis. The lateral pressure, p, is carried almost equally by the two plates of the IG unit in all cases in this example.

It is observed that the location of maximum principal stress moves from the center of the plates (Cases A and B) to an area near the corner of the plates (Cases C and D) as p increases from 10 psf (0.48 kPa) to 100 psf (4.79 kPa). Distributions of boundary reactions for all four cases are illustrated in Figs. 5 and 6. High corner reactions are observed in each case. In Fig. 5, the boundary reaction decreases gradually as it moves from the center of the boundary (edge) toward the corner and becomes a negative value at the corner. A significant difference in the distribution pattern of boundary reactions is observed in Fig. 6 compared to Fig. 5. In Fig. 6, the boundary reaction increases from the center of the boundary (edge) to a large value close to the corner. At the corner, the boundary reaction becomes negative in a very short distance.

Example 2: This example employs an IG unit 30 x 48 in. (0.762 x 1.219 m) in size. The spacer thickness and initial pressures (airspace pressure and external barometric pressures) are as given in Example 1: 0.5 in. (12.7 mm) and 14.7 psi (101.4 kPa), respectively. The glass plates of the IG units for Cases A and B are equal in thickness and are equal to 0.5 in. (12.7 mm). For Cases C and D, the outer plate (plate #1) and leeward plate (plate #2) have thicknesses of 0.225 in. (5.72 mm) and 0.180 in. (4.57 mm), respectively. A value of 10^3 psi (6.895 MPa) is used for k_2. The results of the analyses and assigned values for k_1 and p are shown in Table 2.

The location of maximum principal stress stays at the center for all cases in this example. The distributions of boundary reactions for Cases A, B, C, and D are shown in Figs. 7 and 8. The reactions in this example do not rise as sharply at the corners as those cases in Example 1.

It was assumed in the model analyses presented above that the load sharing between the plates of the IG units is the same as that determined for IG units with simply supported (nondeformable supports) plates. To

TABLE 1 -- Lateral Pressure, Spring Constant and Results of Cases in Example 1

	P (psf)	k_1 (psi)	%*	Plate #	W_{max}/t	α_{max} (psi)	Location of σ_{max}
Case A	10	10^5	50.31	1	1.397	1038.7	**
				2	1.385	1029.7	**
Case B	10	10^4	50.31	1	1.398	1038.9	**
				2	1.385	1029.7	**
Case C	100	10^5	50.87	1	5.428	5030.3	***
				2	5.339	4880.6	***
Case D	100	10^4	50.87	1	5.433	5037.0	***
				2	5.338	4877.4	***

k_2 = 1000 psi
* - Percentage of load carried by windward plate
** - At center
*** - Around corner area
(1 psi = 6.895 kPa; 1 psf = 0.0479 kPa)

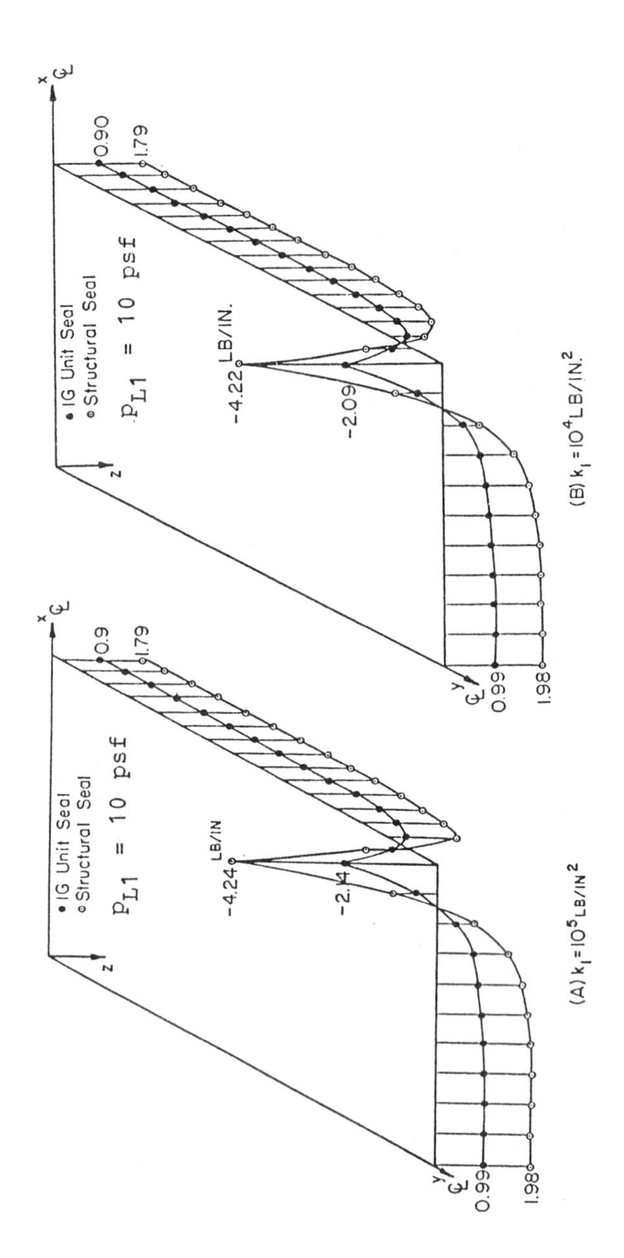

FIGURE 5 -- Boundary Reactions: Case A and B, Example 1

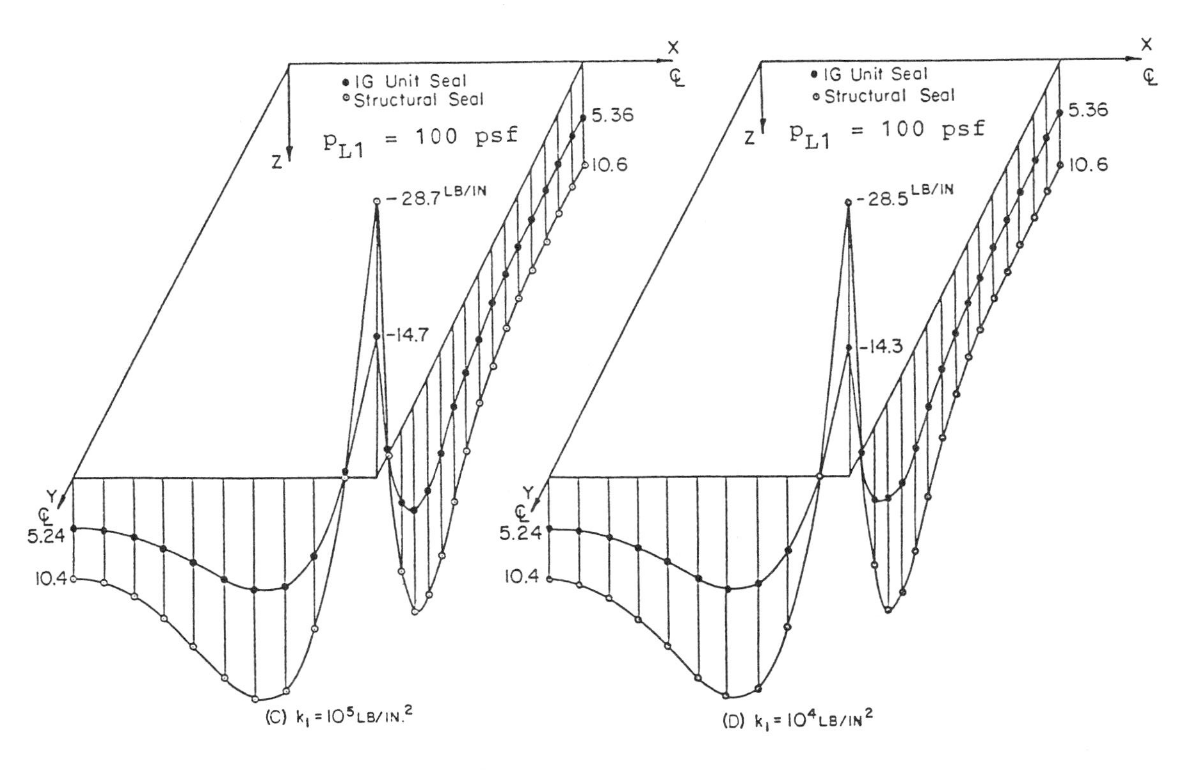

FIGURE 6 -- Boundary Reactions: Case C and D, Example 1

TABLE 2 -- Lateral Pressure, Spring Constant and Results of Cases in Example 2

	P (psf)	k_1 (psi)	%*	Plate #	W_{max}/t	α_{max} (psi)	Location of σ_{max}
Case A	100	10^5	69.66	1	0.086	951.8	**
				2	0.052	453.1	**
Case B	100	10^4	69.66	1	0.087	956.5	**
				2	0.051	450.5	**
Case C	100	10^5	66.57	1	1.260	3654. 3	**
				2	1.476	2714.1	**
Case D	100	10^4	66.57	1	1.264	3656. 7	**
				2	1.476	2713.5	**

k_2 = 1000 psi
* - Percentage of load carried by windward plate
** - At center
(1 psi = 6.895 kPa; 1 psf - 0.0479 kPa)

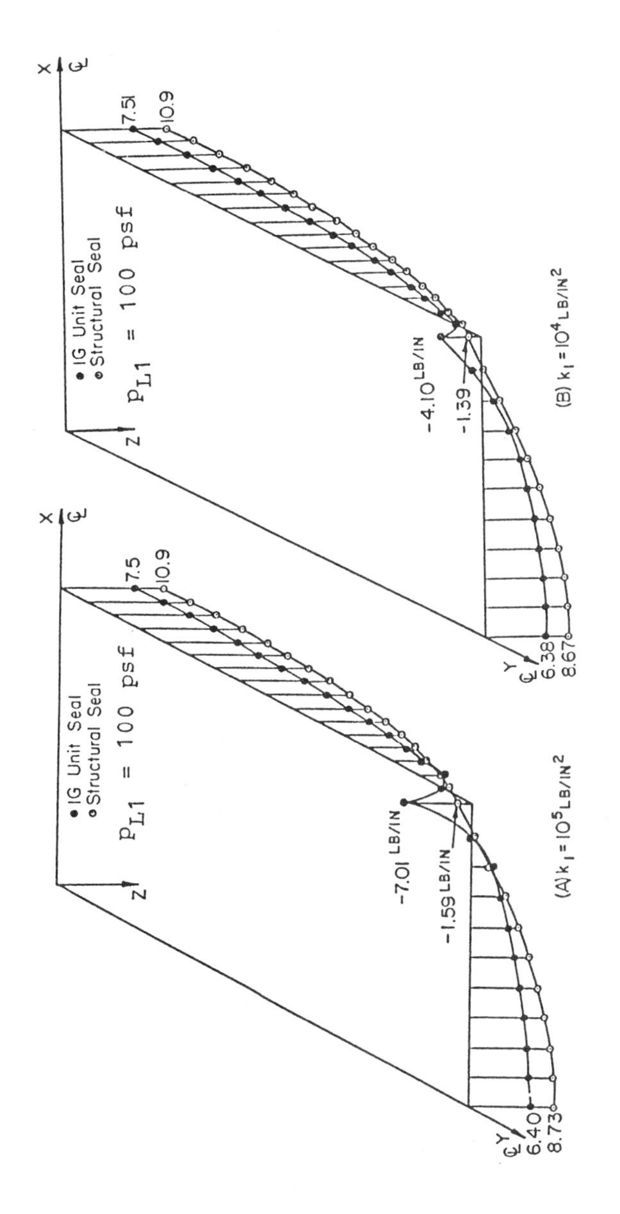

FIGURE 7 -- Boundary Reactions: Case A and B, Example 2

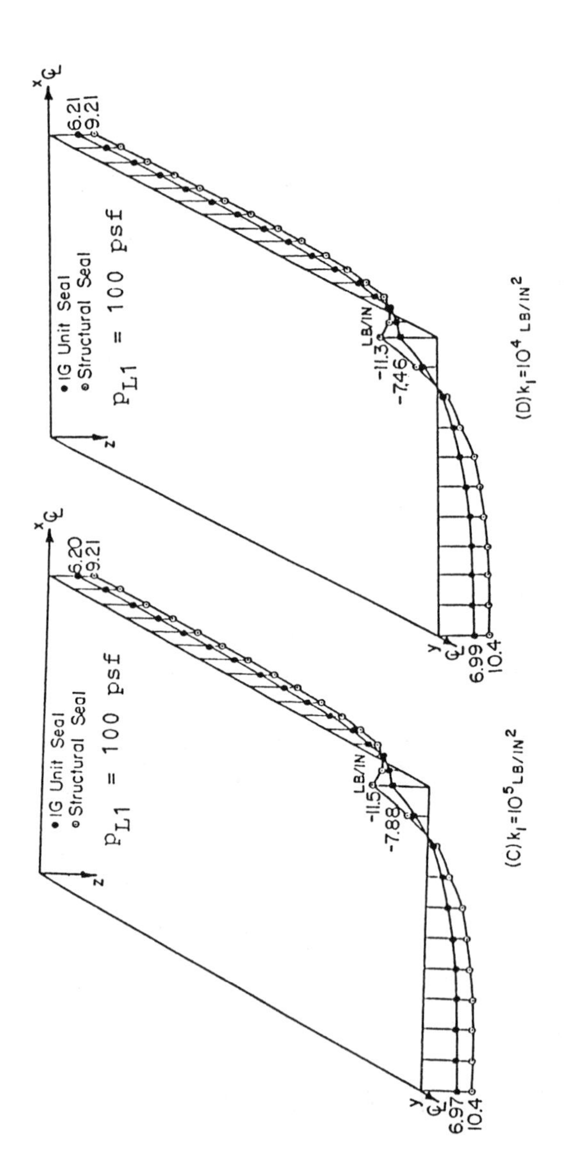

FIGURE 8 -- Boundary Reactions: Case C and D, Example 2

TABLE 3 -- Volume of Change of Airspace

	Example 1		Example 2	
Initial Volume of Airspace	2880 in.^{3}	2880	720	720
$k_1 = k_2 = \infty$	6.76 in.^{3}	65.32	10.16	11.20
$k_1 = 10^5$ psi	7.72(A)	65.36(C)	10.40(A)	11.16(C)
$k_1 = 10^4$ psi	8.48(B)	71.24(D)	11.40(B)	11.00(D)

$k_2 = 1000$ psi

(1 psi = 6.895 kPa; 1 in.^{3} = 16.387 cm^3.)

check the adequacy of this assumption, calculations of volume changes in the airspace as functions of seal stiffnesses were performed. The volume changes in the airspace for the eight cases are presented in Table 3. All quantities shown in Table 3 represent reductions in the volume of the airspace.

For $k_1 = 10^5$ psi (689.5 MPa), it is seen that the airspace volume change is almost the same as that for simply supported conditions. But for $k_1 = 10^4$ psi (68.95 MPa), there is a small difference. However, these differences in volume change are so small when compared to the total initial volume of airspace that their effects on load sharing are negligible.

SUMMARY AND CONCLUSIONS

The structural glazing system model has been used to analyze example IG units with various cases of loading and geometry. It is important to note that boundary reactions close to the corner are always opposite in direction to the applied load. Also, distribution patterns of the boundary reactions are significantly different between linearly behaved and nonlinearly behaved plates [9]; at the same time they are relatively insensitive to small changes in the value of k_1 or even k_2. These reaction patterns will be particularly useful in the design of IG unit seals and structural seals in a structural glazing system.

In the mathematical model presented here, it is assumed that the forces in the sealants are only in the direction of the applied forces. Truly, the sealants are capable of having edge moment and shear restraints; the mathematical model can be extended to consider these reactions also.

ACKNOWLEDGEMENTS

Research results reported herein were developed as part of a program of research on structural glazing systems being conducted at the Glass Research and Testing Laboratory at Texas Tech University under the direction of Dr. Joseph E. Minor, Director. Support for the model of the structural glazing system was provided by the Critical Engineering Systems Division of the National Science Foundation under Grant No. ECE-8500620. The support of Dr. Michael P. Gaus and Dr. Eleonora Sabadell is gratefully acknowledged.

REFERENCES

[1] Quade, David, "New Technologies in Insulating Glass," Glass Magazine, March, 1986.

[2] Solvason, K.R., "Pressures and Stresses in Sealed Double Glazing Units," Technical Paper No. 423, Division of Building Research, National Research Council of Canada, 1974.

[3] Vallabhan, C.V.G., and Chou, Gee D., "Interactive Nonlinear Analysis of Insulating Glass Units," Journal of Structural Engineering, ASCE, 112(6), 1986.

[4] Chou, G.D., Vallabhan, C.V.G., and Minor, J.E., "A Mathematical Model for Structural Glazed Units," Glass Research and Testing Laboratory, Texas Tech University, Lubbock, TX, 1987 (NTIS Acc. No. PB87196192).

[5] Vallabhan, C.V.G., "Iterative Analysis of Nonlinear Glass Plates," Journal of Structural Engineering, ASCE, 109(2), 1983.

[6] Vallabhan, C.V.G. and Minor, J.E., "Experimentally Verified Theoretical Analysis of Thin Glass Plates," Computational Models and Experimental Methods, June, International Conference for Computational Methods in Engineering, Southampton, England, 1984.

[7] Timoshenko, S. and Woinowsky-Krieger, S., Theory of Plates and Shells, McGraw-Hill Book Company, Inc., New York, NY, 1959.

[8] Ugural, A.C., Stresses in Plates and Shells, McGraw-Hill Book Company, Inc., New York, NY, 1981.

[9] Chou, Gee. D., "Analysis of Insulating Window Glass Units," Ph.D. Dissertation, Dept. of Civ. Engrg., Texas Tech University, Lubbock, TX, 1986.

James R. Bailey, Joseph E. Minor, and Richard W. Tock

CHANGES IN SEAL SHAPES OF STRUCTURALLY GLAZED INSULATING GLASS UNITS

REFERENCE: Bailey, J. R., Minor, J. E., and Tock, R. W., **"Changes in Seal Shapes of Structurally Glazed Insulating Glass Units,"** Building Sealants: Materials, Properties, and Performance, ASTM STP 1069, Thomas F. O'Connor, editor, American Society for Testing and Materials, Philadephia, 1990.

ABSTRACT: A structural glazing system bonds glass, stone, or other materials to the structural frame of a building using structural silicones. The most complex structural glazing system uses insulating glass (IG) units with a structural silicone holding all four sides of an IG unit to a building frame. This complex structural glazing system is known as a four-sided structurally glazed IG unit. Researchers at Texas Tech University conducted tests on samples representing four-sided structurally glazed IG units. Each sample underwent applied simulated wind pressures while instruments measured changes in structural seal and IG seal shapes. This paper describes changes in the shapes of the structural seals and the IG seals of these test samples.

KEYWORDS: silicone, structural glazing, insulating glass unit, curtain wall, sealants

INTRODUCTION

Structural glazing has become a popular method for finishing building exteriors [1]. About one in three new glass curtain walls is structurally glazed. Most structural glazing systems use mechanical connectors and structural silicones to fasten glass products to a frame (mullion). Other structural glazing systems, however, use structural silicones alone to fasten all sides of a glass product to a frame. No mechanical connectors hold the glass product in place. Only the structural silicone holds the glass product in place. This design is known as a four-sided structural glazing system.

Dr. Bailey is a research associate of Civil Engineering, Texas Tech University, Lubbock, TX, 79409; Dr. Minor is the Thomas Reese Professor and Chairman of Civil Engineering, University of Missouri-Rolla, Rolla, MO, 65401; Dr. Tock is a professor of Chemical Engineering, Texas Tech University, Lubbock, TX, 79409.

Four-sided structural glazing systems provide a building with a smooth exterior finish. Four-sided structurally glazed IG units are very popular since these systems also improve building insulation [2]. Four-sided structurally glazed IG units, however, are the most complex type of structural glazing system. Thus, researchers at Texas Tech University began a test program involving four-sided structurally glazed IG units [3]. The goal was to measure changes in the shapes of the structural seal and the IG seal of samples representing four-sided structurally glazed IG units.

Most designers use the Trapezoidal Rule to estimate stresses in a structural silicone seal. Use of the Trapezoidal Rule suggests that the distribution of normal strains along the side of a four-sided structurally glazed IG unit also is trapezoidal in shape. Thus, maximum normal strain will occur near the center along the side of a structurally glazed IG unit while normal strain at the corner will equal zero. The purpose of tests which measure changes in structural seal and IG seal shapes is to provide data to determine the distribution of seal strains.

TEST PLAN

Researchers selected eight sample designs using information provided by curtain wall designers and makers of structural glazing systems (Figure 1). These eight sample designs represent a broad range of four-sided structurally glazed IG units. One of two structural silicones, a medium modulus Dow Corning 795 (DC795) or a high modulus Dow Corning 983 (DC983), forms the structural seal which fastens an IG unit to a sample frame. Structural seal sizes are 6.35 mm x 6.35 mm or 19.1 mm x 9.52 mm (0.25 in. x 0.25 in. or 0.75 in. x 0.375 in.). A foam rubber backer rests between an IG unit and a sample frame for quality control during application of the structural silicone.

Two fully-tempered 1.52 m x 2.44 m x 5.71 mm (60 in. x 96 in. x 0.225 in.) glass plates separated by a 14 mm (0.55 in.) spacer form the IG units. Using fully-tempered glass reduces the chance of glass failure during testing. Test samples use two different IG seal sizes (minimum, maximum). Only the width (depth) of the secondary seal differs between the two IG seal designs. Otherwise, each IG seal design uses the same primary seal (PIB), secondary seal (DC982), and spacer. Smooth aluminum tubing forms the sample frame. Designers select aluminum for a secondary frame (mullion) since structural silicones form very strong bonds with aluminum.

Researchers designed a test facility to apply pressure or partial vacuum to one or both sides of a test sample. The facility has a platform, outer chamber, inner chamber, support frame, and control panel (Figure 2). A test sample fastens to the platform. A space forms behind the IG unit when fastening the sample frame to the platform, resulting in an inner chamber. An outer chamber fastens to the platform, covering a test sample. A control panel directs air through pipes to the inner and outer chambers which increases or decreases pressure on one side or both sides of a test sample. Initial tests used maximum effective pressures of 4.79 kPa (100 psf) to represent a practical upper limit for most design wind loads.

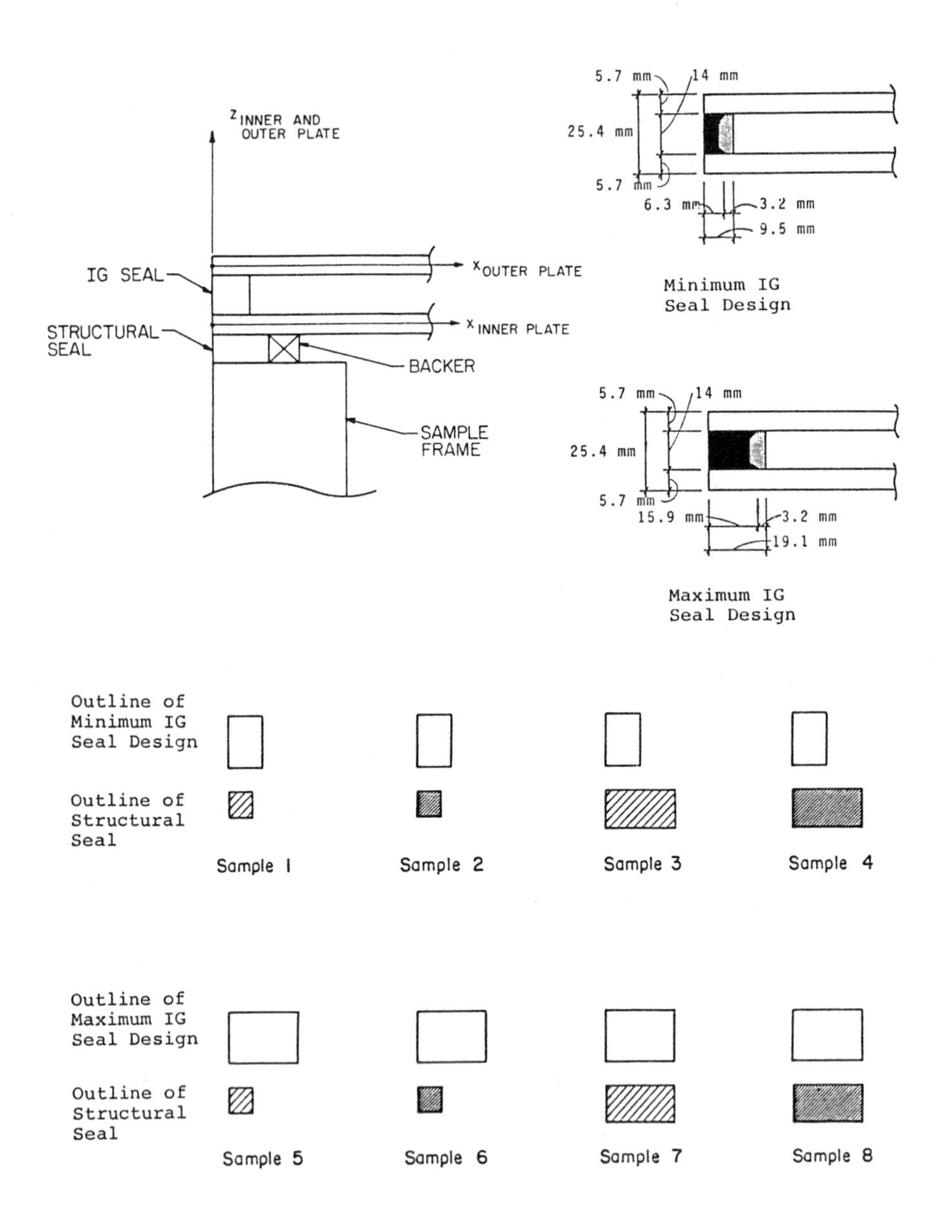

Notes: Smaller structural seal is 6.35 mm x 6.35 mm (0.25 in. x 0.25 in.).
Larger structural seal is 19.1 mm x 9.52 mm (0.75 in. x 0.375 in.).
Light shaded structural seal is a medium modulus sealant (DC795).
Dark shaded structural seal is a high modulus sealant (DC983).

FIGURE 1 -- Selected Test Samples

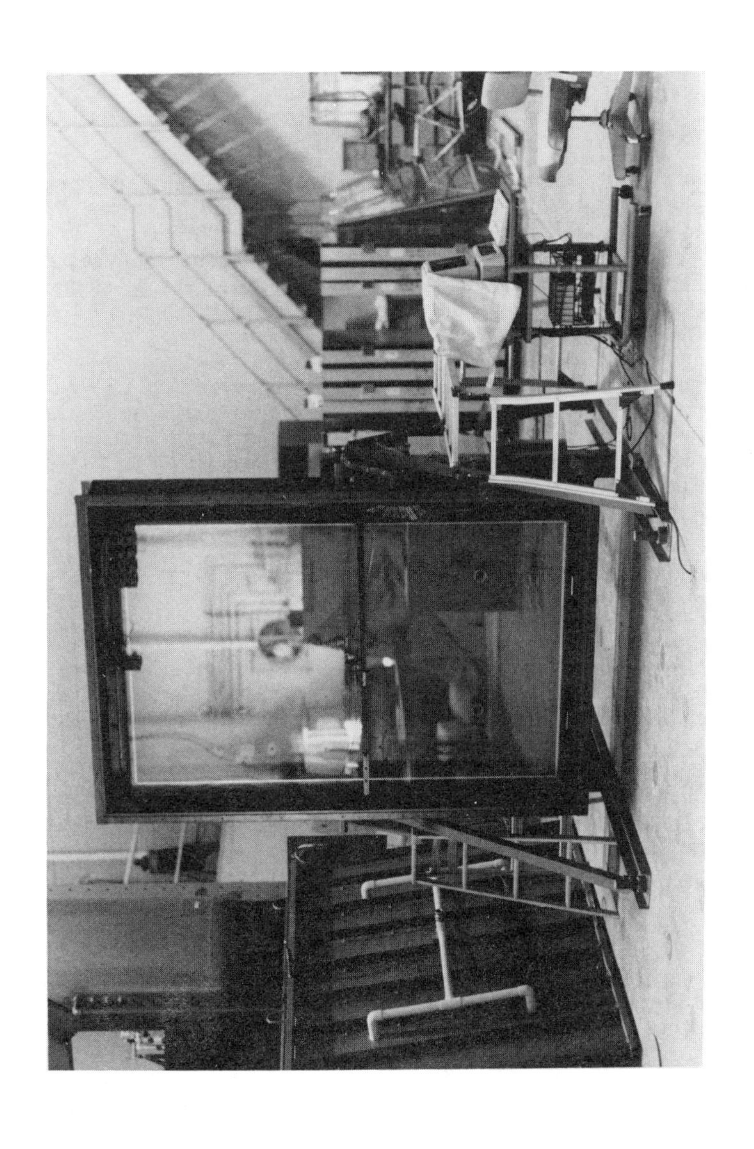

FIGURE 2 -- Test Facility (Cover Off)

Brackets position instruments to measure displacements of the inner and outer IG plates (Figure 3). Other instruments measure inner chamber, IG air space, and outer chamber pressures during a test. A microcomputer receives signals from these instruments and records displacements and pressures. These displacements define positions of the IG plates relative to the sample frame at a given pressure. Researchers use these IG plate positions to determine changes in the shapes of the structural seal and the IG seal for a given test sample.

SELECTED RESULTS

All eight test samples underwent decreasing applied pressures adjacent to the outer IG plate. These applied pressures (partial vacuum) ranged from 4.60 to 4.93 kPa (96 to 103 psf). Pressures decreased evenly to a maximum value in five seconds, then returned to zero in five seconds. Researchers observed changes in the shape of the structural seal relative to the sample frame and changes in the shape of the IG seal relative to the inner IG plate.

The change in angle between two lines intersecting at 1.57 rad (90°) defines shear strain in a material. Knowing in-plane seal displacements and seal sizes allows solving of shear strains. Shear strains in the structural seals did not exceed 0.087 rad (5.0°). Most shear strains in the structural seals were less than 0.044 rad (2.5°). Shear strains in both IG seals were within 0.09 rad (5.2°).

The change in length divided by the original length defines normal strain in a material. Knowing out-of-plane seal displacements and seal sizes allows solving of normal strains. Maximum normal strains in the structural seals of Sample 1 and Sample 5 were 16 percent of seal thickness (6.35 mm). Maximum normal strains in the structural seals of the other six test samples were within 9 percent of seal thickness (6.35 mm or 9.52 mm). Maximum normal strains of the IG seals, however, were within 2 percent of IG seal thickness (14 mm).

Edge slopes of the inner and outer IG plates ranged from less than 0.0087 rad (0.5°) at the sample corner to about 0.07 rad (4°) at the center of either sample side. These edge slopes began to increase linearly from the plate corner, then leveled toward the center of the plate edge. Edge slopes of the inner and outer IG plates differed little at a given location along the side of a sample.

Comparing structural seal and IG seal shapes of samples using the minimum IG seal size (Sample 1 through Sample 4) provides useful information on the response of these samples to simulated wind pressures (Figure 4). Bailey et al [3] provides data for determining the structural seal and IG seal shapes of Sample 1 through Sample 4. Data used for comparing seal shapes (Figure 4) came from the first test of each sample at near 4.79 kPa (100 psf) applied pressures. Changes in un-deformed structural seal and IG seal shapes are magnified five times for illustration when comparing seal shapes.

The structural seals deform in several directions. Structural seal shapes vary unevenly from one location to the next location along the short side of Sample 1 through Sample 4. Other tests reveal that

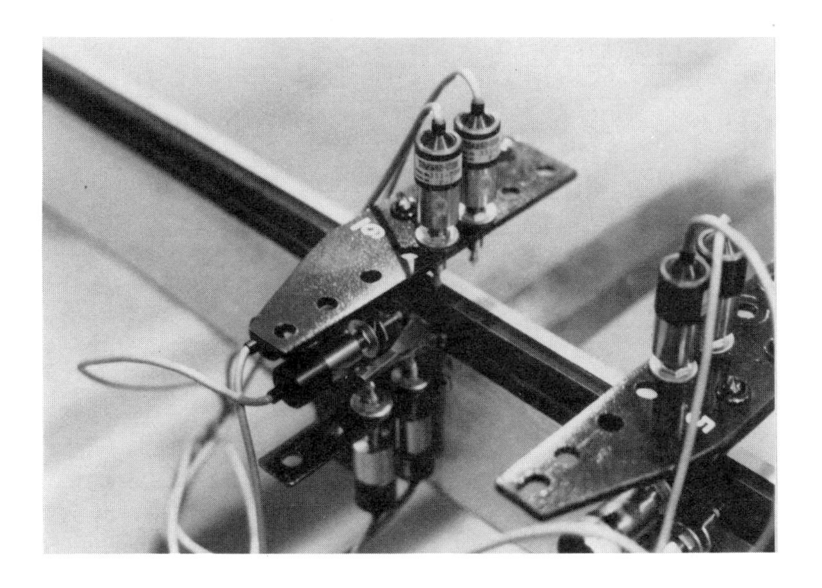

(a) Fastened to Sample Frame

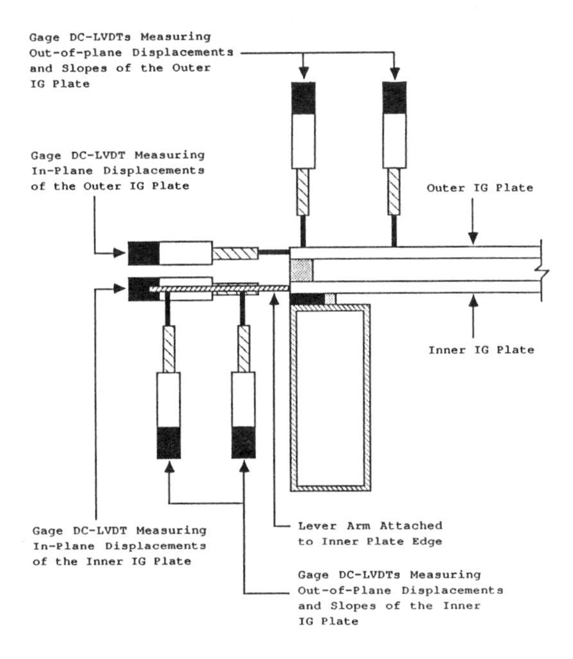

(b) Profile

FIGURE 3 -- Brackets Measuring IG Plate Positions

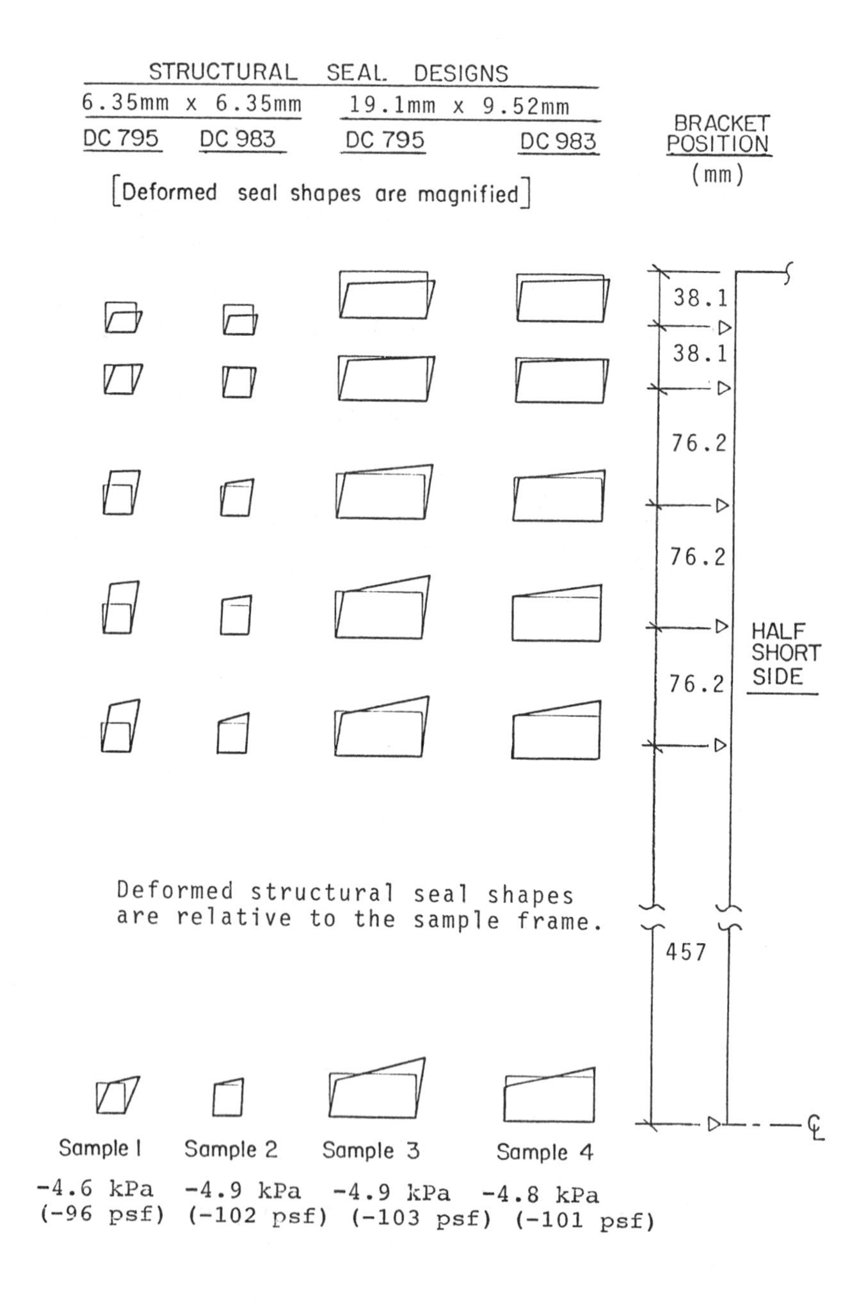

(a) Structural Seals

FIGURE 4 -- Deformed Shapes of the Structural Seals and the IG Seals of Samples Using the Minimum IG Seal Size

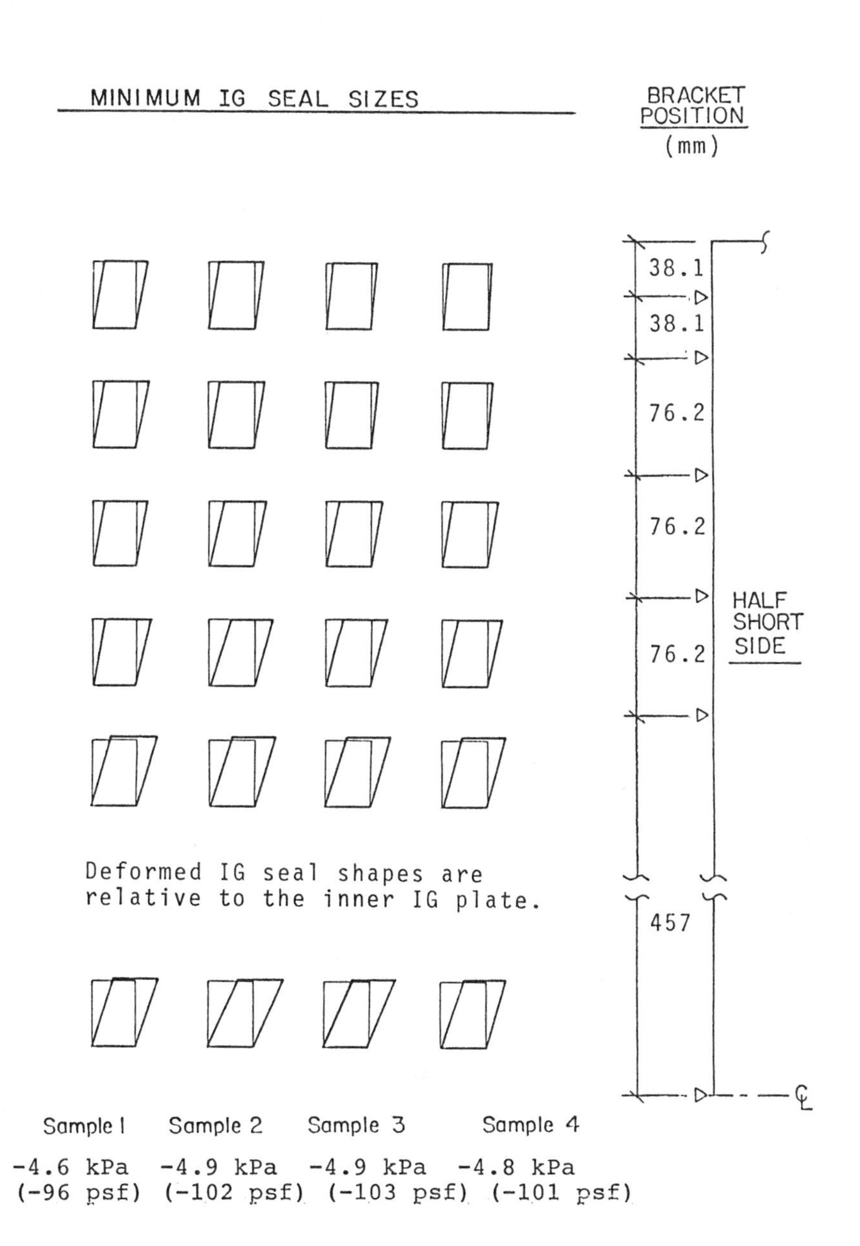

(b) IG Seals

FIGURE 4 -- (Continued)

structural seal and IG seal shapes also vary unevenly from one location to the next location along the long side of Sample 1 and Sample 8.

Comparisons of the structural seals of Sample 1 through Sample 4 also show that larger out-of-plane displacements did not always occur at the center of the short side (Figure 4). Maximum normal strains often occur between the center and the corner of the short side of Sample 1 through Sample 4. Normal strains at the corners do not equal zero. These observations are more obvious with samples having the smaller structural seal size (Sample 1 and Sample 2). Similar observations also result when comparing the structural seals of Sample 1 through Sample 4 along the long side.

Models predicting structural seal and IG seal stresses show maximum normal stresses occurring between the center and the corner along the side of a sample [4]. Thus, test results and model predictions imply that normal strain distribution along the side of a four-sided structurally glazed IG unit is not trapezoidal in shape.

Comparing test results to estimated strain values from an assumed stress distribution is difficult because converting stress to strain requires a relationship between stress and strain. Some materials have linear elastic stress-strain relationships which allow simple conversions of stress to strain. Structural silicones, however, do not always behave as a linear elastic material. Several factors affect the stress-strain relationship of a structural silicone [5]. Thus, uncertainties remain concerning the accuracy of the Trapezoidal Rule when estimating stresses or strain in a structural silicone seal.

Comparisons of IG seal shapes along the short sides of Sample 1 through Sample 4 show that changes in IG seal shapes result mostly from the shearing action of rotating IG plates (Figure 4). This shearing action is most apparent at the center of the short side of Sample 1 through Sample 4. The center of the short side of Sample 1 through Sample 4 also is the location where maximum slopes of the inner and outer IG plates occur along the short side. Similar observations result when comparing IG seals along the long side of Sample 1 through Sample 4. Similar observations also result when comparing samples using the maximum IG seal size (Sample 5 through Sample 8).

Researchers repeated tests for each sample to see if the same data resulted. Researchers applied the same load for each repeated test. Repeatable data is important. Non-repeatable data means that sealant properties changed between tests. Sample 2 through Sample 4 and Sample 6 through Sample 8 data proved repeatable at near 4.79 kPa (100 psf) applied pressures.

Sample 1 data proved non-repeatable at 4.79 kPa (100 psf) applied pressures. Sample 1 is the least stiff sample of the eight test samples. The deformed shapes of the structural seal along the short side of Sample 1 differed notably between tests at 4.79 kPa (100 psf) applied pressures. Inner IG plate displacements actually exceeded instrument ranges at some locations along the short side of Sample 1 during repeated tests. The IG seal of Sample 1, however, showed similar deformed shapes between tests at 4.79 kPa (100 psf) applied pressures.

Sample 5 has the same structural seal design as Sample 1. Thus, researchers decided to reduce applied pressures of Sample 5 to 2.39 kPa (50 psf) since Sample 1 data proved non-repeatable. When Sample 5 data proved repeatable at 2.39 kPa (50 psf) applied pressures, researchers then tested Sample 5 once at applied pressures of 3.59 kPa (75 psf) and 4.79 kPa (100 psf). Small differences result when comparing seal strains of Sample 5 at pressures of 4.79 kPa (100 psf) to seal strains of the first test of Sample 1 at pressures of 4.6 kPa (96 psf).

CONCLUSIONS

Researchers at Texas Tech University applied simulated wind pressures on eight test samples representing four-sided structurally glazed IG units. The goal was to measure changes in the shapes of the structural seal and the IG seal of these eight test samples. The purpose of these tests is to provide data for determining the distribution of seal strains.

Test results show that structural seals undergo mostly normal strains due to out-of-plane displacement and rotation of the inner IG plate edge. Changes in IG seal shapes, however, result mostly from the shearing action of rotating IG plates.

Structural seal shapes vary unevenly from one location to the next location along both sides of a test sample. Maximum normal strains in a structural seal do not always occur at the corner or at the center of either side of a test sample. Maximum normal strains often occur between the center and the corner along the side of a test sample. Normal strain in a structural seal also exists at the corners of a test sample. Test results show that normal strain distribution of a structural seal along the side of a test sample is not trapezoidal in shape.

Results from the research work described in this paper hopefully will improve understandings of the response of four-sided structurally glazed IG units to wind pressures. This research work, however, does not address all concerns involving structural glazing systems [6]. Other research topics include effects of dynamic loads and extreme temperatures on deformed seal shapes and full-scale failure tests.

Structural glazing provides a new design concept to the architect. Glass makers, sealant suppliers, engineers, and contractors need to improve their understandings of structural glazing response to wind pressures. Improved understandings will help assure owners, tenants, and the public of the soundness of structurally glazed buildings.

ACKNOWLEDGEMENTS

The authors thank the National Science Foundation (Grant ECE-8500620), Washington, D.C., Cardinal IG, Inc., Minneapolis, Minnesota, and Olden and Co., Dallas, Texas, for their support. The authors base this paper on research completed at the Glass Research and Testing Laboratory of Texas Tech University.

REFERENCES

[1] Swanson, John G., "Structural Glazing Continues to Grow in the Market Place," Glass Digest, August, 1987, p. 108.

[2] Farrell, John and Schoenherr, Bill, "Silicone Structural Glazing With Insulating Glass," Glass Magazine, November, 1985, p. 86.

[3] Bailey, James R., Minor, Joseph E., and Tock, Richard W., "Experiments Involving Structurally Glazed Insulating Glass Units," Glass Research and Testing Laboratory, Texas Tech University, January, 1989.

[4] Chou, G. David, Vallabhan, C.V.G., and Minor, Joseph E., "A Mathematical Model for Structurally Glazed Insulating Glass Units," Glass Research and Testing Laboratory, Texas Tech University, April, 1987.

[5] Tock, Richard W., Dinivahi, M.V.R.N., and Chew, Choon Hoi, "Visco-elastic Properties of Structural Silicones," Advances in Polymer Technology, Vol. 8, No. 3, pp 317-324.

[6] Bailey, James R. and Minor, Joseph E., "Structural Glazing Tests Show Wind Pressure Effects, "Glass Digest, October 15, 1989, pp 68-76.

Lawrence D. Carbary

STRUCTURAL SILICONE SEALANT REQUIREMENTS FOR ATTACHING STONE PANELS FOR EXTERIOR APPLICATIONS

REFERENCE: Carbary, L. D., "Structural Silicone Sealant Requirements for Attaching Stone Panels for Exterior Applications," Building Sealants: Materials, Properties, and Performance, ASTM STP 1069, Thomas F. O'Connor, editor, American Society for Testing and Materials, Philadelphia, 1990.

ABSTRACT: This paper attempts to discuss the requirements for an adhesive attachment for stone panels 1.25-3 cm in thickness used in exterior applications. Adhesion requirements are discussed and tensile adhesion testing is suggested through various environments such as heat and humidity cycling, freeze thaw cycling and acid rain exposure. Structural perimeter seals, stiffeners and bedding applications are discussed. ASTM C-880 is discussed with modifications to properly identify if the stone used on a project is proper for the exterior application. Limitations through fire are discussed and recommendations to the design professional are suggested.

KEYWORDS: Structural silicone, freeze thaw test, QUV weatherometer, ASTM C-880. Tensile adhesion, natural stone, thin stone, structural perimeter seal, stiffener, structural bedding.

INTRODUCTION

As thin stone becomes more popular among architects for use as an exterior cladding material the technology for cutting and attaching thin stone will progress. The ASTM Stone Symposium held in New York City in March of 1987 was the first such symposium to discuss thin stone, however most of the discussions were dealing with stone cladding materials down to 3cm (1 1/4") in thickness. Most 3cm (1 1/4") stone has a kerf cut at the perimeter, and supported by metal anchors. Other stone is supported by dovetail anchors, drilled holes with metal pins epoxyed into the stone, and various other means of providing a mechanical anchorage.

Mr. Lawrence D. Carbary is a Technical Service Specialist at Dow Corning Corporation, 3901 S. Saginaw Road, Midland, MI 48686.

This paper intends to address adhesive anchorages on stone measuring from 1.25 - 2cm (1/2" to 3/4") in thickness. Typical mechanical anchorages can not be attached into thin stone panels 1-1.6m (3-5') wide and 1-2m (3-6') tall used in exterior applications. An attempt is made to identify the requirements for a silicone adhesive using the knowledge which has been obtained with the use of structural silicone sealants used to attach glass panels to aluminum and steel frames and the use of weatherseal silicone sealants used to seal natural stones.

Silicone sealants have been used successfully to weatherseal natural stones for better than 20 years. The sealants have retained their elastomeric and adhesive properties through natural weathering as predicted by weatherometer testing.[1] Weatherseals are subjected to higher working stresses and strains through thermal cycling than structural silicone adhesive applications. A silicone sealant used as a weatherseal between large spandrel panels of natural stone can undergo annual movement cycles of up to ± 25% to ± 50% of the original joint width. This depends upon the size of the panels joint width, temperature differentials and if the panels are subject to live load floor slab deflections. Silicone sealants have been extremely successful when used in these applications. Structural silicones in a typical glass to aluminum application are designed such that differential thermal expansion may put the sealant into a 15-25% shear movement and windload forces will not elongate the seal more than 15% depending upon the modulus of the silicone used. In fastening stone adhesively with structural silicone sealants, both of these attributes must be combined. One may ask what is the driving force behind this new technology? Thin stone is less expensive than thick stone. A conversation with a contractor purchasing stone for an aluminum curtainwall yielded a statement that a cost savings of $4/ft.^{2} could be realized by purchasing 2cm (3/4") granite rather than 3cm (1 1/4") granite. Depending on the project windload the weight could be cut by 1/3 and aluminum extrusions could also be down sized to support the lesser load. Yet the problem of attaching the stone and retaining the original specified sizes must be addressed.

Bortz, Erlin and Monk stated that after reviewing forty stone veneer buildings in a decade that "the ability of an anchor to engage its resistance as soon as the lateral load is applied is crucial to thin veneer structural success."[2] Adhesive applications lend themselves to these kinds of resistances. Benevengo referenced sealants used as kerf sealants to prevent freeze thaw damage at the kerf.[3] Gere referenced sealants used as kerf sealants to prevent stress concentrations at the anchor points.[4] It has been the authors experience that kerf sealants are used to hold the stone in place on truss supported stone panel systems not only for live loading on the building, but retaining the panel in its original position during transportation to the jobsite on the back of a truck.

During the last five years structural silicone sealant has been used in combination with mechanical devices to attach natural stone materials to buildings. All to date have been successful. Based upon past performance of weatherseal and recent past performance of actual structural applications, the industry will proceed with this new technology. The industry must move with this new technology with new methods of quality control rather than old methods.

IDENTIFICATIONS OF STRUCTURAL APPLICATIONS

Structural silicone attachment to thin stone for exterior applications must withstand the environmental forces found in typical spandrel areas. Extreme heat and humidity coupled with freeze thaw conditions make this environment the worst possible. The silicone adhesive must be capable of withstanding this environment while maintaining its ability to resist windload, live load deflections and seismic forces. Structural silicones used in glass to metal applications have withstood these forces in spandrel areas since the early 1970's. Concentration on stone testing under spandrel environments will be discussed later.

Aluminum curtainwalls which utilize thin stone provide a natural application of a perimeter seal of structural silicone between stone and aluminum, an application identical to structural attachment of glass. Trapezoidal loading of the stone must be considered for this application. The industry standard equation for predicting the amount of silicone based upon trapezoidal loading theory uses one half the shortest panel side, the projected windload and the sealant design strength to determine the sealant bite.

$$\text{Sealant Bite} = \frac{1/2\ SP \times W}{DS}$$

SP = shortest panel side
W = projected windload
DS = sealant design strength

The sealant design strength has been equal to or less than 137 kPa (20 psi) and should never exceed the design stress as recommended by the manufacturer, however the design engineer should closely examine the stress/strain curves for the particular adhesive intended to be used on actual substrates and determine a lower design stress or accept the manufacturers recommended allowable stress.

Because the stone is opaque, the attachment of a stiffener may be appropriate to reduce the effective panel size or support span. This application will be determined by the structural engineer as 2cm (3/4") stone strength is less than 3cm (1 1/4") stone strength. Preliminary mock-up testing on 2cm (3/4") granite with and without a stiffener showed a higher ultimate stone failure on the sample which used a stiffener when compared to a sample which did not use a stiffener.

Adhered thin veneer panel systems such as the Cygnus system use the structural silicone as a bedding adhesive.[5] Typical applications use dollups of silicone with a prescribed sealant dimension of square inches of silicone per square foot of panel. This sealant dimension is based upon the windload placed upon the panel, the strength of the stone, and a conservative safety factor. Depending on the weight of the stone in question the deadload forces may be supported by the sealant or by a shelf angle. The sealant in all cases carries the windload.

The sealant in the Cygnus application is applied to a galvanized steel deck attached to a steel frame and the stone set into the bed of sealant.[6] The bed of sealant must be at least twice the final sealant thickness so that when the stone is pressed into the bed the silicone sealant will firmly wet out both surfaces. As stone cutting technology will progress, precise stone thickness will be obtained. At this time stone thickness is variable. This additional requirement for the structural silicone in the bedding application must be considered, for all stone thickness tolerances are taken up within the structural silicone. Planning for worst case tolerances is mandatory.

STONE TESTING

Numerous tests can be performed on natural stone to determine if the stone is appropriate as a cladding material. The standard tests include ASTM C-880-89 Flexural Strength of Natural Building Stone, ASTM C-99-87 Modulus of Rupture of Natural Building Stone, ASTM C-97-83 Absorption and Specific Gravity of Natural Building Stone, and ASTM C-170-85 Compression Strength of Natural Building Stone to name a few. Gere reported that ASTM C-97 and ASTM C-99 are the two most important physical properties for which ASTM has established minimum standards.[2] Although ASTM C-880 has no minimum standards, this test with some modifications should be equally if not more important. There are presently no durability tests for stone, but ASTM C-18 is developing one.

As stone is placed in spandrel areas of a building where the environment has extreme heat and humidity along with freeze thaw cycles, the ASTM C-880 can be modified to more closely represent the actual applications.

Flexural strength should be tested according to the ASTM method with representative project stone at the specified project thickness. This testing should include testing dry and wet, to predict strength difference and durability.[4] Testing for flexural strength should be done both from the top and bottom to simulate performance under negative and positive windload. Testing should also be done parallel and perpendicular to the rift [4] or parallel or perpendicular to the saw marks left from slabbing.[7] Other variations may include testing after 300 cycles of freeze thaw as specified by ASTM C-666 if the project is in a freeze thaw environment. Skolnik proposed spraying acid water over samples which represented the acid rain exposure of twenty years and visually inspecting the samples for degradation. [8] An appropriate incorporation of this onto a modified ASTM C-880 test would be to spray the stone samples with this solution to evaluate the flexural strength after acid exposure. Jack Stecich of WJE proposed 300 freeze-thaw cycles in 0.01M sulphorous acid would represent stone degradation in an acid rain environment. [17]

The equation for the calculation of flexural strength is shown below.

$$\Theta = \frac{3WL}{4bd^2} \quad \text{Where}$$

Θ = Flexural Strength MPa (psi)
W = Maximum Load N (lb_f)
L = Span mm (in.)
b = Width mm (in.)
d = Depth mm (in.) [9]

The flexural strengths are calculated for the various above mentioned conditions and the data is analyzed. A safety factor for the stone is chosen based upon the scatter of data and the worst case condition. At that point the attachment method and span can be determined.

As is evident, this is an elaborate testing program with much time and effort devoted to stone selection long before construction of a project begins. This battery of tests will describe the stone and determine the kind and number of anchorages required to support the stone for the application.

SEALANT TESTING

Structural silicone sealant glazing of glass has provided the silicone industry with various test methods to determine the suitability of a product used for the application and a list of requirements needed for structural application. The most controversial test method for determining adhesion is ASTM C-794-80. This peel adhesion test is variable up to ± 100% between laboratories [10] and Klosowski states that it is not a good sealant test.[11] However this test is required in most architectural specifications and sealant manufacturers do perform this test hundreds of times each year in support of structural silicone sealant projects. An appropriate test method would be to utilize the tensile adhesion test [12] with a representative substrate and evaluate the stress strain curves of the samples. A major consulting firm in the U.S. specifies a test which incorporates the actual joint configuration for a job to be pulled to three times the design load after seven days water immersion. Testing takes place at three temperatures, 160°F, 72°F and 0°F after the seven day immersion.

The Cygnus panel system requires a tensile adhesion test on each stone substrate to be used on a project by project basis. The joint design used is as specified in ASTM C-719 and the joints are pulled to destruction after room temperature cure and seven days water immersion. Initial qualifications are completed using 2400 hours in a QUV weatherometer in addition to the room temperature and water immersion testing. Weatherometer results do not show significant degradation in strength and modulus compared to water immersion.

Testing with a medium modulus silicone using the tensile adhesion test procedure gave the following test results. Numbers given are average values of 18 test samples:

Tensile Adhesion Results
Granite To Galvanized Steel
Medium Modulus Silicone Sealant
12.5mm X 12.5mm X 50mm Joints (1/2" X 1/2" X 2")
Pulled at 50mm/min. (2 in./min.)

	Ultimate Stress	Ultimate Strain	Stress @ 25% Strain
28 D/RT Cure	0.500 MPa (73.07 psi)	72.8%	0.323 MPa (47.1 psi)
28 D/RT Cure + 7 D/H_2O	0.433 MPa (63.20 psi)	74.3%	0.269 MPa (39.24 psi)
28 D/RT Cure + 2400 hrs QUV	0.460 MPa (67.19 psi)	77.1%	0.266 MPa (38.88 psi)

Testing with a high modulus silicone sealant using the tensile adhesion test procedure gave the following test results. Results are an average of three samples:

Tensile Adhesion Results
Granite to Galvanized Steel
High Modulus Two Part Silicone Sealant
12.5mm X 12.5mm X 50mm (1/2" X 1/2" X 2")
Pulled at 50mm/min. (2"/min.)

	Ultimate Stress	Ultimate Strain	Stress @ 25% Strain
14 D/RT Cure	1.18 MPa (172.5 psi)	126%	0.440 MPa (64.23 psi)
14 D/RT Cure + 7 D/H_2O	0.856 MPa (125.0 psi)	93.3%	0.374 MPa (54.58 psi)
14 D/RT Cure + 2400 hrs QUV	1.22 MPa (178.6 psi)	64.0%	0.708 MPa (103.3 psi)

The QUV weatherometers used in these studies were cycled for four hours at 40°C and 100% RH to four hours at 60°C with UV light provided by UVB 313 bulbs. Since all materials were opaque, the weatherometer is used as a heat and humidity cyclic tester. This is intended to simulate the hot and humid spandrel environment found in buildings.

Tensile adhesion testing through freeze-thaw is important and has not been performed in the past. Working weatherseals with silicone sealant between stone panels is common in the freeze-thaw environments of the U.S. Testing was performed on joints of both high and medium modulus silicones between galvanized steel and granite. The tensile adhesion test pieces were submitted to Hauser Labs in Boulder, Colorado to be cycled 250 times according to ASTM C-666. The test pieces were constantly immersed in water and cycled between -18°C (0°F) and 5°C (40°F) averaging seven cycles per day. The samples were then allowed to equilibrate for a week at room temperature and the following results were obtained. Results are averages of three samples:

Tensile Adhesion Results
Granite To Galvanized Steel
Medium Modulus Silicone Sealant
12.5mm X 12.5mm X 50mm (1/2" X 1/2" X 2")
Pulled at 50mm/min. (2"/min.)

	Ultimate Stress	Ultimate Strain	Stress @ 25% Strain
Control 3 month/RT	0.550 MPa (80.3 psi)	81%	0.314 MPa (45.8 psi)
After 250 Freeze/Thaw Cycles 7 D/RT	0.512 MPa (74.8 psi)	96.0%	0.247 MPa (36.1 psi)

Tensile Adhesion Results
Granite To Galvanized Steel
High Modulus Two Part Silicone
12.5mm X 12.5mm X 50mm (1/2" X 1/2" X 2")
Pulled At 50mm/min. (2"/min.)

	Ultimate Stress	Ultimate Strain	Stress @ 25% Strain
Control 14 D/RT	1.18 MPa (172.5 psi)	126%	0.440 MPa (64.23 psi)
Control 2400 hrs QUV	1.22 MPa (178.6 psi)	64.0%	0.708 MPa (103.3 psi)
2400 hrs QUV + 250 Freeze Thaw Cycles + 7 D/RT	1.33 MPa (193.8 psi)	61.0%	0.753 MPa (109.9 psi)

These first results show excellent physical property and adhesion retention through 250 cycles of freeze-thaw testing.

The high modulus silicone sealant was removed from the QUV weatherometer for the freeze-thaw test.

As thin stone has been sealed with various silicone sealants the potential for staining on the stone exists. Testing for staining according to ASTM C-510 is a minimum exposure test.[13] A proposed stain test is to make a joint between actual stone substrates according to the ASTM C-719 joint design, let it cure for three weeks and compress it 50% at 70°C (160°F). This will accelerate any fluid migration from the sealant into the stone. However, this test does not address dirt pick up of exterior sealant joints nor does it address water absorption differences in the stone.

A proposed test for acid rain exposure would be much like the test proposed by Skolnik.[8] Tensile adhesion testing to represent job site substrates immersed in acid water for seven days should be adequate. The concentration of the acid should be representative of the project location. Silicone sealants show excellent physical property retention [14] when immersed in weak acids, however the bond strength to substrates which is vulnerable to acid attack should be evaluated on a project by project basis. Past history shows no failure of silicone sealants on stone substrates used in weatherseal applications attributed to acid rain attack.

STRUCTURAL JOINT DESIGN

Structural silicone perimeter seals should follow the same design criteria as was set up for glazing systems previously mentioned. Again the manufacturer-recommended design stress should not be exceeded. The stress strain curves of the sealant should be evaluated to determine the anticipated joint deflection under windload. Smith reported most stone specifications limit the deflection of the stone supporting elements to the lesser of L/360 or 1/8".[15] This adds a new dimension to structural silicone sealant joint design as a design stress may not be as important as a design strain. Structural perimeter seals should be of uniform dimensions for the entire length of the joint as this will provide symmetrical loading pattern. Trapezoidal loading must be considered with perimeter seals.

Flexural strength testing of stone may dictate that a perimeter seal application be accompanied by stiffeners to effectively reduce the panel size and span. The structural joints should be uniform to produce a symmetric loading pattern. What is suggested here is that a continuous silicone sealant attachment around the perimeter of a stone panel with or without stiffeners will produce a continuous symmetric loading pattern and can be substituted for discontinuous metal anchors placed into kerfs at the quarter or fifth points. Bortz et al report that as stone materials are brittle in nature they are very sensitive to stress concentration in kerfs or other reentrant corners.[2] Adhesive attachment with a continuous flexible adhesive will minimize stress concentration and maximize stress distribution.

Structural bedding applications can be linked to the anchor pullout tests reported by Kafarowski.[16] U type pins are conventional anchors in precast concrete which attach stone to the concrete panel and are placed every 0.19 - 0.28 m^2 (2 - 3 $ft.^2$). Pullout loads for this type of anchorage showed 318 Kg (700 lbs.) with epoxy and 136 Kg (300 lbs.) without epoxy. Although strain is not reported, an aluminum pad could be attached to the backside of the stone with a silicone adhesive to provide the necessary stress strain characteristics found in this type of anchorage. When the requirements are identified for anchorages, structural adhesives can be designed to provide the necessary strength required while distributing the stress over a larger area.

An anchor which provides a 318 Kg (700 lb.) stress at a strain of 3 mm (1/8") could be designed with an adhesive. An adhesive with a modulus characteristic of 514 kPa (75 psi) at 50% strain could be used in a dimension of 78 mm X 78 mm X 6 mm (3 1/16" X 3 1/16" X 1/4") to meet the above requirement. The stress is distributed over 60.2 cm^2 (9 1/3 in^2) rather than in a single 5mm (3/16") diameter pin. Drilled holes into the stone are not required.

This type of joint must also be designed with the manufacturers design stress in mind. The above adhesive may have a 137 kPa (20 psi) design stress assigned by the manufacturer. Assuming a straight line linear interpolation, a strain of 13% is present at the 137 kPa (20 psi) stress. The wind or seismic load should not subject this adhesive to more than 84.5 Kg (186 lbs.) of load. Adhesive anchorages can be designed to work within the parameters of a given system.

SEALANT APPLICATION

The sealants which have been discussed so far have been silicones. Silicone sealants are virtually unaffected by the temperature and humidity cycles found on the exterior of a building. Silicone sealants remain as flexible structural adhesives and have a twenty year history of success. The excellent physical property retention through weathering and the success of silicone sealant weatherseals on natural stones have provided a path for the use of structural silicone sealants on stone.

Structural silicone sealant is proven successful on glass anodized aluminum and painted aluminum substrates. Vast amounts of literature and thousands of projects globally, too numerous to reference here, are testimony that structural silicone sealant technology is a viable method of glass to aluminum attachment.

Silicone adhesives must be applied to clean, dry, dust free and frost free surfaces. The sealant industry has recommended the two rag solvent wipe as a surface preparation for the glass to metal applications. Natural stone has been found to require water washing with high pressure water or bristle brushes to adequately clean the stone.[6] Slurry from the stone cutting process is not rinsed off of the stone at the cutting site and dries to a hard layer on the unfinished backside of the stone. Silicone adhesives will easily remove this layer of dried slurry at a very low tensile stress when the stone is exposed to a wet condition. An effort must be made to remove this residue before a structural silicone application. Water washing has proven to be the most practical method of removing the residue. However, the panels must dry at room conditions overnight before sealant application. Storage space for the cleaned panels must be planned for during the production of a panel or unitized curtainwall system.

Peel adhesion testing will determine if a primer is required to obtain adhesion to the substrates. This testing should also be conducted at the fabrication site or plant. It is specified and relatively quick and simple. Tensile testing may also be run to verify the stress strain relationship of the system.

Silicone primers are adhesion promoters and should be applied to clean dry surfaces and are not a substitute for cleaning. Stone that is previously cleaned and allowed to collect dust should be recleaned with a solvent wipe to remove and lift any dust before priming.

Perimeter seal and stiffener applications require the sealant to be extruded from a caulking gun. The bite will be predicted by the trapezoidal loading equation. The bite must not be less than the glue thickness and the glue thickness should not be less than 6mm (1/4"). As the structural sealants today are extremely thixotropic and nonslump, a minimum 6mm (1/4") glue thickness allows the sealant to flow into the cavity and fill the void up to 18 mm (3/4") deep. A good rule of thumb is that structural silicone sealant pumped into a cavity will not flow deeper than three times the glue thickness. It is true that four times the glue thickness can be achieved under controlled and supervised conditions, but sealant is wasted and aesthetics can be affected. Joints with an aspect ratio of greater than 4:1 between opaque substrates may not get totally filled as the supervision relaxes.

Sealant application for structural bedding applications must also be quality controlled. Sealant applied to one substrate and the other substrate pressed into place is a blind application. Quality Assurance testing of the proposed sealant beads placed between glass or clear plastic plates is mandatory to monitor joint sizes throughout the project. Substrates must be pressed together before the sealant has a chance to skin over.

FIELD TESTING/VERIFICATION

The project which uses structural silicone sealant attachment to natural building stone must have the adhesive application documented throughout the project. Silicone sealants will readily bond to themselves. A cut out of silicone sealant to determine the adhesion characteristics of a specific joint seal can be repaired. As a check for adhesion, a simple hand pull test may be run after the sealant is fully cured (usually within 14 to 21 days). The hand pull test procedure is as follows:

> Make a knife cut horizontally from one side of the joint to the other.

Make two vertical cuts (from the horizontal cut) approximately 2" long, at the sides of the joint.

Grasp the 2" piece of sealant firmly and pull at a 90° angle or more.

If adhesion is acceptable, the sealant should tear cohesively in itself before releasing adhesively from the substrate.

Sealant may be easily replaced in the test area, by merely applying more sealant in the same manner as it was originally installed (assuming good adhesion was obtained). Care should be taken to assure that the new sealant is in contact with the original, and that the original sealant surfaces are clean, so that good bond between the new and old sealant will be obtained.

An adhesion test should be performed for each 1/2 day of production or a specified amount of linear feet of applied adhesive. Documentation must include the date of application, substrate, batch of sealant, location on the project, and names of people responsible for the application. This kind of testing is readily performed on perimeter seal and stiffener applications.

Structural bedding applications may not allow for the above type of destructive testing. A loading performance test may be more appropriate, or fabrication of representative samples prepared specifically for load testing. Suggested loading is 1.5 to 2 times the design load. Load testing may be accomplished by placing a suction cup or series of cups on the panel and pulling on the cups with a spring scale to the specified force per unit area. A project with a 1' X 1' (0.3m X 0.3m) panel under a 25 psf (122.Kg/m^2) specified windload could be checked with a suction cup pulling 50 lbs. on that tile. Small prefabricated representative samples could be adhered together and pulled up to 1.5 to 2.0 times design load as part of the quality process. The idea is to verify that the adhesive performs beyond the requirements for the project. Again documentation throughout the entire project is mandatory.

The purpose of the documentation is to provide a map of the project which identifies sealant, substrate, production date and crew for the structural sealant application. If a problem were ever to arise, a systematic approach of tracking the problem could be used. The success of structural silicone application to natural building stone is contingent upon the quality of workmanship.

LIMITATIONS

The structural silicone sealants mentioned in this paper have excellent physical property and adhesive property retention with time in the environments found in a building structure. Although skepticism can always be found when discussing adhesive anchorage advantages over mechanical anchorages, mechanical anchorages are not always fool proof as noted by Bortz et al.[2] However, the adhesives discussed here do not retain their physical properties through fire exposure. Building code officials place a large emphasis on systems performance after fire exposure as public safety is part of their charter. A high rise building using nothing but adhesive attachment may not become a reality but an adhesive attachment with a few mechanical anchorages used for panel support in case of fire may be a more realistic approach.

While combinations of mechanical and adhesive attachment may allow a greater intuitive comfort with the sytem, one must realize that incorporating a rigid mechanical attachment into a rigid system sacrifices seismic performance. Displacements due to seismic loads can be designed into structural silicone sealants. If fire clips are designed into a system they should not engage the panel until long after the sealant has surpassed its design strains.

Structural adhesives should not be used beyond the manufacturer recommended design stresses. The future may bring standardization to silicone sealants. Until a global standardization and agreement on test methods occurs, the individual manufacturers are the current sources of recommendations for their products.

As mentioned previously, workmanship is crucial to the success of adhesive attachment used with natural stone. Proper supervision and quality control is mandatory. Although a comparison of field application and shop application has not been discussed, many feel that shop application is the only proper method to utilize for structural silicone sealants.

CONCLUSIONS AND RECOMMENDATIONS

This paper has discussed the various types of applications for structural silicone sealants for natural building stone, stone testing, sealant testing, joint design considerations, sealant application considerations, field testing and documentation. Not all stones will be appropriate for exterior applications nor will all structural silicone sealants be appropriate for all stones.

It has been the author's experience that stone is chosen and tested at the building owner and architectural level approximately one year before a contractor begins erection. During this period the battery of ASTM tests are run and the thickness, method of attachment and span dimensions are determined for the chosen stone. During this testing period is when the adhesive should be evaluated. Tensile adhesion tests should be run along side with all stone property tests to observe the physical properties of the adhesive. Staining and dirt pickup evaluations could also be observed. Tensile adhesion testing should be run dry, after water immersion, after freeze-thaw, and after acid exposure. The stress strain characteristics of the adhesive should be evaluated as the physical properties of the stone are being developed. Additional samples could be fabricated to study long term tensile effects of the adhesive under actual aging conditions for additional safeguards.

Granite has been the most common stone tested and used. Some soft limestones show poor peel or tensile adhesion results when measured by cohesive failure of sealant at break. Other stones have not been tested with this new concept. It is crucial to the success of a project designed with adhesive attachment of stone that the adhesive/stone combination is proper. The adhesive/stone combination should be tested and verified before stone is purchased. All parties involved will find themselves in an awkward position if the project stone is on site with no attachment method.

Structural silicone sealant attachment clearly has certain advantages over conventional attachments used with natural building stone. One is uniform distribution of stress over a large area of stone which will minimize the potential for cracking and breaking associated with concentrated stresses in thin stone. Structural silicone sealants are proven performers in the field of structural attachment of glass and metal. A methodic scientific approach can be taken to realize the full potential of this new technology.

FUTURE WORK

At the time of this writing a mock-up is being built to evaluate various modulus silicone sealants in a perimeter seal structural application and stiffener application on stone. Results are to be evaluated after initial cure and two years in a northern freeze-thaw climate.

Continuous tensile adhesion testing on various granites and limestones is commencing with emphasis on physical property determination of the adhesive after accelerated weathering. A further understanding of natural building stone for this application will continue as long as this type of technology can provide advantages over conventional attachment.

REFERENCES

[1] Hilliard J. R., Parise, C. J., and Peterson, C. O. Jr., "Structural Sealant Glazing," Sealant Technology In Glazing Systems, ASTM STP 638, American Society For Testing And Materials, Philadelphia, 1977, pp 67-99.

[2] Bortz, S. A., Erlin, B., and Monk, C. B., Jr., "Some Field Problems With Thin Veneer Building Stones," New Stone Technology, Design, And Construction For Exterior Wall Systems, ASTM STP 996, B. Donaldson, Ed., American Society For Testing And Materials, Philadelphia, 1988, pp 11-31.

[3] Benovengo, E.A. Jr., "Design Criteria For This Stone/Truss Wall System," New Stone Technology, Design And Construction For Exterior Wall Systems", ASTM STP 996, B. Donaldson Ed., American Society For Testing And Materials, Philadelphia, 1988, pp 57-65.

[4] Gere, A. S., "Design Considerations For Using Stone Veneer On High Rise Buildings", New Stone Technology, Design And Construction For Exterior Wall Systems", ASTM STP 996, B. Donaldson Ed., American Society For Testing And Materials, Philadelphia, 1988, pp 32-46.

[5] Loper, W., and Obermeier, T., "Thin Stone Veneers - A Steel/Silicone Diaphragm System", New Stone Technology, Design,

And Construction For Exterior Wall System, ASTM STP 966, B. Donaldson, Ed. American Society For Testing And Materials Philadelphia, 1988, pp 137-140.

[6] Carbary, L. D. and Schoenherr, W. J., "Structural Silicone Sealants Used To Adhere Stone Panels On Exterior Building Facades", New Stone Technology, Design, And Construction For Exterior Wall Systems, ASTM STP 966, B. Donaldson, Ed., American Society For Testing And Materials, Philadelphia, 1988, pp 160-165.

[7] Heinlein, M., "Selection, Purchase, And Delivery Of Building Stone - The Obstacle Course," New Stone Technology, Design, And Construction For Exterior Wall Systems, ASTM STP 966, B. Donaldson, Ed., American Society For Testing And Materials, Philadelphia, 1988, pp 47-53.

[8] Skolnik, A. D., "Testing For Acid Rain", Progressive Architecture, July 1983.

[9] ASTM Standard Test Method For Flexural Strength Of Natural Building Stone C-880.

[10] ASTM Standard Test Method For Adhesion-in-Peel Of Elastomeric Joint Sealants C 794-80.

[11] Klosowski, J. M., "Sealants In Construction", Marcel Dekker Inc., New York, NY, 1989 pg 99.

[12] Klosowski, J. M., "Sealants In Construction", Marcel Dekker Inc., New York, NY 1989, pg 107.

[13] Klosowski, J. M., "Sealants In Construction", Marcel Dekker Inc., New York, NY 1989, pg 106.

[14] Fluid Resistance Guide To Silastic Silicone Rubber. Published by Dow Corning Corporation.

[15] Smith, G. H., "Exterior Wall Systems Performance And Design Criteria: Should These Vary With Different Types Of Cladding Systems - Glass Fiber Reinforced Cement, Stone, Metal, Glass Panel, Frame, Or Veneer?" New Stone Technology, Design And Construction For Exterior Wall Systems, ASTM STP 996, B. Donaldson, Ed., American Society For Testing And Materials, Philadelphia, 1988, pp 155-159.

[16] Kafarowski, G., "Stone On Precast Concrete Or Steel In Wall Design And Construction". New Stone Technology, Design, And Construction For Exterior Wall Systems, ASTM STP 996, B. Donaldson, Ed., American Society For Testing And Materials, Philadelphia, 1988, pp 105-118.

[17] Stecich, Jack. Paper presented at the 1989 Fall Symposium. "Exterior Claddings on High Rise Buildings", sponsored by Chicago Committee on High Rise Buildings, November 13, 1989.

Jean Iker and Andreas T. Wolf

COMPARISON OF EUROPEAN AND U.S. TESTING/QUALIFICATION PROCEDURES FOR STRUCTURAL GLAZING SILICONE SEALANTS

REFERENCE: Iker, J., Wolf, A.T. "Comparison of European and U.S. Testing/Qualification Procedures for Structural Glazing Silicone Sealants", Symposium on Building Sealants: Materials Properties and Performance, ASTM STP 1069, Thomas F. O'Connor, editor, American Society for Testing and Materials, Philadelphia, 1990

ABSTRACT: The paper reports on the state-of-the-art test methods and qualification procedures for SSG sealants currently applied in Europe and compares those to the ASTM C-24 proposal. The test data generated show that the 50°C water immersion required by some European building code authorities results in a much lower success rate in the adhesion test than the 1 week room temperature immersion applied by the silicone sealant manufacturers in the U.S. The paper also describes the efforts to consolidate the different national test and requirement standards into one European standard and directive on structural glazing by 1992.

KEY WORDS: Structural Glazing, CEN, UEATC, ASTM C-24, AVIS Technique, standardisation work, adhesion test

INTRODUCTION

Structural Glazing is a method of attaching glass, metal, or stone panels to the curtain wall of a building using a silicone sealant/adhesive to hold the panels in place. Although the Structural Glazing concept has been in existence in the United States for more than 20 years, interest in this type of construction had not grown dramatically until the 1980's.

Ing. Iker is Technical Service Specialist and Dr. Wolf is Technical Service & Development Section Manager at Dow Corning S.A., Parc Industriel, 6198 Seneffe, Belgium

Structural Glazing has changed the face of U.S. cities during the last 10 - 15 years and the same trend can be observed in Europe since the introduction of Structural Glazing there 5-6 years ago. It has been estimated [1] that in 1987 25 to 40% of all new glass curtain wall construction in the USA has been structurally glazed. In Europe, the growth potential of this glazing technique is still enormous, as architects only consider Structural Glazing for approx. 5 to 10% of all new glass curtain walls. It may well be that in 5-8 years time, Structural Glazing will become the fastest growing glazing system in Europe, as has already been reported for the USA in 1987 [2].

Although Structural Glazing is considered for very prestigious projects through-out Europe, the level of acceptance of this glazing technique with the building code authorities of the various European countries differs widely. It is therefore essential that the technical expertise in implementing this glazing method keeps pace with the market development to ensure the highest quality structural glazing systems.

NEED FOR STANDARDISATION

Structural Glazing, being an innovative construction technique, is still very often regarded as a dangerous technique in the sense that a failure of the silicone adhesive may cause damage to goods and people. It is of the utmost importance therefore to understand and control the risks involved in this application and to limit them to an acceptable level. The way this control is implemented in today's industries is usually via norms and standards. The glass and sealant manufacturers have already demonstrated the benefit of such an approach in a related application, the insulating glass manufacture. Through development and application of standard test methods, the quality - and simultaneously the performance - of insulating glass units has significantly improved over the last 10 to 20 years in the United States and in Europe. The improved quality resulted in higher acceptance levels of insulating glass with architects, specifiers and building owners. Using a similar approach for Structural Glazing therefore looks quite promising, and this is what is on the way on both sides of the Atlantic Ocean.

A closer look at the application is necessary to effectively analyse what can be standardised and what is difficult to cover in a standard. During the erection of a structural sealant glazing project, three phases can be identified, which require our attention:

(a) the pre-production phase, which consists of designing a system capable to perform under the known service conditions

(b) the material choice in terms of glass, aluminium surface finish, and the silicone structural glazing sealants

(c) the production of the SSG modules in-shop and their installation on site

The adhesion of the selected structural glazing sealant is the key to the success of the project, even if the value of the sealant represents only a small percentage of the value of the total facade, as it controls the integrity of the whole project.

Contrary to the material properties of the sealant, which today can be easily tailored for the different applications, adhesion is a complex science and not much is really understood in terms of its mechanisms and controlling parameters. It is therefore very difficult to design a sealant for good and long-lasting adhesion to <u>all</u> construction substrates. This is what makes the adhesion test such an important feature in the structural glazing project approval procedure.

Standard test procedures to evaluate the adhesion of sealants to substrates exist in any European country. These tests were initially designed for standard sealant applications. If adhesion represents such a critical element in the structural glazing application, it seems logical to set higher requirements in the existing adhesion standards to satisfy the need for a higher safety level. One way to achieve this would be to increase the severity of the condition method that the adhesion test specimens have to undergo by increasing the temperature, humidity, level of UV exposure or just the duration of the condition method itself. Furthermore, the criteria for passing the test can be changed. The requirement standard, instead of tolerating a 10 or 20% loss of adhesion, can demand 100% cohesive failure only.

Although there are infinite ways to increase the severity of the adhesion tests, only a few of them will give a good correlation with the practical experience in SSG projects. For countries, where sufficient experience of the structural glazing technique is available, a group of experts should be able to define a set of adhesion and other performance standards to ensure the high safety margin required for this application.

DEVELOPMENT OF EUROPEAN STANDARDS FOR STRUCTURAL GLAZING

In Europe, Structural Glazing was introduced to the market only 5 to 6 years ago, and is therefore still regarded as a new technique. The building code authorities allow the construction of structurally glazed curtain wall facades on an "exceptional" basis; currently there is no "general approval" of this technique. The exceptional approval requires a project-by-project approval of the facade by the building code authorities. Once a general approval has been issued for a construction technique, the curtain wall company is free to design and construct facades, as long as they stay within certain design limits. Obtaining a general building code approval in Europe may take decades, as can be illustrated by the example of heat tempered glass, which still has not received a general approval in Germany for use in curtain walls since its introduction in the mid 1960's. Some countries have started to work on test and requirement standards, which will form the basis of a future general approval. The country that has made the furthest progress on this is France, which has set up an "AVIS Technique" for structural glazing, which will be discussed in some detail later.

On a European basis, two institutions play a key role in developing technical requirements for structural glazing: The first one is the UEATC, the "Union Européenne pour l'Agrément Technique dans la Construction". The UEATC is an association of institutes that issues "agréments" (approvals) in the Western European countries. An "agrément" is a favourable assessment of fitness for purpose for construction products, processes or equipment, with the aim of encouraging the mutual recognition of "agréments" amongst the UEATC members. The second body is the CEN, the European Committee for Standardisation, a body that consists of the 16 national standardisation institutions of the European Community and of the European Free Trade Association, with the main objective of preparing standards contributing to European harmonisation.

Whilst CEN is principally concerned with established construction processes and products and their use, the UEATC is oriented more towards innovative construction processes and products, whilst accepting that the innovative products of today are the established products of the future. CEN and UEATC have agreed on a certain workload distribution: CEN concentrates primarily, insofar as construction products are concerned, on the development of test methods, specifications and certification for existing products, while UEATC is involved with the same aspects of recently introduced products, and also systematically deals with the constructions in which these products will be used. The activities of CEN and UEATC are therefore far from being at variance with one another. They are in actual fact complementary and both institutions will continue to work towards an enhancement of their collaboration.

In order to avoid any misunderstanding which may arise from the fact that both organisations are working in the field of construction products, they have adopted the following resolutions:

* The UEATC will not become involved with assessing products for which standardisation is in progress.
* CEN recognises that innovative products fall within the scope of the UEATC.
* CEN and UEATC will consult one another as to the choice of subjects, which they intend to include in their respective work schedules.
* CEN will actively consider the standardisation test procedures evolved by the UEATC.
* CEN agrees to give preference to the UEATC directive, whenever relevant, as a basis for the formulation of European standards.

In summary, the UEATC will primarily deal with innovative techniques for which normalisation is in progress and the document issued by UEATC will be considered as a pre-standard and used as working document in the CEN activities.

With regard to structural glazing, the UEATC has been working on a directive for the last 3 years. The draft document (Directives UEATC de base pour l'agrément des vitrages extérieurs collés) has already been reviewed in two public meetings of the general assembly and the final directive is to be published before the end of 1990. The UEATC document highlights the general considerations necessary for controlling

Structural Sealant Glazing production. The basic outline of this document is as follows:

* General introduction describing the application and the position of UEATC versus innovative techniques.
* Quality in terms of safety, habitability, durability, maintenance, use and production.
* Performance evaluation as far as tests and specifications are concerned.
* Quality assurance and follow-up.
* Content of the "Agrément".

Within the CEN Technical Committee 129, Working Group 16 has the charter to develop a European standard for structural glazing. As the UEATC directive is not yet available, the CEN129/WG16 has started its work based on a different document prepared by the French. This document essentially comprises the tests developed for the "AVIS Technique" by the French test institute CSTB (Centre Scientifique et Technique du Bâtiment : Guide pour la présentation des demandes d'avis techniques vitrages extérieurs collés.). During the last few meetings of the CEN129/WG16 considerable discussion has arisen on the question, how "innovative" Structural Glazing still is in Europe, and whether it should currently only be dealt with as a UEATC directive instead of preparation for a European standard. Germany and Belgium questioned the wisdom of standardising a technique on a European level, which has not yet even spread to all the European countries. The German delegation insisted that more experience should be gained on this application, before the state-of-the-art expertise could be cast into an European standard. The decision on whether to continue or to hold the development of a CEN standard on structural glazing can only be taken at the European Commission level. As both the UEATC directive and a potential future CEN standard will be strongly influenced by the test procedures originally developed by the CSTB for the AVIS Technique qualification, they shall be compared to the draft documents issued by the ASTM Committee C-24.

COMPARISON OF FRENCH AVIS TECHNIQUE AND ASTM C-24 DRAFT DOCUMENTS

One fundamental difference between both papers is that the ASTM proposal fixes limits whilst the French AVIS technique does not. Some of the limits in the ASTM proposal being : a sealant should not sag by more than 4.8 mm, sealants should have a Shore A hardness between 20 and 60°, a sealant should not lose more than 10 % of its original weight or show any cracking when tested in accordance with C 792, the tack free time of a sealant should be less than 3 hours. Furthermore, for tensile/adhesion samples, aged or unaged, each one must have a minimum average tensile strength of 50 PSI at break with a minimum average cohesive failure of 90 %. Due to their lack of experience, the French authorities are not setting limits on the material properties of sealants but every set of results for a given sealant which is submitted to the AVIS Technique Committee is judged separately and its suitability for the desired application evaluated. Both papers split into a section on material characterisation, suitability for the application and durability requirements. In the product characterisation section, a minimum extrusion rate is required by ASTM, while this property is not mentioned by the CSTB. On the other hand, CSTB imposes a thermogravimetric analysis, while ASTM does not. The CSTB requirements are more demanding in the suitability for application section as compression, elastic recovery and shear tests in 2 directions are all required. The durability of the sealant's mechanical and adhesion properties is - of course - the main concern. A good agreement between both documents can be found on the durability testing (see Table 1). The main discrepancy concerns the influence of high temperatures: while ASTM requires only 1 hour at 88° C, the CSTB asks for 7 days 100° C storage. In the French recommendations, tensile/adhesion samples are aged for 7 days at 100° C in a dry environment (oven) and then tested to destruction after 2 days at 25° C and 50 % relative humidity reconditioning.

TABLE 1 -- Comparison of test parameters in CSTB test procedure and in ASTM C24 proposal

Test Parameter	ASTM C24	CSTB
Temperature	1 hour at 88° C 1 hour at -29° C (tested at storage temperature)	7 days at 100° C
Water Immersion	7 days room temp. max. 10 mins. drying time	7 days room temp. 2 days drying time
UV and Humidity	5000 hours ASTM G53	4000 hours NFP 85-516
Static Load	...	0.14 N/mm2 at 50° C

EVALUATION OF THE DIFFERENT ADHESION TEST REQUIREMENTS

As mentioned earlier, a durable adhesion to the curtain wall substrate is one of the key properties of a structural glazing sealant. For a given substrate, let us say anodised aluminium, strong variations of sealant adhesion occur from one anodisation to another. Even within one anodisation type, differences in adhesion occur, depending on the type and the age of the sealing bath. This makes it impossible to qualify substrates as such. Instead a very tedious project-by-project control procedure is required. For larger projects, which are erected over a longer period of time, variances of the quality of the substrate - and therefore the sealant adhesion to this substrate - may occur even within one project. For anodised aluminium, as an example, this may be due to the fact that different lots of the same anodisation type were used on the project or that all the anodised aluminium was derived from the same lot, however, the surface properties of this anodisation may show changes with ageing. To cope with situations like these, constant monitoring of the quality of the substrates is required for large structural glazing projects, besides the initial first-time approval of the substrate.

The control of adhesion on a project-by-project basis is not directly required in any of the current standards or draft proposals, but it is the key to the success of the structural glazing technique. Around the world experts are aware of the necessity of project-by-project testing, however, the way it is being carried out differs strongly from country to country. There are basically two ways to evaluate the adhesion of a sealant to a substrate: one is the peel test, as documented for instance in ASTM C-794, the other is a tensile adhesion test, as described in many national and international standards, such as ISO8339, NFP 83-504, etc. Both methods of evaluating the adhesion of a sealant lead to different results, even if all other test parameters are kept constant. The other parameter that will influence the results of the adhesion test strongly, is the conditioning method that is imposed on the specimen prior to testing. In the literature, different conditioning methods are described: water immersion, high humidity environment, and combination of both with UV exposure or high temperature storage. Water immersion is by far the simplest and most severe way of conditioning peel or tensile adhesion test samples.

The test method generally chosen by the U.S. structural glazing silicone sealant suppliers is a peel adhesion test according to ASTM C-794 carried out after immersing the cured samples for 1 week at room temperature. This test method was introduced to the different European countries 5 to 6 years ago as part of the technical support of our structural glazing sealant commercialisation program. Several European countries were concerned that this test method was not severe enough and the building code authorities required more stringent test methods. We have therefore over the last 3 years added one and two weeks water immersion at 50° C to our project approval test procedures.

In France, the controlling bodies were concerned about the poor reproducibility of the force values obtained in the peel test. This phenomenon had been observed by many other companies and discussed in the various national standardisation bodies. While the failure mode turned out to be very reproducible in the peel test, the force values showed a strong variation. Some countries considered the failure mode the only relevant outcome of the peel test and the requirements on the mechanical properties of the sealant were based on a separate tensile test. France, however, required the tensile adhesion test as its only criterion for project approval. For each project, 9 test specimens have to be prepared for each substrate. All specimens are stored in a standard 23° C / 50% rel. humidity climate for 1 week in the case of two-component sealants and 3 weeks in the case of one-component sealants to ensure proper cure. 3 specimens are tested without any additional conditioning and serve as references. 3 further specimens are tested after 7 days room temperature water immersion and 2 days reconditioning in an indoor climate. The last 3 specimens are tested after 7 days dry 100° C storage and 2 days reconditioning at room climate. The last test is designed to check whether any coating or anodisation additive is migrating to the bond line and is impairing adhesion.

In order that the differences between the various test procedures are more fully understood, we would like to discuss the results obtained in our laboratory for a neutral, two-part structural glazing silicone sealant on different anodised and coated aluminium. During the years 1987 and 1988, 369 different aluminium samples were tested according to the ASTM C-794 peel test method after three different conditioning methods: 1 weeks water immersion at room temperature and 1 week and 2 weeks water immersion at 50° C. The samples were prepared using 5 different cleaning methods, and the sealant was applied both with and without primer. The results obtained are displayed in Fig. 1.

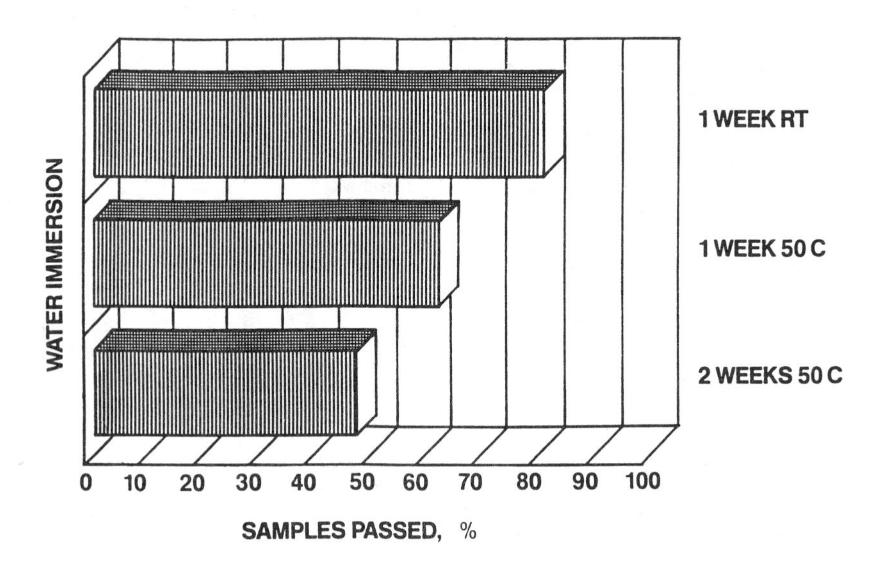

FIG. 1 -- Peel adhesion results on 369 various aluminium substrates tested in 1987/88

As expected, the success rate of samples passing the peel adhesion test (100 % cohesive failure) degrades rapidly with the severity of the conditioning method. While 83% of all aluminium types passed the 1 week RT water immersion test, only 62% and 48% passed the one and two weeks 50° C water immersion, respectively. The same study conducted on 163 samples tested in 1989 showed the same trend, namely 92%, 77% and 60%, respectively (see Fig. 2). If we separate the samples into 2 classes, anodised and coated aluminium, we clearly see that the adhesion success rate is as high as 80% after 2 weeks at 50° C on coated aluminium (see Fig. 3) and as low as 56% and 39%, respectively, on unpigmented and pigmented anodised aluminium (see Fig. 4 and 5) after the same conditioning method.

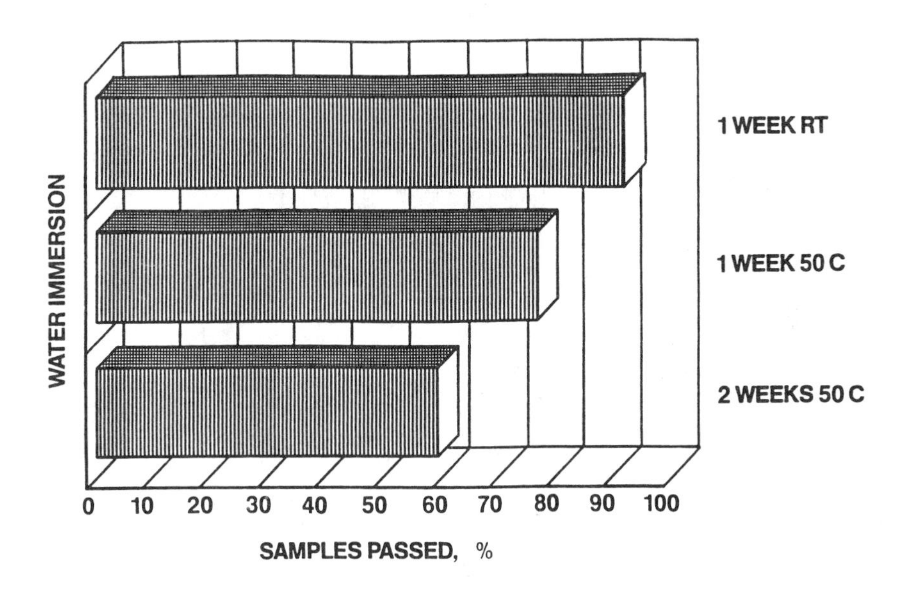

FIG. 2 -- Peel adhesion results on 163 various aluminium substrates tested in 1989

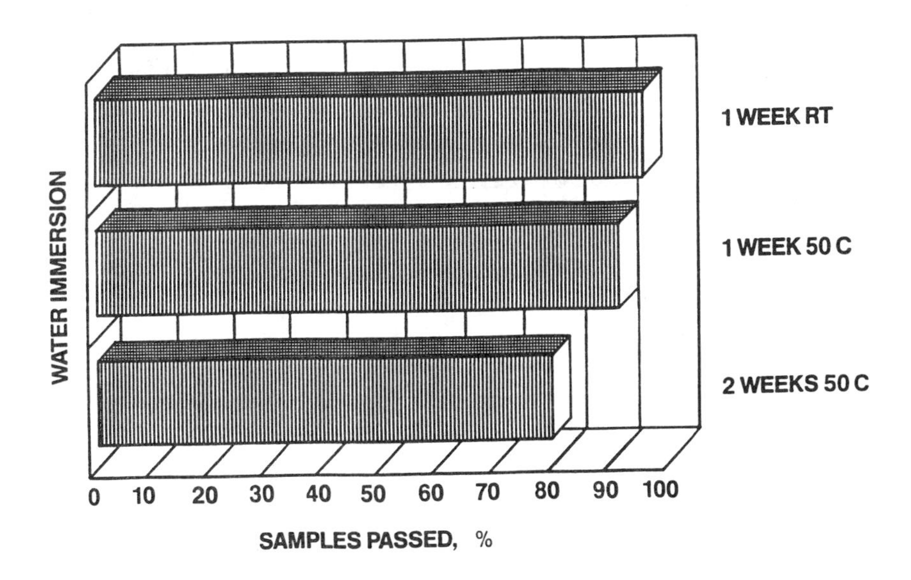

FIG. 3 -- Peel adhesion results on 26 coated aluminium substrates tested in 1987/88

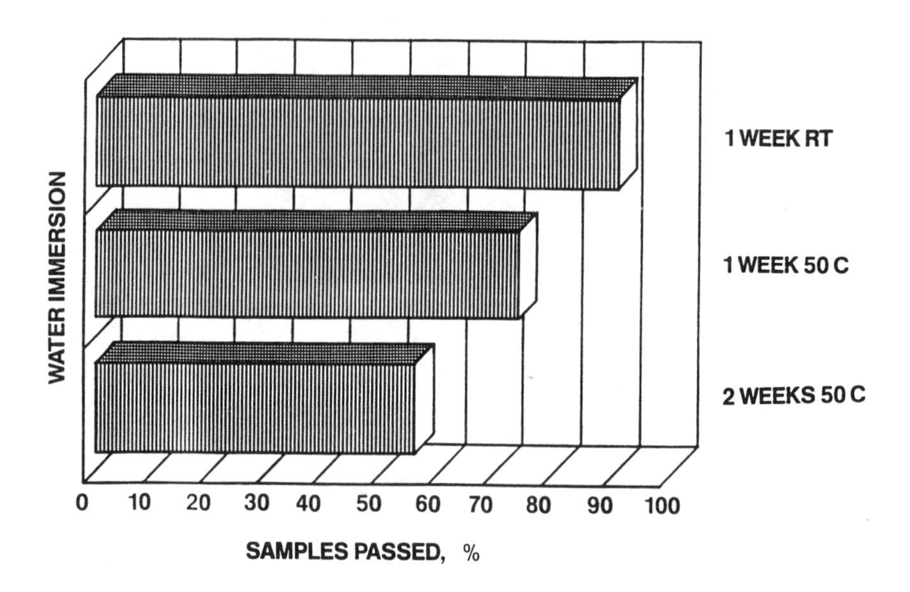

FIG. 4 -- Peel adhesion results on 137 unpigmented anodised aluminium substrates tested in 1987/88

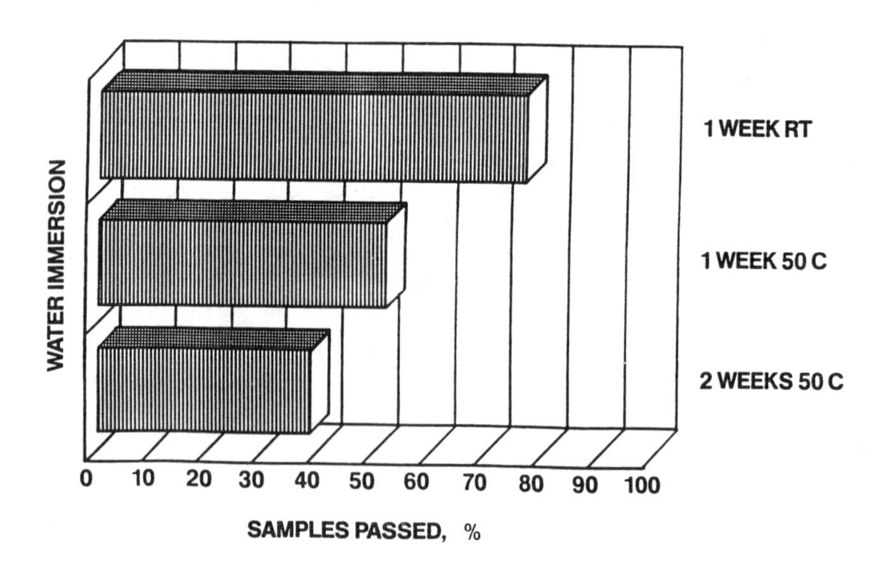

Fig. 5 -- Peel adhesion results on 206 pigmented anodised aluminium substrates in 1987/88

The comparison of the peel test method to the newly imposed French tensile adhesion test produced much confusing data. Samples on which no adhesion was recorded using the peel adhesion test passed the French tensile adhesion test with flying colours. As a simple rule, all samples that passed the 1 week RT water immersion peel test, also passed the French requirements. This can be attributed to the fact that forces at the bond line are lower in the tensile adhesion test samples than in the peel test, due to the dissipation which occurs. Tab. 2 gives a comparison of the failure modes in both test methods.

TABLE 2 -- Comparison of failure modes in peel and tensile adhesion tests for 8 anodisations

Peel Test ASTM C794-80			Tensile Test NFP 85-504/508		
1 week RT water water	2 weeks 50° C water	2 weeks 50° C water	Dry	1 week RT water	1 week 100° C
CF	25% AF	AF	CF	CF	CF
CF	CF	CF	CF	CF	CF
BF	CF	CF	CF	CF	CF
CF	BF	AF	CF	CF	CF
CF	CF	AF	CF	CF	CF
CF	AF	AF	CF	CF	CF
AF	AF	AF	CF	CF	CF
AF	AF	AF	CF	AF	CF

SUMMARY

Structural Sealant Glazing is actively being studied on both sides of the Atlantic Ocean and many standardisation committees are dealing with the necessary and complex task of evaluating the suitability of the sealants for this application.

All standards and proposed regulations include a characterisation of the product, an assessment of the suitability for the application and a durability test on the mechanical and adhesion properties of the sealant. The current test proposals do not include the requirement for a project-by-project test. With the variations in substrate quality observed between different projects or even within one project, we believe that this project approval procedure should become a requirement in future structural glazing standards.

The peel adhesion test is more frequently used in the United States, while European countries are more inclined to the tensile adhesion test. Due to the dissipation of forces occurring in the tensile test sample the stress on the bond line is lower in the tensile test than in the peel test. Water immersion is the simplest and most severe conditioning method, especially if carried out at elevated temperatures. We therefore suggest that a future structural glazing standard should rely on the peel test method after 2 weeks of 50° C water immersion to assess the durability of the adhesion. However, the peel test should only be evaluated for the failure mode. The retention of the mechanical properties of the sealant upon ageing can better be assessed in a tensile test.

ACKNOWLEDGEMENTS

The authors are indebted to Guy Tilmant and Dominique Culot for carrying out the lab peel and tensile adhesion tests.

REFERENCES

[1] Swanson, J.G., "Structural Glazing Continuous to Grow in the Market Place", Glass Digest, Aug. 1987, pp. 108

[2] Sanford, A.G., "High Tech or Black Magic", Glass Magazine, March 1987, pp. 125

Leon S. Jacob and Brad. Johanson

STRUCTURAL GLAZING FROM A CURTAIN WALL DESIGNER'S PERSPECTIVE

REFERENCE: Jacob, L. S. and Johanson, B., **"Structural Glazing from a Curtain Wall Designer's Perspective,"** Building Sealants: Materials, Properties, and Performance, ASTM STP 1069, Thomas F. O'Connor, editor, American Society for Testing and Materials, Philadephia, 1990.

ABSTRACT: This paper endeavours to discuss the various issues that have to be considered by the curtain wall designer. Emphasis has been placed on the relativity and impact of new technology in the use of glass, sealants and other associated materials relative to the function and performance of the curtain wall. The need for acceptance and implementation of the current knowledge and technology is clearly demonstrated. The difference in the level of technology between the Researcher and the Practitioner is clearly identified and widening, specifically in relation to structural silicone design criteria.

KEYWORDS: Structural glazing, silicone, curtain walls, stress analysis, aspect ratio.

INTRODUCTION

Structural glazing may be defined as a system of bonding glass to a building's structural framing, utilising a high strength, high performance silicone sealant designed and tested for this application. This definition identifies the critical ingredients that go to make a sound structural glazing system. What is not specified, but clearly implied, is that Quality Control is necessary to make the concept work.

In this paper it is hoped to review the key issues and factors that confront the designer while working with structural silicone in the design and construction of curtain walls.

Leon Jacob Group Technical Manager, O'Brien Glass Industries, 45 Davies Road, PADSTOW, NSW Australia
Brad Johanson President, Johanson & Associates, 24955 Pacific Coast H'way, MALIBU, California.

Historically, the design and selection of glass type and window systems has been the Architect's responsibility. Structural considerations, though acknowledged, were considered less important due to the relative size of the window and, more importantly, the limited perceived consequences of window glass failure.

Until recently, the design charts used internationally were produced by the primary glass manufacturers.

The advent of structural glazing, the incidence of glass failure through natural and other causes, and the fascination that the Architectural fraternity has for larger windows, has forced glass technology into the 21st century.

Substantial theoretical and experimental work has been carried out at Texas Tech. University which has enlightened the industry on the performance of glass plates. We will consider a few key factors in this paper.

CURRENT LIMITATIONS

Even though structural glazing has caught the imagination of Architects, Builders, and Owners alike, there are some serious pitfalls which exist. These limitations have developed primarily due to the evolution of structural glazing from the old stick and other conventional glazing systems.

It is not sufficient for the industry just to adapt existing technology to suit the current structural glazing systems. It has to incorporate the latest technology, systems and criteria in its designs. For example, the method used in supporting the perimeter of the glass panel is different and will result in different stresses being developed in the plate due to edge constraints.

The method of fixing the unitised curtain wall to the core structure is totally different due to the loading conditions and forces being totally different. These factors must be recognised and incorporated in the design.

More importantly, the use of a sealant to position and restrain the edge of the glass panel under the anticipated loading conditions must be evaluated very critically to ensure that the right sealant in the correct quantity (correct joint dimensions) is used.

PERFORMANCE SPECIFICATION

There are primarily 2 distinct mechanisms used for tendering prior to the construction of a curtain wall.

(1) The pre-designed system.
(2) The performance specification.

With the pre-designed systems, the curtain wall fabricator has already completed the necessary design work, testing and certification to ensure that the curtain wall will perform satisfactorily and in accordance with the stated objectives of existing standards and product specifications.

However, for the situations where a pre-designed system is not specified, the fabricator will have to develop a curtain wall to a set of performance specifications and criteria. For example, generally in Australia, the performance document is used as the nucleus in the development of the total curtain wall system.

The document is primarily aimed at ensuring that designers, contractors, building owners and other interested and involved parties, fully understand the demands for design, waterproofing, Quality Control, specification and future maintenance necessary for the curtain wall.

The document generally includes minimum technology standards that have to be met by the individual components and materials to be used and reference is generally made to existing ASTM and such other standards.

It is essential that one is aware of the following in order to ensure compliance with the performance criteria.

(a) The properties, the quality, and the constraints applicable to materials that are selected.
(b) The work practices that are prevalent in the industry.
(c) The compatibility of the materials that go together to make the curtain wall.

This is a very arduous task which results in the curtain wall designer becoming reliant on the glass maker/supplier, the sealant maker/supplier, and the substrate processing/coating applicator being able to provide him with recent and relevant information.

STRESS ANALYSIS IN GLASS

The growth in the technology and understanding of the behaviour of glass has risen exponentially. Unfortunately, users are lagging the academics. This lag in technology must be taken up by the industry.

Glass panels used in curtain walls are technically thin plates. The deflections experienced under normal design conditions are generally far greater than the thickness of the glass plate.

Consequently, the conventional elastic plate analysis, Timoshenko, S.[1], which primarily predicts bending stresses and deflections proportionately to load, is no longer valid.

In actual fact, "in plane" (membrane) stresses are developed in such panels. This stress may become as large, or larger than, the bending stresses.

The location, the pattern and the magnitude of these stresses tends to move away from the centre of the plate towards the corners as the aspect ratio of the panels tends to 1 (a square plate), Backmenver, R.D.[2].

Texas Tech. University and, more particularly, Vallabhan[3], has developed a unique solution using a finite difference technique to solve the Von Karmon equations that characterise thin plate behaviour.

The International code bodies, such as A.S.T.M., are currently in the process of publishing design charts acknowledging the performance of glass plates in terms of membrane stress development. The significant impact of this acknowledgement is that the industry now has at its use the mechanisms and tools to more accurately determine the appropriate glass thickness relative to the design requirement.

There are numerous computer programmes that have been developed to resolve the large deflection plate equations as defined by Von Karmon. The most popular of them is the finite difference programme developed by Vallabhan.

Features of Membrane Stress in Glass Panels

Aspect Ratio: Aspect ratio is defined as the ratio of the length to the width of the glass panel (b/a - where "b" is the long side and "a" is the short side). It is a key factor in determining the glass thickness for a particular loading condition. The aspect ratio is generally chosen by the Architect and is not subject to change pending strength analysis. Let us consider 3 separate aspect ratios in terms of stress generation and build up.

(a) For a panel with a high aspect ratio (4:1) membrane stresses do not have a significant influence on the stress field. Generally, centre stresses can be approximated by using one-way bending along the "long axis".

(b) For medium aspect ratio (2:1) membrane stresses begin to play an important role. At the centre of the plate tensile membrane stresses increase the total stresses while at the corner of the plate the compressive membrane stress tends to reduce the total stresses. Maximum stresses are of similar magnitude at the centre and the corners.

(c) For plates with low aspect ratio (1:1) the effect is significant on membrane stress. These membrane stresses cause the maximum stress to move towards the corner of the plate.

Thickness: Beason, W.L.[4] proved that the stress fields are different for glass plates with different aspect ratios and thicknesses. The thickness of the glass also contributes to the change in the membrane stresses developed in a plate. The glass designer can, and must, use these variables to the best advantage in designing or, rather, selecting the glass thickness for a particular panel size and loading.

I.G. Units: In a similar fashion, the use of I.G. units for vision panels for curtain walls has also been extensively researched and reasonably well understood. Once again, the tools and the theory are available for the curtain wall designer to optimise the design. Beason W.L., Vallabhan and others.

EDGE REACTIONS

Historically, the industry worldwide has used the trapezoidal method of determining the edge reactions, i.e. the forces that are applied through the edge of the glass panel via the structural silicone on to the supporting frame. It has been argued that the trapezoidal theory is conservative enough for continued use. Minor[5] in his paper talked about edge seal forces. However, Chou, Vallabhan and Minor[6] developed a series of 27 charts that provide the edge forces in glass plates for different operating conditions. A summary of their findings is as follows:-

(a) As the width of a panel is increased, the boundary forces increase at the corners, in other words, as membrane stresses increase the reaction forces at the corners are also correspondingly increased (see figure 1, next page).

(b) The spring constants have a significant influence on the boundary forces. Once again, as membrane stresses are developed, the boundary force increases with any increase in the spring constant (see table 1).

TABLE 1 -- REACTIONS - lb/ins.

Panel / Stiffness	540x360		1080x 720		1620x1080		2160x1440		2700x1800	
	Pos.	Neg.	Pos.	Neg.	Pos.	Neg.	Pos.	Neg.	Pos.	Neg.
K1 = K2 = 300	13	2	42	15	64	25	80	45	100	70
K1=300, K2=2100	17	4	42	20	64	60	80	90	100	120
K1 = 2100 = K2	18	8	42	40	64	80	80	120	100	160

K1 - Stiffness of I.G. units sealant.
K2 - Stiffness of silicone between I.G. units and frame.
Wind pressure - 100 p.s.f.
Aspect ratio - b/a = 1.5

FIGURE 1 -- Forces in Structural Glazing Seals

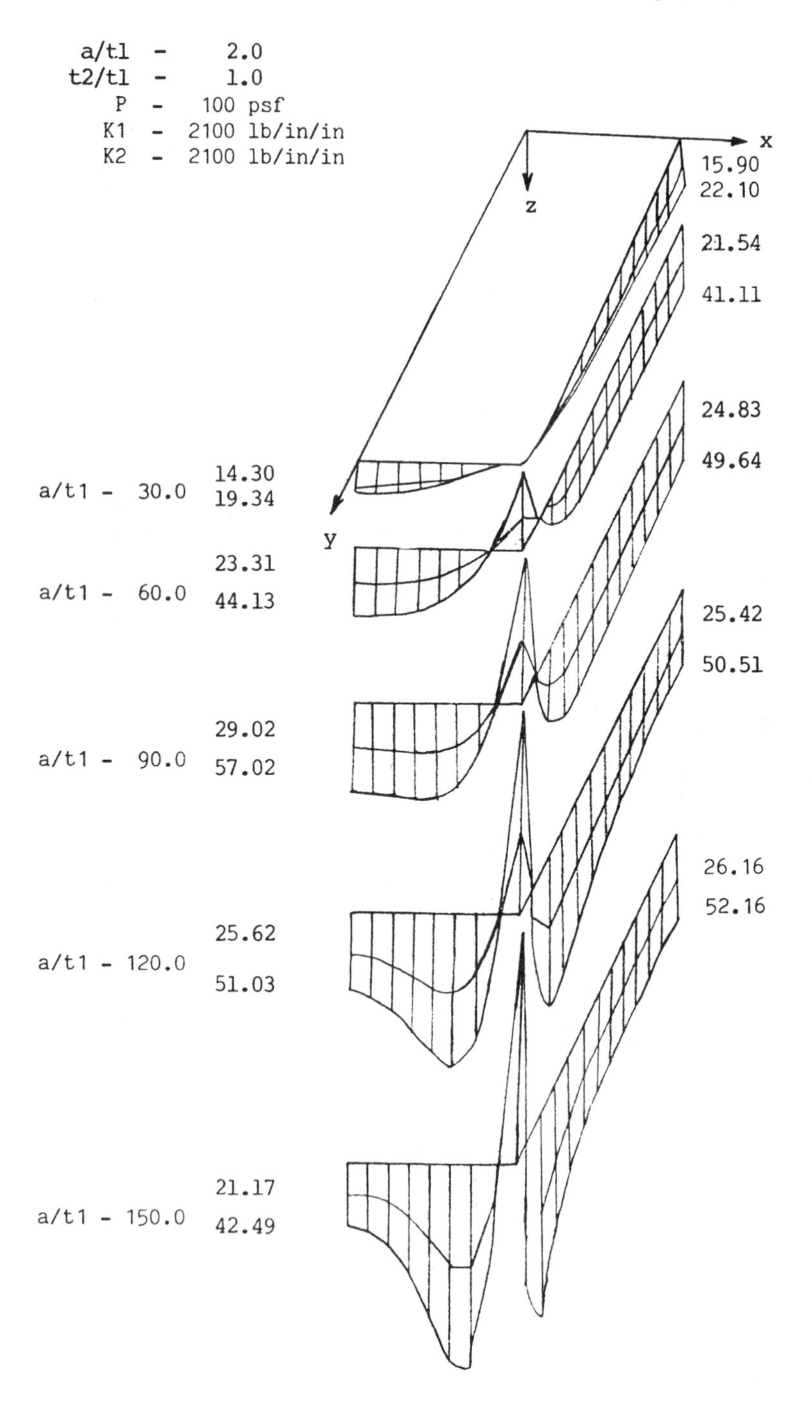

(c) Evaluating figure 1, it can be deduced that as the stresses in the plate change from pure bending (high aspect ratio) to membrane, (low aspect ratio), the location and magnitude of the reaction forces at the edge change significantly.

(d) The boundary reactions for glass plates are as follows.

1. For a/t < 60 - the maximum positive reactions occur at the centre of the boundary.
2. For 60 < a/t < 120 - corner reactions have negative values.
3. For a/t > 120 - maximum positive reactions do not occur at the centre of the boundary. (Here 'a' is half the shorter side and 't' is the thickness.)

Comparison Between the Trapezoidal and the Finite Difference Methods of Determining Edge Reactions

In table 2, a sample comparison is made between the edge reactions expected on a glass plate using the finite difference analysis provided by Chou et al and the conventional trapezoidal method.

TABLE 2 -- Boundary Reactions in lb/ins.

	a/t =30	a/t =60	a/t =90	a/t =120	a/t =150
Centre a	4.84	11.03	14.26	12.76	10.63
Centre b	5.53	10.28	12.25	12.63	13.04
Positive Maximum	4.84	11.03	14.0	22.86	32.86
Corner Negative	1.25	12.14	22.86	31.43	37.14
Trapezoidal	5.12	10.26	15.38	20.52	25.64

b/a = 2.0 (long side/short side)
Design Wind Pressure = 100 psf.
K (stiffness of silicone) = 2100/lb/in/in.
Silicone bite = 1/2 inch.
Glass thickness = 1/4 inch.

It is quite easy to discern that for the larger panels where the membrane stresses dominate, the reactions are substantially higher at the corners (as illustrated in figure 1). Because the edge distance subject to these higher reactions is proportionately small it has been considered acceptable. So the trapezoidal rule continues to be used.

However, if the smaller dimension of the panel is low, e.g. a/t = 30, then the quantity of silicone that is used with the trapezoidal formula is excessive since the bite size is larger than necessary (see Figure 2).

So in summary, it can be stated that correct design using the solutions of Von Karmon plate equations will give you a better design and consequently less structural silicone.

This can affect reasonable cost savings when the total project is considered.

FIGURE 2 -- Structural Silicone Bite

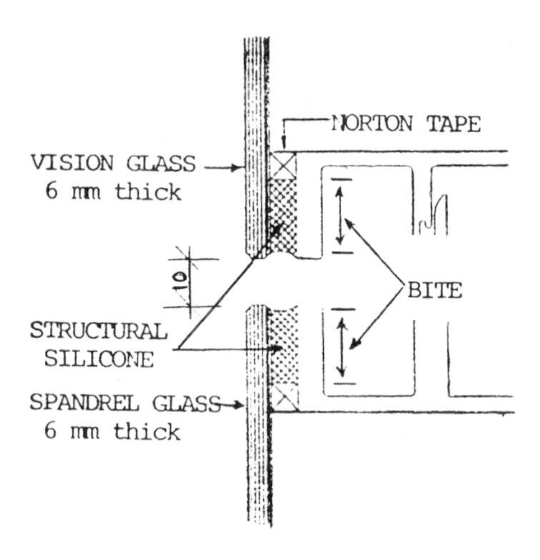

SPANDREL PANELS

During the past few years there has been a significant and positive trend towards the use of granite and natural stone as spandrel panels in the construction of curtain walls. Unlike days of old, stone is now being used in far thinner sheets ranging from 20 to 50 mm in thickness. Fixing of these thinner plates takes on significant proportions when structural glazing is considered.

Various techniques of keying the stone back to the curtain wall frame have been tried. These methods are generally coupled with structural silicone.

The dangers that have to be overcome with this technique are:

1. Poor mechanical fixing.
2. Poor compatibility between materials chosen (i.e. porosity of the stone).
3. Break down in mechanical fixing.
4. Chipping and flaking of the edges of the stone.
5. The quality and strength of the stone.

The natural modification to structural glazing would be to use a higher strength sealant for the spandrel panels. Polyurethane sealants have tensile strength up to 3 times higher than silicone (see Table 3). The only limitation that urethane has is its poor resistance to ultraviolet radiation.

TABLE 3 -- Sealant Strength (MPa)

	Exposure	No. of Samples	x Arithmetic Mean	s Standard Deviation	x - 3s 99.86%
SILICONE	Air	1200	1.126	0.152	0.670
	QUV	200	1.129	0.164	0.637
	Water	300	1.096	0.121	0.733
URETHANE	Air	20	3.455	0.19	2.885
	QUV	20	3.270	0.20	2.670
	Water	20	3.317	0.17	2.807

Given that the spandrel panel is opaque, with careful design it is quite conceivable that the urethane can be used for the structural fastening of stone or granite spandrels. The added strength of urethane will enhance the total design by reducing the reliance on the mechanical fastening of the spandrel panels. The mechanical fasteners will only need to become operative in the event of panel failure with the urethane.

UNITISED GLAZING

The advantages of unitised glazing to on-site glazing are well established especially for structural glazing. However, the advantages which are associated with glass and silicone performance have not yet been fully exploited by the industry.

The use of finite element techniques for both thickness and reaction computations for a thin glass plate can offer benefits in terms of silicone site lines and reduced mullion widths. Quality Control is yet another area which can also reduce the quantity of silicone used and hence the costs for unitised glazing (Jacob 1987)[7]. This additionally effects edge stiffness.

The use of structural silicone for unitised glazing will be enhanced if some of the following aspects are seriously and properly considered.

(a) Structural Interaction Between Different Materials: The designer has a responsibility to ensure that the whole system can and will interact within the tolerance limits for the building. For example, building movement can be generated through loading, creep (as for concrete), settlement, etc. It is important that the designer allows for this expected movement, failing which there could be serious long term problems that could result in glass fracture and ultimate fall out.

(b) Drain Principles: The drain principle or pressure equalisation principle has been generally accepted. Unitised glazing is a method wherein the drain principle can be effectively designed and installed. By-and-large the unitised system lends itself to the pressure equalisation principle. Some important aspects of the drain principle that must be considered by the designer are -

1. Head of Water: The designer must allow for an adequate head of water depending on the climatic conditions prevalent to the geographic location of the curtain wall.
2. Weep Holes: must be adequate in size and correctly positioned with a baffle to ensure that wind driven rain does not blow into the building.
3. Sponges - The use of small sponges across the weep hole is also recommended.
4. Water Track - At the intersection of unitised frames ensure that there is a track for the water to drain to the outside of the building rather than the inside.

(c) Compatibility: Compatibility between the various components used in curtain walls is extremely important. The sealant suppliers and the unitised manufacturer must take responsibility for the choice of materials. More importantly, test reports and data must be provided to the designer to ensure that the right products are selected.

(d) Primers: It is well understood that unprimed adhesion is better than primed adhesion and consequently makes for a safer system. So the designer has the opportunity to specify components, or rather substrates, that do not require priming when the specifications are written. J. M. Klosowski[8] in his treatise on Sealants and Construction identifies numerous secondary stresses that are generated in structural silicone glazing which must be considered. Primarily, they apply to the effects of glass rotation, dead load and thermal movement on the silicone joint.

SILICONE

Silicone is the key ingredient for structural glazing. Its quality, performance and strength are vital. There are numerous problems associated with the application of silicone. Quality control and a commitment to consistency in application is mandatory for the success of curtain wall glazing. The key checks that must be incorporated in terms of the quality of the silicone are :-

1. Material uniformity - consistency, cure rate, viscosity, and gunability.
2. The failure strength of samples made during assembly after the minimum 1 week exposure to water, ultraviolet radiation and ambient air must be consistent.
3. The wetting of the silicone to the substrate. This is a function of the quality of the substrate and the surface

preparation that is given to it prior to silicone application.

The long term performance of the silicone and hardness are also important properties that must be monitored. Silicone is known to harden with age. Hardness values to 60 durometer are not uncommon and are not desirable (see table 4). Figure 2 gives the tensile strength of silicone over a 3-year period. It can be seen that there is some stabilisation in the tensile strength while the hardness has gradually increased.

FIGURE 2 -- Long Term Silicone Strength

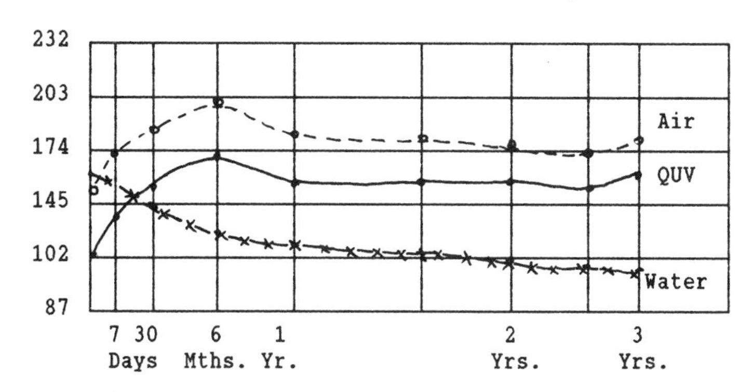

From table 2 it can be seen that the silicone strength is substantial when compared with current design practice. Even though we use an additional factor of safety of 3 to the 99.86% probability value the design stress would be in excess of 0.2 MPa which is approximately 30 psi. This increase in design stress will reduce the quantity of silicone which, in turn, reduces the stiffness or spring constant for the edge restraint. This works in unison with membrane stress analysis to provide a better design.

HARDENING OF SILICONE

Table 4 gives the relationship between Hardness and Time for 2-component structural silicone. These samples were stored under standard laboratory conditions of 23°C and 55% relative humidity. If this pattern of gradual increase in hardness continues, there must be serious concern regarding the effectiveness of the silicone joint to respond to the dynamic forces that will apply on that building. Sway , building racking and, most importantly, seismic loads are potentially serious factors that could cause joint rupture and ultimately result in glass fracture and fall out.

It appears that the sealant manufacturing companies have not addressed this facet of structural silicone.

TABLE 4 -- Shore A Hardness

Sample Set Up Date	No. of Samples	Shore A		
		Maximum	Minimum	Mean
1986	120	80	62	67.9
1987	150	66	55	59.2
1988	110	65	44	56.8
1989	50	52	40	47.2

SUBSTRATES

The quality, consistency and type of treatment that is applied to the aluminum frame, which is the structural member on to which silicone is applied, is vitally important in determining the long term adhesive qualities of the silicone and hence the life of the curtain wall. Numerous coatings have been tried with varying degrees of success. The question that must be asked is - "What is the adhesive and the cohesive strength of the surface coating that is used on the aluminum?"

For instance, if a powder coat is to be used, because of the thickness of the coating, it is quite conceivable that the total curtain wall is reliant on the cohesive strength of the powder coat.

Alternatively, it is reliant on the adhesive strength between the powder coat and the aluminum. The only method of evaluating these factors is consistent testing and quality control to ensure uniformity of product.

Experience has shown that the anodised substrate is potentially the most consistent. Even though PVF2 coatings offer better adhesion, they are less reliable and less desirable because of their intrinsic softness.

The alodyne substrate, on the other hand, attains extremely good adhesion but is not an aesthetically pleasing surface.

QUALITY CONTROL

The primary goal of structural glazing, from both an application and performance point of view, is "zero defects".

To this end quality control is vitally important during the assembly process for the unitised glazing.

However, there is always the possibility that a lite of glass will require replacement due to bad handling, surface damage or accidental glass failure. It would be a good idea for the curtain wall designer to consider the methodology and provide the mechanism which will facilitate the cut out and subsequent reglazing of another lite given the fact that a one-component silicone will need to be used for the replacement.

QUALITY REQUIREMENTS IN THE FACTORY

Next to design, the assembly of the curtain wall (unitised glazing) has the most impact on its structural performance.

It is the designer's responsibility to ensure that the specification demands strict adherence to a Quality Control programme which must be instituted for the curtain wall manufacturer. The benefits of such a programme are to ensure that -

(a) The operators maintain a sufficiently high level of "cleanliness" and "work consistency" to ensure that the expected tensile strength in the structural silicone is always attainable.
(b) The Quality Control of the two-component structural silicone is consistently acceptable.

The use of a stringent Quality Control programme during the assembly of unitised structural glazing not only ensures that the complete project will perform its designed function but also provides the designer with the facility for being more precise in the determination of the structural silicone bite (refer Jacob 1987).

Changes made in the field from the specification such as the use of alternatives, could be the most significant cause of failures in the construction industry. For example, substituting one gasket for another may seem insignificant. However, it is nearly impossible to judge how it will affect roll-out through glass deflection under load without mock up testing.

CONCLUSION

Although conservative design in the glass industry is still needed, we now have the tools and the theories to facilitate a more rational and correct design using glass and its associated material.

The work done in Lubbock, Texas on glass strength has contributed significantly to the selection and use of glass in buildings to the extent that international code bodies, such as ASTM, have adopted those recommendations.

The attainable design stress for 2-component structural silicone must also be reviewed in light of any Quality Control problem and our general awareness of the forces that come to bear on the silicone joint.

From a structural glazing point of view, there is a significant trend towards improved quality control in the assembly of unitised panels which further enhances the technological awareness. It is important for this ground swell to be promoted by the design professional who can implement these requirements in terms of improved design and quality control programmes. By setting minimum tensile strength standards for all production units it will then facilitate a more accurate design and ensure a higher degree of confidence in the system. This will ultimately reduce the quantity of silicone used, and hence control the spring constant which, in turn, renders the glazing more precise. It is quite conceivable that glass and curtain wall technology will develop and mature into an exact science for everyone's benefit.

However, the first step in this evolution must be the recognition that the technology now exists which will require a qualified professional to do the structural analysis. This is necessary to ensure that the system and the material, namely glass and framing, and the total system is structurally adequate for the anticipated life time of that facade.

REFERENCES

[1] Timoshenko, S. & Woinowsky-Krieger, Theory of Plates & Shells. McGraw Hill Book Co.

[2] Backmenver, R. D., Visualisation of Stress Fields in Thin Glass Plates. Texas Civil Engineer, Vol.51, No.9, Nov., 1981.

[3] Vallabhan, C. V. J., Finite Difference Program, 1983.

[4] Beason, W. L., Failure Prediction Model for Window Glass, Inst.of Disaster Research, Texas Tech.Uni., Lubbock, May 1980.

[5] Minor, J.E., Developments in the Design of Architectural Glazing Systems, 1st Nat'l Structural Eng. Conference-1987, Melbourne, Aust.

[6] Chou, Vallabhan & Minor. A Mathematical Model for Structurally Glazed I.G. Units, Glass Research & Testing Lab. - Texas Tech. Uni. - 1987.

[7] Jacob. L. J., Strict Q.C. Allows New Thinking About Structural Glazing, Glass Digest, Feb.5, 1988.

[8] Klosowski, J. M., Sealants & Construction, Marcel Dekker Inc., New York.

Movement Capacity of Sealants

Franklin W. Shisler III, Jerome M. Klosowski

SEALANT STRESSES IN TENSION AND SHEAR

REFERENCE: Shisler III, F. W., and Klosowski, J. M., "Sealant Stresses in Tension and Shear", Building Sealants: Materials, Properties, and Performance, ASTM STP 1069, Thomas F. O'Conner, ed., American Society for Testing and Materials, Philadelphia, 1990.

ABSTRACT: This paper addresses the question, "Are the stress vs strain or modulus characteristics of various generic sealants the same in shear as they are in tension?"

The ASTM tests and Federal Specifications for sealants address the joint movement ability of sealants when they are stressed in tension and compression. Nowhere does a test or specification relate to the movement ability in shear directions. Many joints are made of dissimilar materials so that with temperature changes these materials change dimension by differing amounts. Also, some joints are stressed as one substrate moves in or out of the plane of the wall or floor. The above mentioned cases are examples of shear movement. The authors stressed a variety of different generic sealants in the tension mode as well as the two shear modes (longitudinal and transverse). They show the relationship of the tension movement to the shear movements for sealants from each of the generic types.

KEYWORDS: Joint movement, modulus, sealant, shear, stress, tension.

Mr. Franklin W. Shisler III is a TS&D Specialist at Dow Corning Corporation, 3901 S. Saginaw Road, Midland, MI 48686; Mr. Jerome M. Klosowski FASTM is a Scientist at Dow Corning Corporation, 2200 W. Salzburg Road, Auburn, MI 48611

SUMMARY

Sealant joints between materials having different thermal expansion and movement characteristics may be subjected to tension, compression, longitudinal shear, and transverse shear as illustrated in Figure 1. The modulus, or stress vs elongation characteristics of a sealant in tension and compression are readily established using standard tests. Because there are no tests or specifications relating to sealant joint movement in shear, allowable shear movement characteristics have been calculated using the Pythagorean Theorem and the sealant's tension elongation or modulus characteristics established by test methods such as ASTM C-719 [1]. This calculation technique is illustrated in Figure 2 and discussed later. Such calculations inherently assume that the elongation or modulus characteristics of a sealant are the same in shear as they are in tension. Therefore, in order for this to be a valid calculation technique, the modulus characteristics of the sealant must be such that the shear stress, for both longitudinal and transverse shear, is equal to or less than the tensile stress at the same test elongation.

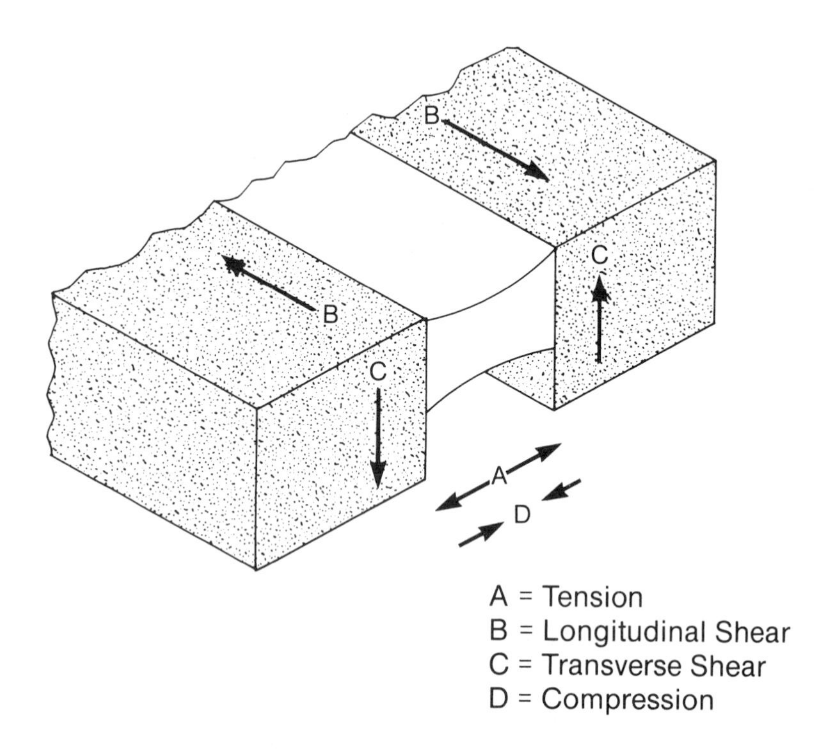

FIGURE 1 — Types of Joint Movement

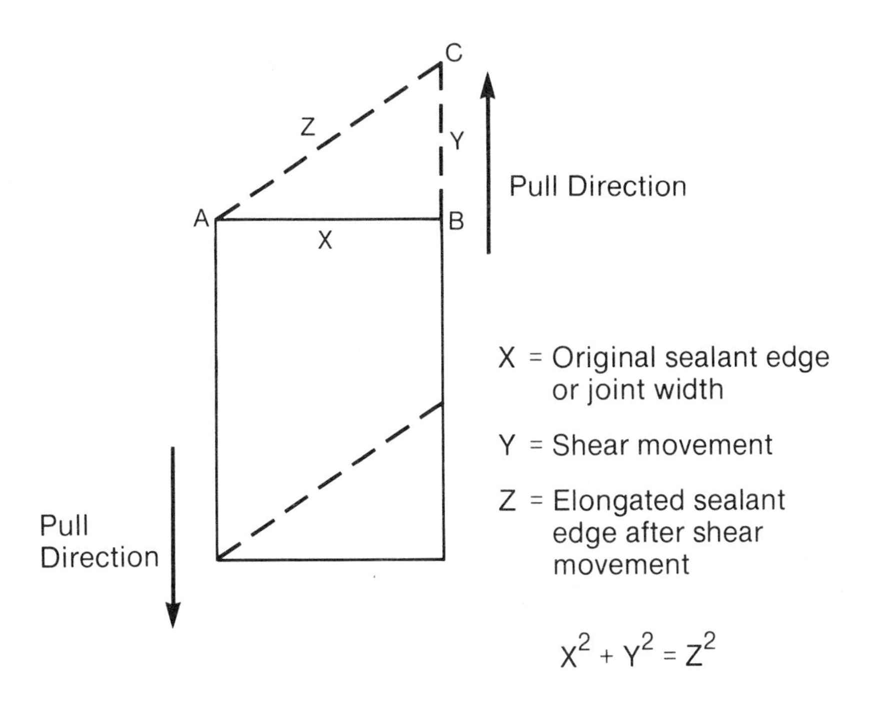

FIGURE 2 — Sealant Movement In Shear

This paper describes the experimentally determined modulus characteristics of several different sealants which were subjected to tension, longitudinal shear, and transverse shear movements. It also assesses the validity of using the Pythagorean Theorem, together with a sealant's tension modulus properties, to calculate shear movement in either shear direction.

The intent of the work presented in this paper is to simply show that, for the variety of sealants tested, allowable shear movement, in either shear direction, can be calculated using the sealant's allowable tension movement. From a practical standpoint, the only data known about most sealants is their tension modulus properties or movement ability in tension. This is the data architects and applicators understand and use. Therefore, this is the data the authors used for their calculations.

Three silicone sealants with different modulus of elasticities were tested. All three silicone sealants exhibited a greater elongation in shear than they did in tension when they were pulled to the same stress value. If this is true for all silicone sealants, this is surely one of the reasons for the great success of silicone joints. The two different modulus polyurethane sealants tested exhibited essentially the same elongation in both shear and tension when pulled to the same stress value. Therefore, based on the modulus characteristics of the sealants tested, under the conditions evaluated, the Pythagoren Theorem can be used, together with the sealant's tension modulus properties, as a valid technique for calculating allowable shear movement. This calculation technique provides conservative results, or a built in safety factor, when applied to the silicone sealants. This built in safety factor appears to be a function of the sealant's modulus properties and does not exist for the polyurethane sealants tested.

The authors recognize the excellent work of Dr. L. Bogue Sandberg and Theresa Ahlborn who studied simultaneous application of tension and shear. They found strong indications that the interaction relation between ultimate tensile and shear displacements is linear as follows [2].

$$\frac{\Delta t}{\Delta t_u} + \frac{\Delta s}{\Delta s_u} = 1$$

where

Δt = tensile displacement

Δt_u = ultimate tensile displacement in a pure tension mode

Δs = shear displacement

Δs_u = ultimate shear displacement in a pure shear mode

The authors also recognize the work of Dr. L. Bogue Sandberg and Jerome M. Klosowski which shows that the shear modulus in the elastic range, the range studied and discussed in this paper, is remarkably constant; while the tensile modulus (modulus of elasticity) is sensitive to width-depth ratio and strain or elongation level [3].

Both of these are excellent works showing the forces at work when failure occurs and the sensitivity to design of the combined stresses on a sealant. Particularily important is the combined stress work in the above cited paper of Sandberg and Ahlborn since almost no joint moves in pure tension or pure shear, but some combination of tension, longitudinal shear, and transverse shear. This is a complex set of forces to understand.

Sandberg and Ahlborn's work shows that tension and shear forces are interactive, and where significant shear stress is expected, the designer should reduce the tensile stress (e.g. design larger joints) to insure the combined stresses on the joint, the sealant, and the bond will not be excessive. This paper shows how to calculate allowable shear movement from a known allowable tension movement. This is a fundamental first step in joint design. Calculating joint movement as a function of interacting tension and shear is not covered in this paper. Table 1 and Figure 3 provide a quick reference for those wishing to use the Pythagoren Theorem calculation technique.

TABLE 1 -- Tension vs Shear Movement In Either Shear Direction For A 12.7 mm (0.50 in.) Wide Sealant Joint

Joint Movement In Tension or Incremental Sealant Edge Elongation (Distance Z in Figure 2)				
- Percent Of Joint Width	10%	25%	50%	100%
- mm	1.3	3.2	6.4	12.7
- in.	(0.05)	(0.13)	(0.25)	(0.50)
Total Elongated Sealant Edge (Distance Z In Figure 2)				
- mm	14.0	15.9	19.1	25.4
- in.	(0.55)	(0.62)	(0.75)	(1.00)
Allowable Shear Movement (Distance Y In Figure 2)				
- Percent Of Joint Width	46%	75%	112%	173%
- mm	5.8	9.5	14.2	22.0
- in.	(0.23)	(0.38)	(0.56)	(0.87)

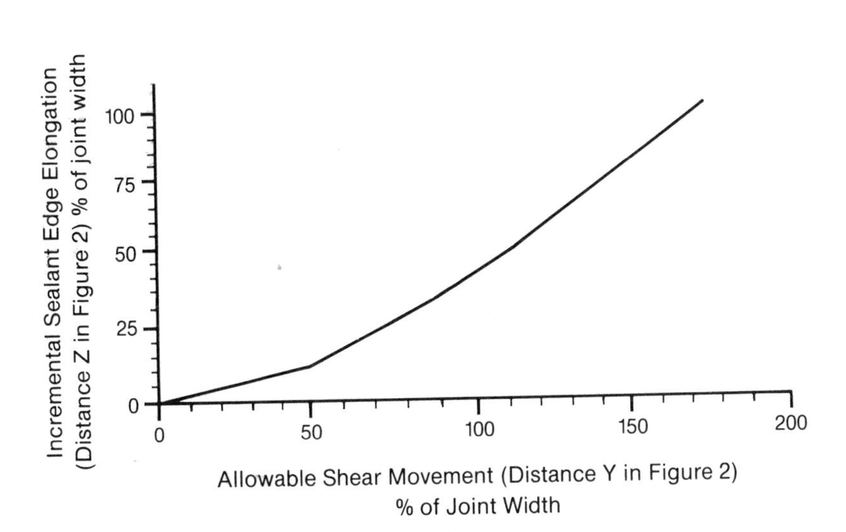

FIGURE 3 — Tension vs. Shear Movement

DISCUSSION

The objective of this study was to examine the relationship between tension modulus, longitudinal shear modulus and transverse shear modulus for a variety of different sealants. Seven sealants consisting of four generic types were selected for evaluation. These included: three silicone sealants (high, medium, and low modulus), two polyurethane sealants (high and medium modulus), one polysulfide sealant and one acrylic sealant. Unfortunately, even after over two months of curing at laboratory conditions, the polysulfide and acrylic sealants had not cured sufficiently. Because of this, the data generated when the polysulfide and acrylic sealants were tested was inconsistent and not valid. Consequently they had to be dropped from this study.

Each sealant was made into three different test joint configurations: a tension (TA) joint, a longitudinal shear joint, and a transverse shear joint. The sealant dimensions of each joint configuration were the same, namely 12.7 mm X 12.7 mm X 50.8 mm (1/2 in. X 1/2 in. X 2 in.). These are the same sealant joint dimensions specified in ASTM C-719. The difference among the three joint configurations is the direction in which the sealant joint is stressed. These joint configurations are best defined in Figures 4, 5, and 6 which show the tension (TA), longitudinal shear, and transverse shear joints respectively. These figures also show the pull direction in which each sealant joint is stressed as referenced in Figure 1.

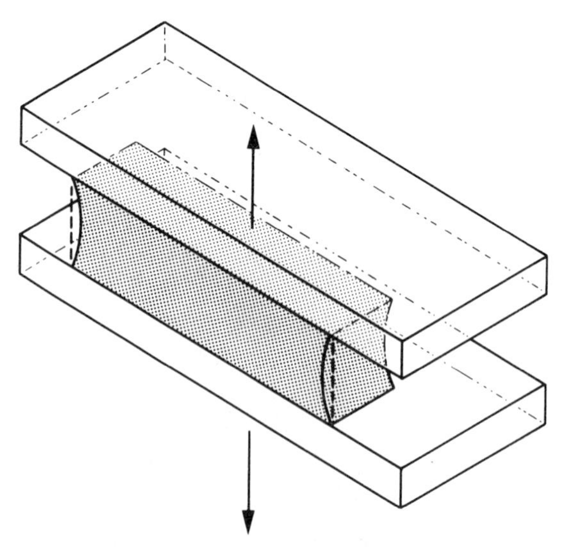

FIGURE 4 — Tension (TA) Joint

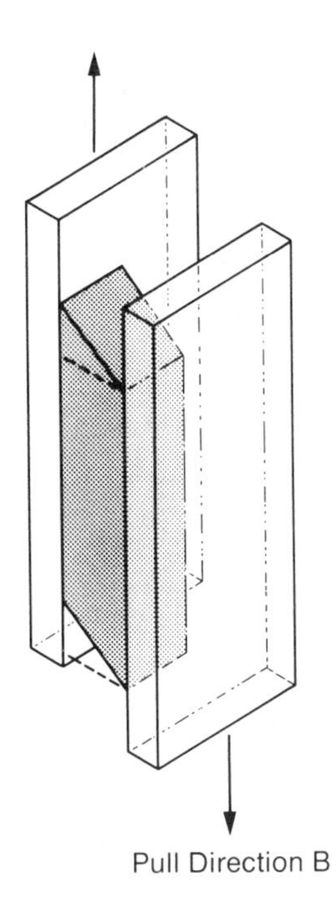

FIGURE 5 — Longitudinal Shear Joint

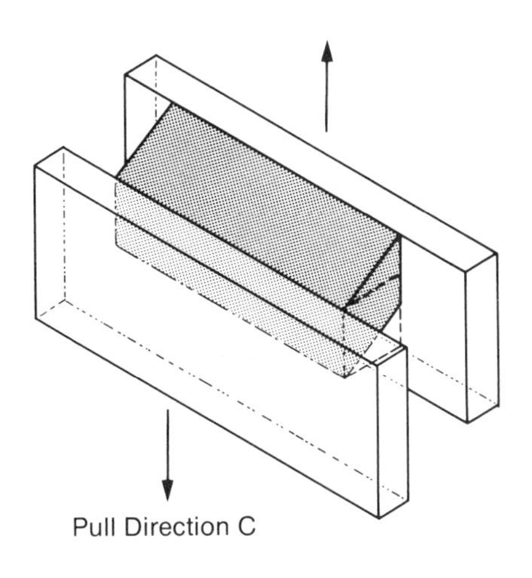

FIGURE 6 — Transverse Shear Joint

Three test samples for each joint configuration were made and tested. The results presented in this paper are the arithmetic averages of the test values generated for these three test samples at each joint configuration. Clear glass was the substrate used for constructing all the samples. The samples were cured and tested at laboratory conditions set at 25°C (77°F) and 50% RH.

Testing was initiated by pulling the tension (TA) joints at 2 inches (50.8 mm) per minute until the test elongation for the sealant being evaluated was obtained. The corresponding TA joint tension or tensile stress was then recorded. Test elongation is defined as the sealant manufacturer's maximum recommended joint movement. Figure 7 shows a tension (TA) joint being pulled.

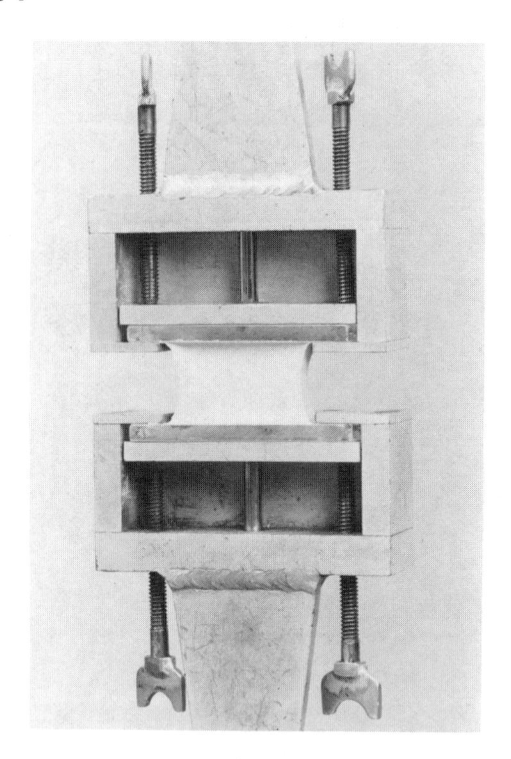

FIGURE 7 -- Tension (TA) Joint Being Pulled

Once the average tensile stress required to pull the tension (TA) joint to its test elongation was determined, identical joints were pulled in shear. Each joint was pulled in shear until both of the following criteria were met. Criteria number one, the joint was pulled in shear until its shear stress value was numerically equal to the average tensile stress value obtained previously. Criteria number two, if the elongated sealant edge had not reached its test elongation value when criteria number one was met, the pull was continued until its test elongation was obtained. The shear stress corresponding to the sealant's theoretical shear movement at its test elongation, as calculated by the Pythagorean Theorem, was recorded. The sealant elongation corresponding to a shear stress value equal to the average (TA) tensile stress value was also recorded. The longitudinal and transverse shear joints were pulled at 2 inches (50.8 mm) per minute. The distance between the glass substrates on both sides of the shear joints was kept constant at 12.7 mm (1/2 in.) by using Teflon spacers as shown in Figures 8 and 9. One might anticipate that this use of Teflon spacers could cause a drag or resistance that would give an increased shear stress value. This was not the case. Any resistance caused by the Teflon spacers which would increase the measured shear stress was not detectable and subsequently not significant. This lack of interference is attributed to the low friction properties of the Teflon spacers and the low normal forces between the Teflon spacers and glass substrates.

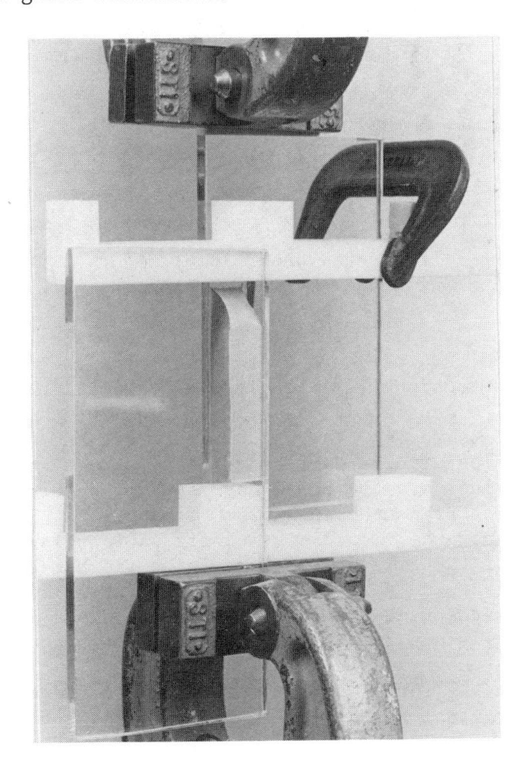

FIGURE 8 -- Longitudinal Shear Joint
(Joint Width Maintained by Teflon Spacers)

FIGURE 9 -- Transverse Shear Joint
(Joint Width Maintained by Teflon Spacers)

Table 2 shows the average values of stress and associated elongation, expressed as a percentage of the original sealant joint width, for the tension (TA), longitudinal shear, and transverse shear joint configurations tested. To make the data in this table easier to review, the values of stress and associated percent elongation are listed side by side regardless of which of these values was the one experimentally determined. The percent elongation presented in Table 2 for both the longitudinal and transverse shear joint configurations is the percent elongation of the original sealant joint width measured along the hypotenuse of the sheared sealant (distance Z in Figure 2). It is not the percent elongation based on the shear movement of the sealant in the plane of the pull direction (distance Y in Figure 2). The shear movement in the plane of the pull direction is much greater than the corresponding incremental percent elongation of the original sealant width measured along the hypotenuse. This is further clarified by referring to Figure 2, Figure 3, Table 1, and the following discussion.

In Figure 2 the solid lines show the sealant joint

TABLE 2 -- Sealant Stresses In Tension and Shear

Sealant	Typical High Modulus Silicone		Medium Modulus Silicone		Low Modulus Silicone		High Modulus Polyurethane		Medium Modulus Polyurethane	
Test elongation recommended by sealant manufacturer, %	25%		50%		100%		25%		50%	
Stress and elongation at various test conditions	Stress kg/cm^2 (lb/in^2)	Elong. %	Stress kg/cm^2 (lb/in^2)	Elong. %	Stress kg/cm^2 (lb/in^2)	Elong. %	Stress kg/cm^2 (lb/in^2)	Elong. %	Stress kg/cm^2 (lb/in^2)	Elong. %
Tension (TA) joint										
Tensile stress at test elongation	2.39 (34)	25%	3.87 (55)	50%	2.25 (32)	100%	4.22 (60)	25%	1.62 (23)	50%
Longitudinal shear joint*										
Shear stress at test elongation	1.90 (27)	25%	3.30 (47)	50%	1.41 (20)	100%	4.50 (64)	25%	1.55 (22)	50%
Elongation at TA joint stress	2.39 (34)	43%	3.87 (55)	65%	2.25 (32)	430%	4.22 (60)	21%	1.62 (23)	53%
Transverse shear joint*										
Shear stress at test elongation	1.83 (26)	25%	3.02 (43)	50%	1.48 (21)	100%	3.94 (56)	25%	1.48 (21)	50%
Elongation at TA joint stress	2.39 (34)	44%	3.87 (55)	73%	2.25 (32)	355%	4.22 (60)	28%	1.62 (23)	56%

**The elongation referred to under both shear joint headings was measured along distance Z in figure 2.*

configuration as installed. The dashed lines show the sealant joint configuration after it has been pulled in shear. The A to B distance (X) represents the original sealant edge or joint width between the two substrates. In these tests the "X" distance was kept constant as explained earlier. When the joint is pulled in shear, the sealant stays adhered to the substrate and point B moves to point C. The B to C distance (Y) is the shear movement in the plane of the pull direction. The A to C distance (Z) is the hypotenuse and represents the now elongated sealant edge distance. The Pythogorean Theorem, which states $X^2 + Y^2 = Z^2$, can be used to calculate the joint movement of this configuration. For example, if a sealant having an original joint width "X" is sheared to a 25% sealant elongation, the hypotenuse "Z" is 25% longer than the original sealant width "X" or, Z = 1.25X. By substituting this value into the Pythagorean Theorem we have $X^2 + Y^2 = (1.25X)^2$ or Y = 0.75 X. Therefore, for a 25% elongation of the original sealant edge or joint width measured along the hypotenuse of the sheared joint, the shear movement in the plane of the pull direction is 75% of the original sealant width or three times as much as the 25% incremental elongation of the original sealant edge or joint width. Table 1 and Figure 3 show this relationship for a wide range of joint movements.

RESULTS

A review of Table 2 confirms that for the sealants tested, under the conditions evaluated, the Pythagoren Theorem can be used together with the sealant's tension modulus properties to calculate allowable shear movement. This is based on the fact that the shear stress for both the longitudinal and the transverse shear joints is equal to or less than the tensile stress at the same test elongation (See Note 1).

The longitudinal and transverse shear stresses for each individual silicone sealant tested were essentially the same and were significantly lower than their tensile stress at the same elongation. In other words, the silicone sealants provide a greater movement or elongation capability when pulled in shear than they do when pulled in tension, to the same stress value. Since these silicone sealants perform even better in shear than they do in tension, calculations which use the Pythagorean Theorem together with the tension modulus properties of these sealants, provide conservative results, or what amounts to a built in safety factor. This safety factor seems to be a

Note 1. The one apparent exception to this statement is the longitudinal shear stress value for the high modulus polyurethane sealant which is 4.50 kg/cm^2 (64 lb/in^2) or 0.28 kg/cm^2 (4 lb/in^2) higher than this sealant's 4.22 kg/cm^2 (60 lb/in^2) tensile stress. However, based on the experimental accuracy of the testing procedures used, both the 4.50 kg/cm^2 (64 lb/in^2) and 4.22 kg/cm^2 (60 lb/in^2) stresses represent the same value.

function of the sealant's modulus properties and may vary for different sealants.

For all practical purposes the tensile, longitudinal shear, and transverse shear stresses for each of the two polyurethane sealants tested were essentially the same at the same test elongation. In other words, their movement ability was essentially the same in both tension and shear. Thus, as with the silicones, the Pythagorean Theorem can be used together with the tension modulus properties of these polyurethane sealants to calculate allowable shear movement. However, the results do not exhibit a built in safety factor as do the silicone sealants.

All the testing described in this report was conducted at laboratory conditions set at 25°C (77°F) and 50% RH. Future work should include testing at high and low temperatures.

REFERENCES

[1] Klosowski, J. M., "Sealant Movement In Shear," The Applicator a publication of the Sealant and Waterproofers Institute, Vol. 9, No. 2, Second Quarter 1987, pp 8-9.

[2] Sandberg, L. B. and Ahlborn, T. M., "Combined Stress Behavior of Structural Glazing Joints," Journal of Structural Engineering, Vol. 115, No.5, May 1989, pp 1212-1224.

[3] Sandberg, L. B. and Klosowski, J. M., "Structural Glazing: Behavior Details of Double-Bead Installations," Adhesives Age, Vol. 29, No. 5, May 1986, pp 26-29.

James C. Myers

BEHAVIOR OF FILLET SEALANT JOINTS

REFERENCE: Myers, J. C., "Behavior of Fillet Sealant Joints," Building Sealants: Material, Properties and Performance, ASTM STP 1069, Thomas F. O'Connor, Ed., American Society for Testing and Materials, Philadelphia, 1990.

ABSTRACT: This paper investigates the behavior of fillet (triangular) sealant joints using tensile and cyclic tests and finite element analyses. Fillet sealant joints with diagonal bond breakers have greater movement capability and lower peak stresses than fillet sealant joints with flat bond breakers and than square and rectangular joints. Design of fillet joints with diagonal bond breakers should follow current rectangular joint design procedures, with the required bond breaker size based on the horizontal projected length of the concealed face of the joint.

KEYWORDS: sealant joint design, sealant joint stresses, joint shape, movement capability, bond breakers

Traditionally, sealant joints in exterior walls of buildings are butt joints, optimally with a 2:1 width-to-depth ratio and concave faces. Situations arise where the butt joint cannot be constructed, and designers and installers opt for other joint configurations, such as fillet shapes. Fillet joints are generally triangular, but can include any joints in which the two sides of the joint are approximately perpendicular. When these situations arise, there is little guidance available to the designer regarding the movement capability of fillet joints or the proper sizing of the joint and the bond breaker. An informal survey of five major sealant manufacturers that I conducted in January 1989 reveals that none of the manufacturers have done any testing of fillet joints, and many feel fillet joints have a history of poor performance and should not be used. Little fillet joint design information is available in published literature.

This paper summarizes a master's thesis investigation of the behavior of fillet sealant joints [1]. Readers seeking greater detail than this summary paper provides should refer to the thesis [1]. This research provides a rational basis for fillet sealant joint design. Specifically, it answers the following questions that fillet sealant joint designers face.

Mr. Myers is a staff engineer at Simpson Gumpertz & Heger Inc., Consulting Engineers, 297 Broadway, Arlington, MA 02174

1. What is the movement capability of a fillet sealant joint?
2. How does the magnitude and nature of the stresses in fillet sealant joints compare to square or rectangular joints?
3. What is the best shape for the fillet sealant joint?
4. What size and shape are needed for the bond breaker?

This study uses the following tests and analyses to address the above-noted questions.

1. Tensile tests of fillet and square sealant joints.
2. Cyclic tests of various fillet, square, and rectangular sealant joints using a procedure similar to ASTM C719 [2].
3. Finite element analyses of stress/strain distributions in various fillet and butt joints.

LABORATORY TESTING

This section describes the materials and procedures used in the tensile and cyclic tests, and presents the results of these tests. All testing consists of applying displacements perpendicular to the length of the joint, and parallel to the leg of the fillet containing the bond breaker. This study does not consider the effect of longitudinal movement on the joints. The nominal strain in the fillet joint is defined arbitrarily as the displacement of the joint divided by the length of the bond breaker projected onto the plane of movement, i.e., horizontal projected length.

Description of Test Samples and Procedures

All test samples are 50.8 mm (2 in.) long. Test samples include square and rectangular (2:1 width-to-depth ratio) joints. Thirteen other test joint configurations fall into two main categories (Fig 1):

Group I: Fillets without sealant at the root of the joint. A triangular or quarter-round foam backer fills this void.

Group II: Fillets with a flat bond breaker (release tape) on the horizontal surface beside the root of the joint and filled solidly with sealant.

Table 1 contains details of joint configurations, geometry, and bond breaker sizes.

The primary focus of this study is to compare performance of various joint geometries, as opposed to various types of sealants. The test program includes construction of all fifteen joint types with one primary sealant, a medium-modulus silicone with an ASTM C719 rating of ± 25% on aluminum. Four types of joints (designations A1, B, L, and M in Table 1) with five additional sealants, three polyurethanes and two low-modulus silicones, are included to determine if the results with the primary sealant are applicable to other sealants. All sealants are one-component. Reference 1 contains sealant brand names and details of surface preparation procedures.

The displacement rate for all tensile tests is 12.7 mm/min (0.5 in./min). A custom jig holds the samples in the machine grips.

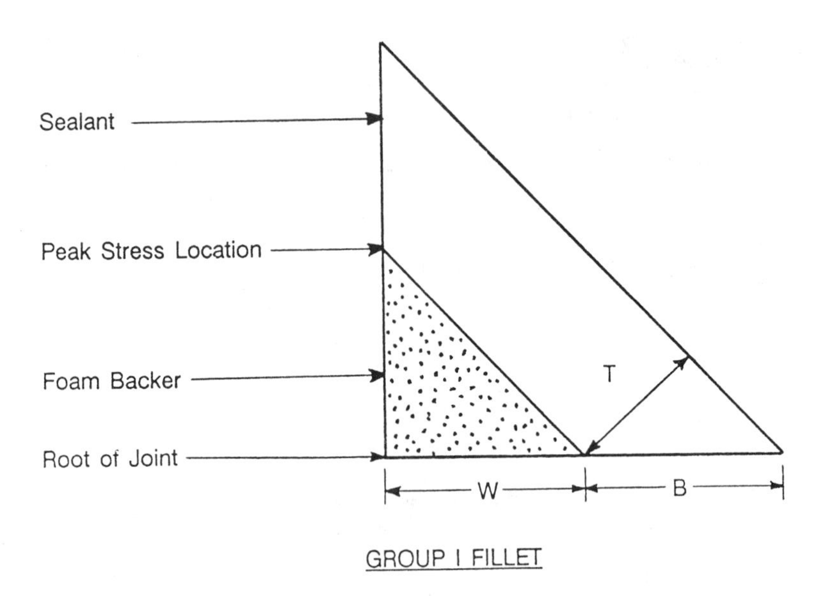

GROUP I FILLET

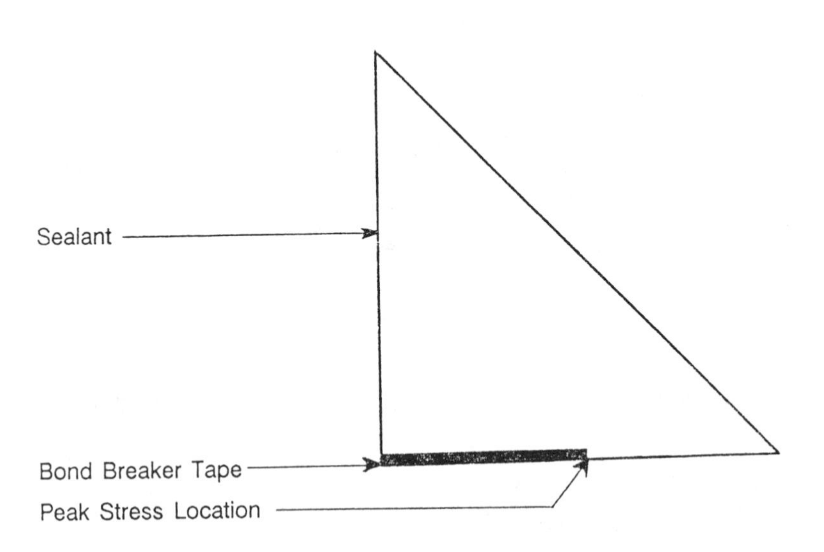

GROUP II FILLET

FIGURE 1. Fillet Sealant Joints

TABLE 1 - - Joint configurations

Designation	Geometry	W[a], mm(in.)	B[a], mm(in.)	Comments
A1	Square	12.7 (1/2)	12.7 (1/2)	
A2	Rectangle	12.7 (1/2)	6.4 (1/4)	Not used in tensile tests.
A3[b]	Hourglass	12.7 (1/2)	6.4 (1/4)	Bond line depth equals W.
A4[b]	Hourglass	12.7 (1/2)	3.2 (1/8)	Bond line depth is one-half of W.
A5[b]	Rectangle	12.7 (1/2)	25.4 (1)	Deep, narrow joint.
B	Group I	9.5 (3/8)	15.9 (5/8)	
C	Group II	. . .	19.0 (3/4)	No release tape.
D	Group II	9.5 (3/8)	9.5 (3/8)	
E	Group I	9.5 (3/8)	9.5 (3/8)	
F	Group II	9.5 (3/8)	9.5 (3/8)	Release tape on vertical and horizontal surfaces.
G	Group I	6.4 (1/4)	12.7 (1/2)	Quarter round backer and concave exposed face.
H	Group II	6.4 (1/4)	9.5 (3/8)	
J	Group I	6.4 (1/4)	9.5 (3/8)	
K	Group I	15.9 (5/8)	9.5 (3/8)	
L	Group II	. . .	9.5 (3/8)	No release tape.
M	Group II	6.4 (1/4)	6.4 (1/4)	
N	Group II	6.4 (1/4)	12.7 (1/2)	Not an equilateral triangle. B on vertical surface is 9.5 mm (3/8 in.)
P	Group II	6.4 (1/4)	12.7 (1/2)	
Q[b]	Group I	6.4 (1/4)	12.7 (1/2)	Quarter round backer and convex exposed face.
R[b]	Group II	9.5 (3/8)	9.5 (3/8)	Convex exposed face.
S[b]	Group I	6.4 (1/4)	12.7 (1/2)	Quarter round backer.
T[b]	Group I	9.5 (3/8)	9.5 (3/8)	Release tape (W) on vertical surface. B applies to vertical surface.

[a] For square and rectangular joints, W is joint width and B is centerline joint depth. For fillet joints, see Figure 1.

[b] These joints are used in finite element analyses only.

The cyclic test procedure is the ASTM C719 (Hockman cycle) procedure [2], except as modified below. The cyclic tests use several strains that are 1.6 mm (1/16 in.) apart for a particular sealant and joint geometry to bracket the likely failure strain. The cyclic tests use a constant frequency (one-half hour per step), rather than a constant displacement rate. There are six room temperature cycles and eight cycles with temperature changes as called for in ASTM C719. Mill finish aluminum angles form the substrates. The cyclic testing involves extending and compressing the joints using bolts with wing nuts and a series of 1.6 mm (1/16 in.) thick shims manually inserted between opposing faces of the aluminum angles, since standard compression-extension machines for sealant joints cannot accommodate the fillet geometry nor the 213 samples to be tested in a timely manner.

Tensile Test Results

Ultimate values refer to values at the point of maximum load. Average nominal ultimate strains range from 235% to 740% for the medium-modulus silicone sealant and from 280% to 550% for one polyurethane sealant. Figure 2 depicts typical load-elongation characteristics for each joint type with the medium-modulus silicone. The polyurethane sealant follows the same trends that the silicone sealant exhibits. Table 2 compares average ultimate strains in the different joints with a polyurethane sealant. Generally, Group I fillets reach larger strains before failing than do Group II fillets.

In general, the square joint requires greater force to reach a given elongation than do the other joints. The Group I joints stretch to a given elongation with less force than do the Group II joints. There is a sharp drop in the slope of the test curves for fillet joints without bond breakers where limited tearing (about 1 mm, 0.04 in.) of the sealant at the root of the joint occurs. Once the tear forms, the joints behave more like that of the other fillets with intentional bond breakers.

In the Group I joints, the strain does not distribute uniformly along the joint based on visual observations of joint deformation in tensile tests, but concentrates on the vertical surface at the peak stress location shown in Figure 1. In Group II fillet joints, the strain concentrates at the edge of the bond breaker tape on the horizontal surface at the peak stress location shown in Figure 1.

In tests with completely cohesive failures, all Group I fillets fail at the vertical face of the joint (perpendicular to direction of movement), while all Group II fillets fail at the horizontal face of the joint. In all cases, the failure begins at the concealed face of the joint and propagates toward the exposed surface. An exception is the joints with the thinnest sealant beads, such as Joint G, which split perpendicular to the exposed face near the middle of the sealant joint.

Cyclic Test Results

The passing strain criterion is similar to that specified in ASTM C719/C920. The results of the manual cyclic tests with square joints agree reasonably well with the manufacturers' ASTM C719 automated test results. Table 2 compares cyclic test results in different joint configurations. The medium-modulus silicone in the square joint passes at $\pm$25% maximum strain. Rectangular joints are tested at strains of $\pm$50% and $\pm$75%, and all failed during the first extension after the initial week-long compression. Most of the Group I fillets of the medium-modulus silicone do not fail at the maximum test strains, $\pm$80% to $\pm$100%. Most of the

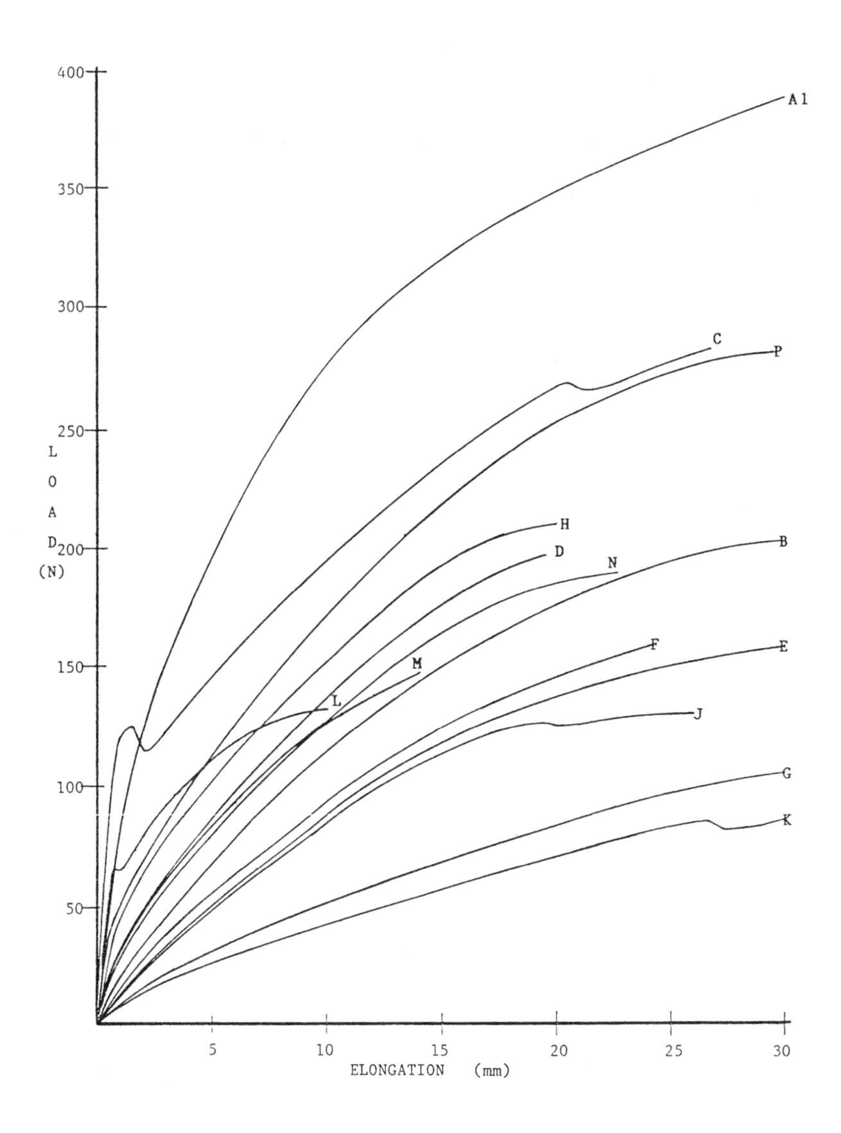

Note: Capital letters refer to joint designations in Table 1. Elongations of joints A1, P, B, E, G, and K extend beyond graph limits and are 38, 30, 38, 32, 50, and 45 mm respectively. Medium-modulus silicone is used.

FIGURE 2. Representative tensile test performance.

TABLE 2 -- Results of tensile and cyclic tests and stress analyses

Joint Designation[a]	Ultimate Nominal Strain in Tensile Test, %[b]	Maximum Passable Strain in Cyclic Test, $\pm$%	Peak Stress Ranking[c]
		Butt Joints	
A5	...	...	10.0
A1	309	25	6.3
A2	...	< 50[d]	4.2
A4	...	...	2.6
A3	...	...	2.6[f]
		Group I Fillet Joints	
B	424	50	2.1
J	460	100[e]	1.8
E	387	83[e]	1.8
K	303	80[e]	1.7
G	550	100[e]	1.0[f]
S	...	...	1.0[f]
Q	...	...	0.9[f]
		Group II Fillet Joints	
R	...	...	6.0
D	248	17	5.9
P	422	50	5.1
M	278	25	5.1
H	338	50	4.9
N	357	25	4.7
T	...	...	3.8
F	332	33	3.4

[a] See Table 1.
[b] Based on average of three tests with a polyurethane sealant.
[c] See text for explanation of rankings; based on linear analyses.
[d] Tests at $\pm$50% and $\pm$75% only.
[e] Joints did not fail at maximum tested strain shown in Table.
[f] Peak stress is in center of sealant, not along bond line.

Group II fillets of the same silicone have maximum passing strains of between ±25% and ±50%. Similar trends hold with the other types of sealants used in the tests.

The square and rectangular joints bulge at the middle when compressed. Compressing the Group II sealant joints creates a bulge in the sealant and at the end of the joint above the bond breaker. The Group I fillets show less distress upon compressing them than do the other joints, but the thinnest ones buckle. Compressing the joints without bond breakers tears the sealant at the root of the joint. All samples return to a width less than the originally formed joint width following the cyclic test. The compression set averages about 50% of the imposed strain. The Group II fillets retain more compression than do the Group I fillets. This trend is more pronounced at imposed strains of 50% or less. Compression set does not increase proportionally to imposed strain; as a percentage of imposed strain, it decreases.

NUMERICAL ANALYSES

Analyses of stresses in sealant joints are complicated by the material nonlinearity of the rubbery sealant materials, the geometric nonlinearity due to large deformations, the incompressibility of the materials, the nonuniformity of the fillet cross-section, and the stress concentrating effect of the bond line. Reference 3 provides background information for readers unfamiliar with the finite element approach.

Assumptions and Procedures of Analyses

Finite element analyses based on linear elastic, small strain and on nonlinear, finite strain (piece-wise linear) models are used to compare the stress and strain distributions in different sealant joints. These analyses are not suitable for predicting absolute values of stresses. While many of the factors discussed below affect the absolute value of the results significantly, the effect on the comparative analyses, which are the purpose of this work, is less significant. The linear analyses use all twenty-two joint configurations in Table 1, and the nonlinear analyses use a square joint (A1) and a Group I (E) and Group II (M) fillet joint.

The analyses are two-dimensional, with horizontal displacements applied to the bonded portion of the horizontal leg of the fillet (Figure 1). The bonded portion of the vertical leg is fixed against horizontal and vertical movement. Plane strain conditions apply to all joints. Poisson's ratio is 0.49. The analyses do not consider the effect of time on the sealant behavior. The behavior of the sealant that is input to the computer is based on the stress-strain curves generated from dogbone tensile tests. The linear elastic analyses arbitrarily use a Young's modulus of 1 MPa (145 psi). The finite element analyses take into consideration the sealant behavior due to presence of the bond line and joint geometry (stress concentrations) that are not present in the dogbone tests.

Corners of the joints are points of stress singularity. The stresses that the program predicts at these locations are dependent on the size of the mesh in this region. To provide a uniform basis of comparison between the peak stresses in different shape joints, each joint has a nearly identical mesh size.

Results of Numerical Analyses

Based on the linear elastic, small strain analyses of joint extension, the peak stresses in a rectangular joint with a 1:2 width-to-depth ratio, i.e., a deep, narrow joint, are about ten times greater than in a Group I fillet joint with a concave concealed face. These are the maximum and minimum peak stress conditions found in the twenty-two joint configurations analyzed. For comparative purposes, these extremes are assigned dimensionless peak stress ranking values of ten and one respectively. Table 2 compares these rankings in each joint configuration. Other Group I fillet joints have stress rankings up to 2.1. Group II fillet joints have stress rankings from 3.4 to 6.0. Rankings for square joints are 6.3. Rectangular joints with a 2:1 width-to-depth ratio have a 4.2 ranking with flat faces and a 2.6 ranking with concave faces.

The relative ranking of the joints remains unchanged through 25% strain with the three configurations used in the nonlinear analyses. The difference between peak strains in different joints remains steady. The predicted average loads are up to 25% above the actual test loads. The location of the peak stress does not change through 25% strain. The stress gradient around the location of peak stress decreases as the joint extends.

Extension of Group I fillets generally creates the sharpest neck down of the sealant (contraction) at the top of the foam backer (see peak stress location in Figure 1). The Group II fillets neck down significantly at both edges of the bond breaker during extension, with the larger portion over the peak stress location shown in Figure 1.

The peak maximum principal stresses during extension lie at the edge of a bond surface in all of the joints, except for some thinner joints with curved surfaces. The peak stress is on the horizontal plane as shown in Figure 1 for the Group II fillets, while it is on the vertical plane as shown in Figure 1 for Group I fillets. For most fillet joints, the stresses along the exposed surface of the joint are sizably less than along the concealed face. The stress gradient near the peak stresses is steep in most joints.

The general trends that the computer analyses predict correlate well with the results of the cyclic tests (see Table 2). The Group I fillet joints have the greatest extensibility and the lowest peak maximum principal stress. The general deformations and locations of greatest deformation predicted by the analyses agree with behavior during testing. The predicted locations of peak stress correlate with the locations of failure initiation.

FILLET SEALANT JOINT DESIGN

Research in the 1960's and early 1970's provided the basis for current design standards for butt sealant joints [4, 5, 6, 7, 8, and 9]. Little guidance has been available to designers faced with specifying fillet sealant joints. Fillet sealant joints have been thought to perform poorly and, consequently, have received little attention from designers and researchers. This research into the behavior of fillet sealant joints shows that these joints can perform well and that improper design and construction (omission of bond breakers) has produced their reputation for poor performance. This research provides guidance to designers on proportioning and movement capability of fillet sealant joints and shows that many of the principles applicable to butt joint design also apply to fillet joint design. As with butt joints, fillet joints need the proper configuration (shape, size, and thickness) and adequate bond breakers to perform successfully.

Joint Shape

The configuration of the fillet sealant joint plays a large role in the performance of the joint. The Group I fillet as a class possess greater movement capability than Group II fillet, square, and rectangular joints. The Group I fillets may perform as well as, or better than, hourglass-shaped joints, since they have lower peak stresses. Hourglass joints are not included in this study.

A key factor in the significant movement capability of Group I fillet joints compared to butt joints is their ability to deform with less stress than the butt joints. The displacement in the sealant in a butt joint is colinear with the joint width. The sealant can only expand perpendicular to the joint width to accommodate joint compression and can only contract perpendicular to the joint width to accommodate joint extension. This lateral expansion and contraction generates stresses, particularly at the bond line which resists this lateral deformation. Unlike with the butt joint, the displacement in the Group I fillet joint is at an angle with the diagonal bond breaker face and the sealant can deform by shear which stresses the sealant less than direct extension and compression.

Movement capabilities and peak stresses in Group II fillets are between those of square and rectangular joints. The Group II fillets develop greater tensile stresses during the compression portion of movement cycles, and tend to have slightly larger compression sets than do Group I fillets. The Group II fillets cannot accommodate joint movement through shear as readily as Group I fillets can.

Group I fillets have a variety of possible configurations, including those where the concealed and exposed surfaces of the joint are concave, convex, or flat (diagonal). Of the twenty-two joints evaluated in this work, the joint with the exposed convex face and concealed concave face (quarter-round backer rod) has the smallest peak stress. Application considerations with this joint are discussed below. Joints with concealed concave faces (quarter-round backer rods) have peak stresses up to one-half of their flat-faced counterparts. This trend is similar to that in butt joints in which the reduced centerline thickness in hourglass-shaped joints improves performance over rectangular joints. The shape of the concealed face in fillet joints plays a much larger role in determining peak stresses than the shape of the exposed face.

Configurations of Group II fillets are limited to variations in the exposed surface. A convex surface increases the peak stress compared to a joint with a flat surface. The difference is slight. The study did not include a Group II joint with a concave face.

Bond Breakers

The type, location, and size of the bond breaker (release tape or foam backer) impact joint performance. Omitting the bond breaker creates unstable conditions, since the sealant tears apart to accommodate movement. The stability of this tear is uncertain in the laboratory, but field experience shows that joints without bond breakers fail prematurely. A bond breaker should always be incorporated. The impact of bond breaker type lies in the resulting joint shape, which is discussed above.

The bond breaker in Group II fillet joints can be along the horizontal or vertical surface or both. Applying the bond breaker to the vertical surface reduces the peak stresses significantly compared to horizontal bond breakers when the joint is extended, due to the "shear effect" discussed in the "Joint Shape" section above. However, the vertical bond breaker cannot accommodate compression of the joint without tearing the sealant along the horizontal bond line.

Bond breakers on both surfaces (Joint F in Table 1) reduce peak stresses over those in joints with a bond breaker on only one surface. However, the increased movement capability and reduction in peak stress do not rival that of the Group I fillets, because extension of Joint F requires a greater volume of sealant to contract laterally than in a Group I fillet which omits the sealant at the root of the joint. This lateral contraction in elastomers is sizable due to their incompressibility.

As with butt joints, Group I fillet joints with wider bond breakers have greater absolute movement capability. Increasing the size of the bond breaker in Group I fillets while holding the sealant thickness constant makes little difference in peak stresses for similar strains (based on a percentage of the bond breaker size). With Group II fillets, the movement capability drops and the peak stress increases with increases in the bond breaker size. Lengthening the bond breaker (dimension W in Fig 1) increases the sealant volume above the bond breaker for equilateral triangles with constant bond width (dimension B in Fig 1) which increases stresses due to incompressibility effects discussed above.

A procedure similar to that currently used for sizing widths of rectangular joints can determine the bond breaker size in fillet joints. Divide the anticipated joint movement by the ASTM C719 movement capability of the sealant material to determine the necessary horizontal projected length of the bond breaker (dimension W in Fig 1).

Joint Size

The thickness of the joint (dimension T in Fig 1) affects the performance of fillet joints. The effect of the thickness is not the same for Group I and Group II fillets. As with butt joints, reducing the thickness of the sealant in Group I fillets reduces the peak stresses and increases movement capability. The trends in peak stresses in Group II joints are not a sole function of the minimum joint thickness at the edge of the bond breaker tape, and depend on other aspects of the joint geometry, such as the volume of the sealant above the tape (incompressibility effects).

Desires of some designers to minimize joint size limit the situations where fillet joints are feasible. The fillet joint is aesthetically inefficient, since both the bond surface (dimension B in Fig 1) and the bond breaker (dimension W in Fig 1) contribute to the sightline of the joint, unlike butt joints. The sag resistance of sealants also limits the maximum size of fillet joints. A maximum overall joint size of 25 mm (1 in.) of horizontal projection (dimensions W plus B in Fig 1) seems to be a reasonable upper limit given the above considerations. This limit will vary with the designer's personal preferences and the sealant's capabilities. With such a limit, the maximum absolute movement capability of a fillet joint using a sealant that passes ASTM C719 at ± 50% is about 8 mm (5/16 in.). This limits the situations in which fillets can be used.

Sealant Materials

Sealants that perform well in proper butt joints, perform well in proper fillet joints. Fillets do not introduce larger or more steeply inclined, i.e., direction relative to bond surface, tensile stresses in areas of significant stress on the bond line or in the material. The ASTM C719 procedure establishes the movement capability classifications of the sealant material.

Application Considerations

The fillet geometry provides installation advantages and disadvantages over that of the rectangular joint. The substrate is more accessible to the applicator for cleaning and preparation.

Presently, foam backing rods of the configurations used for fillets are not commonly available. Consequently, Group I fillet joints with foam backers have not been used significantly and have little track record presently. The rods can be custom-cut from standard rods or custom-fabricated for large applications. A pressure-sensitive adhesive face is the best means for retaining the backer rod. With some joint geometries, it may be possible to force a standard circular rod into a gap in the joint to retain the rod. Another potential area of use is in sealing over existing failed joints and avoiding the significant costs to remove the existing sealant and prepare the previously sealed surfaces. In some cases, covering the face of the existing sealant with tape provides the necessary bond breaker.

Tooling of the joint is not difficult, but requires care to maintain straightness and consistent thickness. Masking the substrate is a convenient means of insuring a straight finished edge and consistent thickness. Joints with exposed convex faces cannot be tooled to produce adequate pressure on the sealant, increasing the likelihood of in-service problems. The thickness of butt and fillet joints depends on the proper position of the foam backer. Thickness in fillets can vary more than in butt joints because there are no substrate edges to guide the applicator in applying and tooling the joint. Some geometries tend to reduce thickness, particularly with exposed and concealed concave faces. The combination of these two surfaces creates a joint that is sensitive to applicator workmanship and has little tolerances.

As with any sealant joint installation, pre-installation testing and mock-ups are essential.

CONCLUSIONS

The results of this research show the following regarding the behavior of fillet sealant joints.

1. The movement capability of properly designed and constructed fillet sealant joints (Group I) exceeds that of square and rectangular sealant joints.

2. The peak stresses in fillet sealant joints range from approximately those of square sealant joints to less than those of 2:1 ratio rectangular joints with concave faces. The character of the peak stresses on the adhesive bond are similar in fillet and rectangular joints.

3. The ultimate nominal strain in fillet joints (with bond breakers) in tensile tests is comparable to and exceeds that of square joints.

4. The degree of compression set that fillet joints retain is comparable to that of square joints at lower strains.

These findings hold for sealants of different chemical composition.

DESIGN RECOMMENDATIONS

The following recommendations for design of fillet sealant joints subjected to movement perpendicular to their length reflect the current knowledge regarding the behavior of sealant joints and may be revised as additional data become available. Incorporation of adequate bond breakers and proper configurations produces fillet sealant joints that can perform successfully. At this time, however, Group I and Group II fillet sealant joints do not have a long track record of performance to substantiate laboratory results, and they should be used with this understanding.

1. Provide a bond breaker in all fillet joints.

2. Divide the anticipated joint movement by the sealant's movement capability per ASTM C719 to determine the appropriate horizontal projected length of the bond breaker (dimension W in Fig 1). See No. 4 and 5 below for limitations.

3. Joints with overall horizontal or vertical dimensions of more than approximately 25 mm (1 in.) should be used cautiously due to sealant material and application constraints.

4. To maximize movement capability, use diagonal bond breakers or similar geometries that omit the sealant at the root of the fillet. Fillets with concealed faces that are concave perform best, such as with quarter round backer rods. Round backer rods are not typically useful, since they require fillet joints with large sightlines to achieve minimum sealant thicknesses. Tool the face of the joint slightly concave. Provide a sealant thickness measured at the point of minimum cross section of 6 mm (1/4 in.) (dimension T in Fig 1).

5. When diagonal bond breakers cannot be used, provide a flat tape bond breaker on the horizontal and vertical surfaces at the root of the joint. If this is not possible, position the bond breaker on the surface parallel to the direction of

movement. Tool the face of the joint concave. Provide a minimum sealant thickness between the edge of the bond breaker and the exposed surface of 6 mm (1/4 in.) (dimension T in Fig 1). The performance of Group II fillets is not as good as Group I fillets and generally lies between that of rectangular joints and square joints, with some configurations slightly below that of square joints.

ACKNOWLEDGEMENTS

Thanks to Simpson Gumpertz & Heger Inc. for use of their laboratory and computer facilities.

REFERENCES

[1] Myers, J. C. Behavior of Fillet Sealant Joints, Master's Thesis, Department of Materials Science and Engineering, Massachusetts Institute of Technology, Cambridge, MA, June 1989

[2] "Standard Test Method for Adhesion and Cohesion of Elastomeric Joint Sealants under Cyclic Movement, C719," Annual Book of ASTM Standards, Vol 04.07, American Society for Testing and Materials, Philadelphia, 1989.

[3] Cook, Robert D., Concepts and Applications of Finite Element Analysis, Second Edition, John Wiley & Sons, New York, 1981.

[4] Tons, E., "A Theoretical Approach to Design of a Road Joint Seal," Highway Research Board Bulletin, No. 229, 1959, pp. 20-53.

[5] Cook, J. P., "A Study of Polysulfide Sealants for Joints in Bridges," Highway Research Board Bulletin, No. 299, 1965.

[6] Catsiff, E. H., Hoffman, R. F., and Kowalewski, R. T., "Predicting Joint Sealant Performance of Elastomers by Computer Simulation - I. Justification of Method," Journal of Applied Polymer Science, 14, 1970, pp. 1143-1158.

[7] Catsiff, E. H., Hoffman, R. F., and Kowalewski, R. T., "Predicting Joint Sealant Performance of Elastomers by Computer Simulation - II. Results in Simple Extension and Compression." Journal of Applied Polymer Science, 14, 1970, pp. 1159-1178.

[8] Catsiff, E. H., "Predicting Joint Sealant Performance of Elastomers by Computer Simulation - III. Simulation of Single- and Multiple-Step Extension of a Stress-Relaxing Material," Journal of Applied Polymer Science, 15, 1971, pp. 1021-1028.

[9] Evans, R. M., "Evolution of Sealant Durability Testing," Adhesives Age, 23, No. 11, 1980, pp. 31-38.

Andreas T. Wolf

EXPERIMENTAL WORK CARRIED OUT IN SUPPORT OF THE DEVELOPMENT OF AN INTERNATIONAL STANDARD ON MOVEMENT CAPABILITY OF SEALANTS

REFERENCE: Wolf, A.T. "Experimental Work Carried out in Support of the Development of an International Standard on Movement Capability of Sealants", Building Sealants: Materials, Properties, and Performance, ASTM STP 1069, Thomas F. O'Connor, editor, American Society for Testing and Materials, Philadelphia, 1990.

ABSTRACT: The paper compares the results of different lab tests, suggested during the development of an ISO standard on sealant durability, with those obtained from outdoor weathering under mechanical strain. The comparison shows that the 70 C compression test, as included in ASTM C-719, is helpful in identifying sealants, which fail due to high compression set. The heat compression test, however, does not rule out sealants that tend to post-cure or embrittle upon ageing. The best correlation between lab tests and outdoor weathering is obtained, if the sealants are allowed to fully cure prior to lab testing.

KEY WORDS: ISO TC59/SC8, durability test method, mechanical cycling, outdoor ageing

Within the International Standardisation Organisation (ISO), Technical Committee 59/Subcommittee 8 is responsible for the development of specifications for sealants. During the first 8-10 years following the establishment of the committee in 1976, priority was given to the drafting of test standards. These activities were highly successful and a wide range of test procedures have now been published as ISO standards or as Draft International Standards (DIS). Over the last 2-3 years, the work of TC59/SC8 has concentrated increasingly on the discussion of requirement standards for facade and glazing joint sealants.

ISO TC59/SC8 DISCUSSIONS ON A DURABILITY TEST METHOD FOR SEALANTS 1979-1981

DIS 9047 [1] on "Determination of Adhesion/Cohesion Properties at Variable Temperatures" was a test standard which experienced a particularly difficult birth. It is a durability test method - comparable to

Dr. Wolf is Technical Service & Development Section Manager at Dow Corning S.A., Parc Industriel, 6198 Seneffe, Belgium

ASTM C-719 - in which the sealant is subjected to cyclical mechanical and thermal stresses in test joints between concrete or aluminium substrates. The idea came from practical experience: a sealant in a construction joint is, after all, compressed in summer (high temperatures) and extended in winter (low temperatures). The first proposal to TC59/SC8 on the testing of the mechanical durability of sealants was submitted in 1979 by the Norwegian delegate [2]. This test method involved much expenditure of time (120-141 days), and no findings were available on how the results achieved with this method could be compared with the practical behaviour of sealants. The proposal was therefore rejected at the next session of TC59/SC8 in London in 1980. The US delegate at this session described the advantages of the ASTM C-719 test method [3]. He placed particular emphasis on the mechanical stress cycle used in the ASTM C-719 method, i.e. the Hockman cycle. According to him, the introduction of the Hockman cycle in the USA had led to excellent correlation between the behaviour of sealants in the laboratory and in practice.

ASTM C-719 specifies an accelerated cure period of 21 days followed by 7 days water immersion of the specimen at room temperature. The specimen is then compressed and kept in this state for 7 days at 70 C. This is followed by 10 mechanical stress cycles in which the specimen is slowly extended and compressed at a constant temperature of 23 C. If the sealant withstands this stress without total failure of all specimens, it is subjected to a further 10 mechanical stress cycles which, however, involve variation of the temperature between -26 C (extension) and + 70 C (compression). The specimens are examined for adhesion and cohesion cracks after completion of the entire test.

The discussions during the ISO TC59/SC8 session in London and the subsequent session in Ottawa in 1981 revolved around the following two questions: what should be the format of the mechanical stress cycle and what should be the minimum number of cycles to ensure good correlation with the practical behaviour of the sealants? The idea of a divided cycle - first at room temperature and then at variable temperatures, as in C-719 - was rejected because most delegations favoured a combined mechanical and thermal cycle. The second question concerned conditioning: should the conditioning only serve to ensure the complete cure of the sealant or should it also serve as stress? The US delegate explained that according to their experience, the ASTM C-719 compression at 70 C was able to eliminate those sealants with high compression sets which fail in practical applications in buildings. Sealants which are deformed irreversibly by compression in the joint during the first summer are subject in the subsequent winters to much higher extension stress than are those with a low compression set. This usually leads to adhesive or cohesive failure of the sealants during the first few winters. The European delegations favoured 28 days conditioning in a standard climate of 23 C and 50% rel. humidity (referred to hereinafter as 23/50) or 28 days conditioning at 23/50, followed by 3 weeks changing climate conditioning. This changing climate conditioning should comprise 3 repetitions of the UEATC cycle [4] - 3 days 70 C, 1 day water immersion at 20 C, 2 days 70 C, 1 day water immersion at 20 C. The German delegation argued that this UEATC conditioning should transform slow curing sealants into a "testable" state without "ageing" the sealant. In contrast, the French delegation suggested reducing the standard 23/50 climate conditioning to 21 days without subsequent UEATC conditioning, this being intended to shorten the required test period.

The following options for conditioning of the specimens were discussed in Ottawa in 1981:

(a) 28 days 23/50
(b) 28 days 23/50 + 3 UEATC cycles
(c) 28 days 23/50 + 3 UEATC cycles + 70 C compression by x%
(d) 28 days 23/50 + 70 C compression by x%

or, alternatively, 21 days 23/50 in each of (a) - (d) above.

The following procedure was agreed with regard to the test cycles:

Cycle A:
1 hour cooling to - 20 C
2 hours at - 20 C with extension by x%
1 hour heating to + 70 C
2 hours at + 70 C with compression by x%
1 hour cooling to - 20 C
17 hours at - 20 C with extension by x%

Cycle B:
1 hour heating to + 70 C
2 hours at + 70 C with compression by x%
1 hour cooling to - 20 C
2 hours at - 20 C with extension by x%
1 hour heating to + 70 C
17 hours at + 70 C with compression by x%

The specimens should not be subjected to any mechanical forces during cooling and heating. Cycle A must be followed by cycle B, with cycle A+B being run five times in total; the total cycle time should therefore amount to 10 days. On the basis of this test cycle, combined with conditioning types (c) and (d), a test programme was agreed as resolution 34 by the delegations present. The French and German delegations expressed their intention to exchange one silicone, polysulphide and polyurethane sealant between the participating test laboratories in order to obtain information on the reproducibility of the test method. However, as resolution 34 was only presented in draft form during the session, different interpretations of the test procedure emerged. France performed the experiments with 21 days 23/50 conditioning and 10 and 25% compression/extension, while Germany applied 28 days and 12.5 and 25% for the same parameters.

The author considered the number of sealants agreed for the round-robin test to be too low to permit a reliable interpretation of the results. Moreover, resolution 34 only took account of a small number of the conditioning types which were discussed in the course of the meetings. He therefore decided to carry out his own experimental work with a larger number of sealants and most conditioning types which had been discussed.

OWN LABORATORY TESTS IN 1981

The official ISO TC59/SC8 round-robin tests usually employ sealants which comply with the national standards of the various countries participating in the test. The quality of the ISO test proposal is then measured according to whether the sealants examined perform similarly in the round-robin test as they do in the various national test procedures. However, this comparison only indicates whether the relevant ISO proposal is as good or as bad as the existing national standard. For his own experimental work, the author therefore selected not only the usual commercial grades but also laboratory formulations with which some difficulties had been observed in various laboratory or practical tests, even though they had passed different national standards.

The experiments on the basis of resolution 34 were thus performed with 10 silicones (8 acetoxy, 1 benzamide and 1 amine cure), 2 water-based acrylics and a one-part polysulphide.

The author selected the following test parameters for his own experimental programme:

Specimens:

12x12x50 mm^3 tensile joints with aluminium and/or concrete as substrate. 3 specimens per sealant and variable test parameters.

Conditioning:

(1) 28 days 23/50 standard climate
(2) 28 days 23/50 + 3 UEATC cycles (21 days)
(3) 28 days 23/50 + 3 UEATC cycles + 7 days compression at 70 C
(4) 21 days 23/50 standard climate
(5) 21 days 23/50 + 3 UEATC cycles + 7 days compression at 70 C

The proposal discussed in Ottawa for 21 or 28 days 23/50 standard climate conditioning followed immediately by 7 days compression at 70 C was not included in the author's own experimental programme.

Test cycle:

Performing the test cycle as described in resolution 34 - with mechanical and thermal stress on 10 successive days - would have involved the laboratory personnel having to work overtime for two weekends. For this reason, and after consultation with the French and German laboratories participating in the official round-robin test, the timing of the stress cycles was determined as follows:

Week 1: Day 1: Cycle A
Day 2: " B
Day 3: " A
Day 4: " B
Days 5-7: 23/50 conditioning without mechanical stress

Week 2: Day 1: Cycle A
Day 2: " B
Day 3: " A
Day 4: " B
Days 5-7: 23/50 conditioning without mechanical stress

Week 3: Day 1: Cycle A
Day 2: " B
Day 3: 3 hours 23/50 conditioning without mechanical stress and testing of specimens

Extension and compression during the test cycle was achieved by means of mechanical vices (see ref. [5]). Due to the manual adjustment of the vices, the rate of the extension/compression movements could not be controlled exactly, but was about 5-8 mm/min. The total amplitude of the extension/compression process was 25% =+/-12.5% or 50%= +/-25%.

Tensile Test

After the test cycle, the specimens were extended to break in a tensile tester operating at a rate of 6mm/min.

Evaluation:

The following properties were evaluated in the tensile test: 100% modulus (N/mm2), tensile strength (N/mm2), elongation at break (%), and failure mode (adhesive/cohesive).

Table 1 illustrates the results of tests performed on five silicone (3 acetoxy, 1 benzamide and 1 amine system) and one polysulphide sealants.

Because of their fast cure and high elasticity, the conditioning type and the amplitude of the extension/compression cycle have only a marginal influence on the mechanical properties of acetoxy silicones. With regard to their different formulations, slight differences can, however, be observed in the behaviour of the individual acetoxy silicones:

The mechanical properties of the silicone sealant Si1, appear to be practically unaffected by the conditioning method; in particular, the differences observed for the 100% modulus lie within the measurement tolerances. The tensile strength appears to increase slightly if the conditioning period is extended. The amplitude of the extension/compression cycle has an even lesser effect on the mechanical properties of Si1: the values obtained for modulus and elongation at break are almost identical for the various conditioning methods. Increased extension/ compression does, however reveal another phenomenon: After 21 days 23/50 standard climate (conditioning 4), adhesion of the sealant to aluminium is still not fully developed, which results in adhesive failure of the specimens in the extension/compression cycle. After the somewhat longer conditioning method 1, all 3 specimens withstand the extension/ compression cycle, but adhesive failure of 1 specimen occurs in the final extension test.

The phenomenon of the build-up of adhesion can also be seen with sealant Si2. All specimens which had only been conditioned in the 23/50 standard climate for 21 or 28 days failed the adhesion test in the extension/compression cycle. Adhesion has apparently properly developed after conditioning 5. By comparing the results after conditioning 3 and 5 with those of conditioning 2, it can clearly be seen that the 70 C compression reduces the 100% modulus of Si2. There is no direct explanation for this phenomenon.

With sealant Si3, there is a slight increase in the modulus with a longer period of conditioning, and also a reduction in the elongation at break after 70 C compression conditioning.

Adhesion of the benzamide silicone, Si4, develops slowly on aluminium. All specimens which were conditioned according to methods 1 and 4 failed in the adhesion test in the extension/compression cycle. Adhesion is obviously achieved earlier on primed concrete. In view of the very slow cure of the benzamide silicone, the 21 days UEATC conditioning obviously had a major effect on the mechanical properties: Comparing the moduli of the concrete specimens shows that the 100% modulus is doubled to tripled after UEATC conditioning.The 70 C compression combined with the extension/compression cycle causes a deterioration of the elasticity

TABLE 1 -- Results of Lab Tests in 1981

Sealant	Type	Substrate	% Ext./Compr.	Prop.	Conditioning Procedure 1	2	3	4	5
Si 1	Acetoxy	Alu	12.5	MOD 100[a]	0.40	0.40	0.42	0.43	0.42
				TS[b]	0.41	0.48	0.52	0.50	0.48
				EB[c]	120	140	160	140	120
				FM[d]	CF	CF	CF	CF	CF
			25	MOD 100	0.38	0.42	0.41	...	0.42
				TS	0.40	0.45	0.48	...	0.50
				EB	110	120	130	...	150
				FM	2CF,1AF	CF	CF	AF	CF
Si 2	Acetoxy	Alu	25	MOD 100	...	0.47	0.30	...	0.29
				TS	...	0.53	0.51	...	0.42
				EB	...	250	230	...	190
				FM	AF	AF/CF	CF	AF	CF
Si 3	Acetoxy	Alu	12.5	MOD 100	0.42	0.45	0.47	0.44	0.45
				TS	0.52	0.67	0.68	0.72	0.67
				EB	170	220	130	230	210
				FM	CF	CF	CF	CF	CF

[a] MOD 100 : Modulus 100, MPa / [b] TS : Tensile Strength, MPa / [c] EB : Elongation at Break, % / [d] FM : Failure Mode

TABLE 1 (Continued) -- Results of Lab Tests in 1981

Sealant	Type	Substrate	% Ext./Compr.	Prop.	Conditioning Procedure 1	2	3	4	5
Si 4	Benzamide	Alu	12.5	MOD 100	...	0.24	0.29	...	0.23
				TS	...	0.29	0.32	...	0.28
				EB	...	180	150	...	160
				FM	AF	AF/CF	CF	AF	AF
			25	MOD 100	...	0.28	...	...	0.20
				TS	...	0.32	...	...	0.20
				EB	...	130	...	...	140
				FM	AF	AF/CF	CF	AF	AF
		Concrete	12.5	MOD 100	0.13	0.23	...	0.13	0.18
				TS	0.17	0.26	...	0.20	0.18
				EB	250	140	...	250	140
				FM	AF	AF	AF	CF/AF	AF
			25	MOD 100	0.09	0.29	...	...	...
				TS	0.14	0.34	...	...	...
				EB	240	150	...	...	...
				FM	AF	AF/CF	AF	AF	AF

TABLE 1 (Continued) -- Results of Lab Tests in 1981

Sealant	Type	Substrate	% Ext./Compr.	Prop.	Conditioning Procedure 1	2	3	4	5
Si 5	Amine	Alu	12.5	MOD 100	...	0.43	0.43	...	0.43
				TS	...	0.45	0.45	...	0.48
				EB	...	160	140	...	180
				FM	AF	CF	CF	AF	CF
			25	MOD 100	...	0.42	0.39	...	0.42
				TS	...	0.47	0.42	...	0.48
				EB	...	170	150	...	180
				FM	AF	CF	CF	AF	AF
		Concrete	12.5	MOD 100	0.45	0.45	0.48	0.40	0.48
				TS	0.50	0.56	0.52	0.40	0.53
				EB	180	270	260	110	170
				FM	CF	CF	CF	CF	CF
			25	MOD 100	0.48	0.44	0.44	0.40	0.50
				TS	0.52	0.49	0.49	0.48	0.58
				EB	180	160	160	210	180
				FM	CF	CF	CF	CF	CF

TABLE 1 (Continued) -- Results of Lab Tests in 1981

Sealant	Type	Substrate	% Ext./Compr.	Prop.	Conditioning Procedure				
					1	2	3	4	5
PS 1	One-part	Alu	12.5	MOD 100	0.42	0.45	0.38	...	...
				TS	0.52	0.50	0.47	...	...
				EB	150	130	150	...	...
				FM	AF	AF	CF	AF	AF
		Concrete	12.5	MOD 100	0.50	0.43	0.38	0.53	...
				TS	0.58	0.47	0.40	0.62	...
				EB	130	140	130	130	...
				FM	AF	AF	AF	AF	AF

of the benzamide silicone. Plastic deformation behaviour occurs in the final extension test, if the specimens are conditioned according to method 5. Almost no difference can be observed between the 100% modulus and the tensile strength at 140% extension.

The build-up of adhesion described already can also be seen with the amine cure silicone Si5. Otherwise, the various conditioning methods and the amplitude of the extension/compression cycle have only a slight effect on the mechanical properties of Si5.

The one-part polysulphide sealant PS1 appears to soften as the conditioning period is extended. This is in contrast to the increased cure normally observed in polysulphides in the course of the UEATC cycles.

THE ISO TC59/SC8 SESSION IN BERLIN 1982

The results obtained by the French and German delegations in the round-robin test and in the author's own experiments were presented at the ISO TC59/SC8 session in Berlin in 1982 [6]. As the French experiments were performed with slightly different test parameters, it was not possible to achieve full comparability. Nevertheless, the significance of the UEATC cycles for a complete cure of the specimens was also illustrated in these tests. The very slow curing, one-part polysulphide, examined in the round-robin test, was still not fully cured to the depth of the specimen after 28 days 23/50 standard climate + 21 days UEATC cycles. The delegations present in Berlin decided to carry out additional round-robin tests in which two alternative types of conditioning would be permitted:

Method A: 28 days 23/50 standard climate + 7 days 70 C compression by x%
Method B: 28 days 23/50 standard climate + 21 days UEATC cycles + 7 days 70 C compression by x%

The intention was to permit both conditioning types in a subsequent ISO test method alternatively, method A being for fast curing and method B for slow curing sealants. The test parameters were clearly defined in resolution 38.

The author decided to perform further laboratory experiments, but this time mainly using organic sealants to avoid a data imbalance in favour of silicone sealants.

OWN LABORATORY EXPERIMENTS IN 1982

These experiments were again performed with conditioning types 1-3 described previously. Conditioning types 4 and 5 were not taken into account, because the 28 days conditioning period received clear preference during the Berlin meeting. Conditioning type 6, corresponding to type A defined in resolution 38, was added.

The following sealants were selected for this series of experiments: 1 alkoxy silicone, 1 one-part and 2 two-part polysulphides, 1 one-part and 1 two-part polyurethane. All sealants were subjected to +/- 25% amplitude in the extension/compression cycle. The results obtained are illustrated in Table 2.

TABLE 2 -- Results of Lab Tests in 1982

Sealant	Type	Substrate	% Ext./Compr.	Prop.	Conditioning Procedure 1	2	3	6
Si 6	Alkoxy	Alu	25	MOD 100[a]	0.52	0.58	0.61	0.54
				TS[b]	0.67	0.75	0.76	0.71
				EB[c]	160	220	210	210
				FM[d]	CF	CF	CF	CF
PS 2	One-Part	Alu	25	MOD 100	0.23	0.22	0.28	0.25
				TS	0.35	0.30	0.38	0.31
				EB	230	160	160	170
				FM	AF	AF	AF	AF/CF
		Concrete	25	MOD 100	0.27	0.21	...	0.26
				TS	0.42	0.26	...	0.37
				EB	260	140	...	220
				FM	AF	AF	AF	AF

[a] MOD 100 : Modulus 100, MPa / [b]TS : Tensile Strength, MPa / [c]EB : Elongation at Break, % /[d]FM : Failure Mode

TABLE 2 (Continued) -- Results of Lab Tests in 1982

Sealant	Type	Substrate	% Ext./Compr.	Prop.	Conditioning Procedure			
					1	2	3	6
PS 3	Two-Part	Concrete	25	MOD 100	0.28	0.21	0.35	0.33
				TS	0.37	0.26	0.53	0.41
				EB	220	160	180	150
				FM	CF	CF	AF/CF	AF/CF
PS 4	Two-Part	Concrete	25	MOD 100	0.15	0.22	...	0.17
				TS	0.31	0.32	...	0.30
				EB	340	160	...	290
				FM	CF	AF	AF	AF
PU 1	Two-Part	Concrete	25	MOD 100	0.25	0.37	0.39	0.37
				TS	0.30	0.46	0.43	0.37
				EB	170	140	150	100
				FM	CF	CF	CF	CF
PU 2	One-Part	Concrete	25	MOD 100	0.29	0.45	...	...
				TS	0.50	0.49	...	...
				EB	480	160	...	...
				FM	AF	CF	AF	AF

The alkoxy silicone sealant Si6 shows a slight increase in the 100% modulus after the UEATC cycles, while the elongation at break also improves simultaneously. It must therefore be assumed that the cure is still not entirely complete after 28 days 23/50 standard climate. The 7 days 70 C compression has practically no effect on the sealant, regardless of whether the sealant is subjected to it after the UEATC cycles or immediately after the 28 days 23/50 standard climate.

In the case of one-part polysulphide PS2, the UEATC cycles have only a slight influence on the 100% modulus, although the UEATC conditioning causes a sharp reduction in the elongation at break. The specimens are obviously able to withstand the 7 days 70 C compression better if not previously subjected to UEATC cycles. Sealant PS2 shows marked deformation after the 70 C compression, regardless of whether or not this was preceded by UEATC conditioning.

In the case of two-part polysulphide PS3, the UEATC cycles result in a reduction in the 100% modulus, and also a reduction of the elongation at break. The 7 days compression at 70 C causes an increase in the mechanical values, apparently caused by post cure.

The UEATC cycles result in a slight increase in the 100% modulus of the soft-elastic two-part polysulphide PS4, while causing a sharp reduction in elongation at break. This appears to be the reason why the specimens withstand the movement cycle after 7 days compression at 70 C, if it immediately follows the 28 days 25/30 standard climate conditioning, but not if they are exposed to the UEATC cycles in-between (method 3).

The effect of the UEATC cycles on the mechanical properties of the two-part polyurethane PU1 is unmistakable: 100% modulus and tensile strength increase markedly, while elongation at break is reduced. Nevertheless, the specimens withstand the extension/compression cycle after all conditioning methods without any problem. The 7 days heat conditioning, immediately following the 28 days 23/50 standard climate, also results in a post cure. Better mechanical properties are achieved, if the sealant is exposed to the UEATC cycles as well.

During the UEATC cycles, the one-part polyurethane PU2 undergoes pronounced post cure. The 100 % modulus rises by over 50%, elongation at break falls by 2/3. The 7 days compression at 70 C finally causes adhesive failure of all specimens in the movement cycle. The reason for this lies in the fact that the sealant undergoes pronounced post cure when under compression, causing it to be almost "frozen" in the compressed state. The maximum extension achievable from this state was then exceeded in the subsequent movement cycles.

Although many of the experimental results obtained in 1981 and 1982 were easy to interpret, the author remained uncertain about which type of conditioning he should favour until only recently. The fact that the results obtained with some sealants after some types of conditioning were better than in the national standards, and worse after other types of conditioning, was of little concern to him. He was much more interested by the question as to which of the conditioning types correlated most closely with the practical behaviour of the sealants. For this reason, he prepared additional specimens in 1981 and 1982 and exposed them to cyclic mechanical stresses in the outdoor climate.

OUTDOOR WEATHERING WITH ENFORCED MOVEMENT OF THE SPECIMENS 1981/2-1988

The best way of determining whether a laboratory test method produces results relevant to practical conditions is to check it in practice. In this case, that meant exposing the specimens to measurable mechanical stresses in an outdoor climate. The mechanical stress here should correspond as closely as possible to that experienced by a sealant in a facade joint. This is usually achieved with weathering racks in which specimens are subjected to continuous movement induced by the thermal extension of the rack components. These racks are, however, relatively expensive and require much space.

For this reason, in the outdoor weathering tests, the author used the same mechanical vices to mechanically stress the sealants as were used in the laboratory experiments. In order to keep the amount of work involved in testing within reasonable limits, adjustments of the joint width could only be made at longer intervals. This means that the diural movements, which occur in a building joint, must be ignored, and the seasonal movements can only be approximated in stages.

Two specimens per sealant and substrate were prepared and exposed to the outdoor weathering after a cure period of 28 days in 23/50 standard climate. The amplitude of the enforced movement was +/-25% of the joint width. Adjustments were made at monthly intervals, the joint width being adjusted by 1/3 of the maximum compression or extension each time. Maximum extension was achieved in January and maximum compression in July. During the months of April and October, the specimens were held at the standard joint width of 12 mm. Adjustments were performed manually at a rate of approx. 6-8 mm/min.

During the outdoor storage, the specimens were stored horizontally on a sieve plate in order to prevent excessive water ingression. Each time the specimens were extended or compressed, i.e. once a month, the sealants were examined for cracks and tears. Finally, in autumn 1988, after respectively 6 and 7 years of outdoor weathering, the experiment was ended and the specimens subjected to a final examination.

For this purpose, the specimen was flexed round the sealant axis twice by 60° - similar to the C-719 procedure - and the extent of the resultant loss of adhesion was given as a percentage of the total adhesion surface (6 cm^2). The specimen was considered to have passed the test, if the loss of adhesion was less than 20% of the total adhesion surface. This loss of adhesion was allowed because the preparation of specimens, involving the use of spacers, always results in poorer wetting of the substrate at the side edges of the sealant. The dimensioning of the sealant joint with a length of only 50 mm also results in considerably higher stresses at the sealant edges during extension or compression. This local stress is lower in longer specimens, e.g. the 100 mm sealant joints used in earlier test standards. In practice, it would be even lower in a real joint of several meters length. A 20% loss of adhesion in a test joint does not therefore imply failure in practice.

The situation was clear in most cases - after 6 or 7 years weathering either both specimens had failed, or neither of the two had failed. Only in the case of one silicone and one polysulphide sealant did one specimen fail while the other did not. Table 3 compares the evaluation of outdoor conditioning with the results of the extension/compression cycles after conditioning types 3, 5 and 6. Let us first consider the

TABLE 3 -- Comparison of Performance in Outside Weathering with Lab Results

Sealant	Substrate	Outside Weathering	Time to Failure, years	Performance in Lab Cycling After Conditioning 3	5	6
Si 1	Alu	passed	...	passed 25 %	passed 25 %	...
Si 2	Alu	passed	...	passed 25 %	passed 25 %	
Si 3	Alu	failed	3,5	passed 12.5 %	passed 12.5 %	...
Si 4	Alu	passed	...	passed 12.5, failed 25 %	passed 12.5, 25 %	...
	Concrete	passed	...	failed 12.5, 25 %	passed 12.5, failed 25 %	...
Si 5	Alu	passed	...	passed 12.5, 25 %	passed 12.5, 25 %	...
	Concrete	1 failed, 1 passed	6	passed 12.5, 25 %	passed 12.5, 25 %	...
Si 6	Alu	passed	...	passed 25 %	...	passed 25%
PS 1	Alu	failed	2	passed 12.5 %	failed 12.5 %	...
	Concrete	1 failed, 1 passed	5	passed 12.5 %	failed 12.5 %	...

TABLE 3 (Continued) -- Comparison of Performance in Outside Weathering with Lab Results

Sealant	Substrate	Outside Weathering	Time to Failure, years	Performance in Lab Cycling After Conditioning		
				3	5	6
PS 2	Alu	passed	...	passed 25 %	...	passed 25 %
	Concrete	failed	4,5	failed 25 %	...	passed 25 %
PS 3	Concrete	passed	...	passed 25 %	...	passed 25 %
PS 4	Concrete	failed	6	failed 25 %	...	passed 25 %
PU 1	Concrete	passed	...	passed 25 %	...	passed 25 %
PU 2	Concrete	failed	3	failed 25 %	...	failed 25 %

sealants which failed the weathering test: Si3(2), Si5(1), PS1(1), PS2(2), PS4(2), and PU2(2).

Si3 withstood the extension/compression cycle after conditioning types 3 and 5, but still failed the practical test because a considerable hardening of the sealant occurred during outdoor conditioning. The reason for this lies in the formulation of this silicone sealant: Instead of the silicone fluid normally used as a plasticiser, this formulation contains over 12% of a non-silicone plasticiser. Due to the high boiling point and excellent compatibility, this organic extender evaporates from the silicone sealant only very slowly. Because of this, the heat conditioning had only a slight effect on the mechanical properties and also resulted in only slightly higher shrinkage. On outdoor exposure, in contrast, the first specimen failed after three and the second after five years. The indentation hardness rose from its original 18 to 28 Shore A during the five years.

In the case of the amine cure silicone Si5, adhesive failure of a concrete specimen occurred after 6 years of outdoor conditioning. However, the loss of adhesion occurred between the primer and the concrete substrate. The cause must therefore lie in the primer itself or in the quality of the priming work carried out on this specimen. The latter would seem probable, because no problems were discovered with regard to adhesion to concrete in the extension/compression cycles after conditioning types 3 and 5.

The two aluminium specimens of the one-part polysulphide PS1 both failed after 2 years, and one concrete specimen failed after 5 years. Both substrates had been treated with a silane primer before application of the sealant. The adhesion of this sealant was, nevertheless, inadequate, as was also indicated by the laboratory experiments, where adhesive failure usually occurred either in the extension/compression cycle or in the final extension test. Sealant PS1 demonstrated marked hardening after outdoor conditioning. The indentation hardness rose from its original value of 12 to 35 Shore A and cracks formed in the sealant surface perpendicular to the direction of extension/compression. If initially only a sufficient adhesion is combined with a hardening of the sealant due to weathering, adhesive failure is bound to occur over time.

The concrete specimens of the one-part polysulphide sealant PS2 suffered adhesive failure after 4 or 5 years. The sealant had been applied - in accordance with the manufacturer's instructions - without primer. This sealant also consistently suffered adhesive failure in the laboratory experiments in the extension/compression cycle (after conditioning 3) or in the subsequent extension test.

The concrete specimens of the two-part polysulphide sealant PS4 suffered adhesive failure in the last year of outdoor weathering. The sealant had, however, already shown pronounced surface deformations in preceding years, similar to the chewing-gum effect observed in plastic sealants. In addition, in the summer months the sealant surface expelled (through sweating) a material component - probably the plasticiser. It should be added that this formulation contained large quantities of plasticiser, which could also be seen in the initially low modulus and the low elastic recovery. Hardening had occurred over time - the Shore A hardness rose from an initial value of 6 to 18 - resulting in the adhesion being unable to hold the tensile forces when the sealant was extended.

As expected from the laboratory experiments, the one-part polyurethane sealant PU2 underwent pronounced post cure. Both concrete specimens suffered adhesive failure after 3 years weathering. The Shore A hardness rose from an initial value of 12 to 35 over the 3 years.

CONCLUSIONS FROM LABORATORY AND OUTDOOR WEATHERING EXPERIMENTS

If the results of the weathering tests are compared with those of the laboratory experiments, the closest correlation is found if the extension/compression cycles are performed after conditioning 3. Of the total of 16 sealant/substrate combinations, the behaviour in the weathering test is predicted correctly in 10 cases by the laboratory examination after conditioning 3. The prediction is incorrect in 6 cases. The reason for this is that conditioning 3, combined with the extension/ compression cycle, does not ensure sufficient ageing of the sealant. Long-term effects, which play a major role in practice, do not come into effect here.

The experiments performed by the author would suggest that carrying out the extension/compression test after conditioning 3, i.e. after extensive cure of the sealant, permits the closest correlation with practical results. Nevertheless, the following effects were not covered by the laboratory process:

- migration of plasticisers/extenders
- hardening of the sealants as a result of the ageing process

In the case of benzamide silicone, the laboratory process appears to overestimate effects which arise from the very slow cure of this sealant.

PROGRESS OF WORK IN ISO TC59/SC8 1982-88

In view of the central importance of a durability test method for a sealant classification scheme, ISO TC59/SC8 continued vigorously to pursue its work in this field. Further round-robin tests were performed in 1983, in which it became ever clearer how much work was involved with the extension/compression cycle used at the time. This point had already been criticised by the Canadian and American delegates in 1981 and 82. In 1984, the German delegation therefore proposed a new test cycle with only one compression or extension per day, which involved much less work and was suitable for automation at a later date. The number of test cycles was also reduced, thereby allowing the final tensile test to be performed at the end of the second week of the total test cycle. Conditioning types A and B as per resolution 38 were also altered, with the adoption of a 7 days water immersion after the 23/50 standard climate or the UEATC cycles. The water immersion, however, meant an even higher number of conditioning types. As a result, a total of 8 different conditioning types were discussed at the ISO TC59/SC8 session in Philadelphia in 1985, these being derived from various combinations of 23/50 standard climate, UEATC cycles, water immersion and compression at 70 C. Some delegations regarded the large number of possible conditioning types as confusing. In 1987, for example, the British delegation submitted the following comment on discussions in the ISO TC59/SC8: "We object in principle to the large number of alternative procedures allowed, without any criteria to guide the user as to which procedure is appropriate for particular sealants and service conditions". Further round-robin tests showed only slight differences in the behaviour of specimens which had been conditioned in the 23/50 standard climate and UEATC cycles and those which had received additional water immersion and compression at 70 C. For this reason, the number of conditioning types was reduced again to two in draft proposal DP9047 issued in 1987, the water immersion and 70 C compression alternatives being removed. The final extension test was replaced by a visual inspection of the specimens after the extension/compression cycle. This draft proposal was published as Draft International Standard DIS9047 in 1988.

RECOMMENDATIONS FOR A FUTURE ISO DURABILITY NORM

The experiments performed by the author confirm the importance of 70 C compression conditioning for achieving close correlation between extension/compression tests in the laboratory and the practical behaviour of sealants. He therefore recommends a compulsory inclusion of 7 days conditioning under compression at 70 C in the future ISO durability norm . This should not necessitate any new round-robin tests, as a sufficient amount of data has already been gathered both with the old and with the new, simplified test cycle. The author would also suggest that ISO TC59/SC8 should give serious consideration to the ageing behaviour of sealants and to providing appropriate test methods for dealing with this.

ACKNOWLEDGEMENTS

The author would like to express particular thanks to Mr. Alsleben and Mr. Hanke who helped him perform the laboratory experiments in 1981 and 82.

REFERENCES

[1] DIS 9047 "Determination of Adhesion/Cohesion Properties at Variable Temperatures", International Standardisation Organisation, 1988

[2] "Building Sealants, Ability to Accommodate Movements", ISO TC59/SC8 Doc. N52E, Nov. 1979

[3] "Standard Test Method for Adhesion and Cohesion of Elastomeric Joint Sealants Under Cyclic Movements", ASTM C-719, 1972

[4] "UEATC Directive for the Assessment of Building Sealants" Union Europeene pour l'Agrement Technique dans la Construction, 1976

[5] Karparti, K.K. "Device for Weathering Sealants Undergoing Cyclic Movements", Journal of Coatings Technology, Vol. 50, No. 641, June 1978, pp. 27-30

[6] ISO/TC59/SC8 Doc. N93E, May 1982

Thomas F. O'Connor

DESIGN OF SEALANT JOINTS

REFERENCE: O'Connor, T. F., "Design of Sealant Joints," Building Sealants: Materials, Properties and Performance, ASTM STP 1069, Thomas F. O'Connor, Editor, American Society for Testing and Materials, Philadelphia, 1990.

ABSTRACT: Often, thermal movement and other factors are inadequately considered in the design of sealant joints. Deficient design results in premature failure of the sealant and usually water damage to the building and its contents. Performance factors that affect sealant joint design are described, including: thermal movement; moisture movement; live and dead loads; creep, elastic frame shortening, and shrinkage of concrete structures; and material, fabrication, and erection tolerances. Three examples, including calculations, will be presented of joints that accommodate movement: (1) expansion joints in masonry walls, (2) joints that occur between different building systems, and (3) multi-story curtainwalls with floor line expansion joints.

KEYWORDS: construction tolerance, expansion joint, performance factor, sealant joint design, thermal movement

Architects and other design professionals, for aesthetic reasons, have desired to limit the width of sealant joints on the exterior walls of buildings. Frequently the admonition has been "I want to see a 6.4 mm (1/4 in) wide joint". Analysis of the performance factors that affect a sealant joint is necessary to determine if a 6.4 mm (1/4 in) wide joint will be effective in maintaining a durable seal against the passage of air and water. If the performance factors are not understood, and

Mr. O'Connor is a Vice President and Architectural Consultant at Smith, Hinchman & Grylls Associates, Inc., 455 W. Fort Street, Detroit, MI. 48226

included in the design of the joint, then failure is a distinct possibility.

Failure can result in increased building energy usage, water infiltration, and deterioration of building systems and materials. Infiltrating water can cause spalling of friable building materials such as concrete, brick, and stone; as well as, corrosion of ferrous metals and rotting of organic materials, among other effects. Deterioration is often difficult and very costly to repair, with the cost of repair work usually exceeding the original cost of the work.

The objective of this paper will be to: (1) describe some of the performance factors that are considered in sealant joint design, and (2) provide sample calculations for sealant joint width. The selection and properties [1] and installation [2] of sealants will not be described.

PERFORMANCE FACTORS

The following describes some of the performance factors that influence sealant joint design.

Material Anchorage

The type and location of various wall anchors has an impact on the performance of a sealant joint. Large precast concrete panels with fixed and moving anchors, brick masonry relieving angle deflection, and metal and glass curtainwall fixed and moving anchorages are examples of anchorage conditions that must be evaluated when designing sealant joints for movement. These anchors have an effect on the length of wall material or the deflection characteristics to be included in the design of the sealant joint width.

Thermal Movement

Walls of buildings respond to ambient temperature change and solar radiation by either increasing or decreasing in linear dimension. This change also causes a change in the width of a sealant joint opening, producing a movement in the sealant. Determining realistic wall or material temperatures, to establish the expected degree of movement, can be challenging. The author uses reference [3] which lists winter and summer design dry bulb air temperatures for many cities, to help in establishing wall or material temperatures for use in joint width calculations.

The coldest winter wall surface temperature, Tw, is established using the winter design dry bulb air temperature. This is realistic since the exterior wall surface will be within a few degrees of the air temperature, depending on the degree of insulation of the wall. Reference [3] lists 99 and 97.5 percent design dry bulb temperatures. Using the 99 percent value, which represents a total of 22 hours during the winter at or below the listed temperature, seems appropriate. Steady state heat flow calculation to determine the surface temperature does not seem necessary for most uses.

The hottest summer wall surface temperature, Ts, is established as follows. The starting point is reference [3] which lists summer 1, 2.5, and 5 percent design dry bulb air temperatures. Using the 1 percent value to establish the hottest summer air temperature, Ta, which represents a total of 30 hours during the summer at or above the listed temperature, seems appropriate.

Walls also receive solar radiation which warms the wall surface. Wall surface radiative heat gain, XS, must be included with Ta when establishing Ts, which is found using equation 1 [4] [5].

$$Ts = Ta + XS \qquad (1)$$

where
Ts = hottest summer wall surface temperature,
Ta = hottest summer air temperature,
XS = wall surface radiative heat gain,
and
S = wall material solar absorption coefficient,
X = 56 (100) for low heat capacity materials,
= 42 (75) for high heat capacity materials,
= 72 (130) solar radiation reflected on low heat capacity materials,
= 56 (100) solar radiation reflected on high heat capacity materials.

Constant X (shown in both metric and inch-pound units) should be used with the following guidelines. Low heat storage capacity materials or walls are represented by well insulated metal panel curtainwalls; and, high heat storage capacity materials or walls are represented by precast concrete or brick masonry. If light colored or other reflective adjacent surfaces reflect solar radiation to the wall surface, then constants that include this effect should be used in lieu of the previous constants [5].

Equation 1 is for Easterly, Westerly, and Southerly facing walls that receive direct solar radiation. Even Northerly facing walls receive some direct solar radiation during the summer, either in the morning or late in the afternoon. During the day, when there is no direct solar radiation on a Northerly facing wall, there is an indirect solar warming effect caused by diffuse radiation reflected onto the wall from particles suspended in the atmosphere. The increase to the hottest summer air temperature due to this effect is relatively small, probably in the range of 6 to 17°C (10 to 30°F), depending on locale, orientation, wall materials, and degree of insulation. Since most sealant joint widths are established based on the worst exposure, with that joint width repeated for all the building exposures, the calculation of a solar radiation component for a Northerly facing exposure is usually not necessary.

Table 1 lists some values for S to use in equation 1 [4] [5] [12].

TABLE 1 -- Solar Absorptivity Coefficients

Material	Coefficient, S
Aluminum, clear finish	0.60
Aluminum paint	0.40
Mineral Board, uncolored	0.75
Mineral Board, white	0.61
Brick, light buff (yellow) color	0.50-0.70
Brick, red color	0.65-0.85
Brick, white	0.25-0.50
Concrete, uncolored	0.65
Copper, tarnished	0.80
Copper, patina	0.65
Galvanized Steel, unfinished	0.90
Galvanized Steel, white finish	0.26
Glass, clear, 6mm (1/4 in.)	0.15
Glass, tinted, 6mm (1/4 in.)	0.48-0.53
Glass, reflective, 6mm (1/4 in.)	0.60-0.83
Marble, White	0.58
Surface Color, Black	0.95
Surface Color, Dark Grey	0.80
Surface Color, Light Grey	0.65
Surface Color, White	0.45
Tinned Surface	0.05
Wood, Smooth	0.78

The use of reference [3] to establish Tw and equation 1 for Ts will provide temperatures for sealant joint width calculations. Judgement, based on past experience and an evaluation and understanding of the wall construction, must still be used by the design professional in establishing these values.

In addition to the temperature extremes the material will experience, the linear coefficient of thermal movement must also be determined. The amount of linear thermal movement is usually expressed, for each material used in construction, as an increase or decrease per unit length per degree change in temperature of the material. Reference [2] includes a table that lists the linear coefficient of thermal movement for many of the materials commonly used in construction. For construction that is a composite of materials, an appropriate coefficient of thermal movement must be determined for the composite assembly.

Moisture Induced Movement

Some materials respond to changes in water or water vapor content by increasing in dimension when water content is high and decreasing in dimension when water content is low [6]. Materials susceptible to this reversible effect include; some natural building stones, concrete, face brick, and concrete block. For some materials there is also an irreversible change in dimension. For sealant joints, the dominant effect on a reversible change in joint width is usually due to temperature change of a wall surface material. The inclusion of reversible moisture induced movement to thermal movement may not be a truly additive effect. Moisture content tends to decrease with a rise in wall surface temperature and increase with a drop in wall surface temperature, thereby producing movements that are somewhat compensating but not necessarily occurring simultaneously. The net sealant joint movement due to thermal and moisture effects, may be difficult or impossible to determine so some judgement must be used by the design professional when reversible moisture induced movement is considered.

Some materials that are known to exhibit moisture induced reversible movement, Mr, as well as irreversible, Mi, are listed in Table 2 [7]. Reversible movement is based on the likely extremes of in-service moisture content and irreversible movement on the period from manufacture to maturity.

TABLE 2 -- Moisture Induced Movement

Material	Movement, Percent Reversible Mr	Movement, Percent Irreversible Mi
Concrete, gravel aggregate	0.02-0.06	0.03-0.08 (-)
Concrete, limestone aggregate	0.02-0.03	0.03-0.04 (-)
Concrete, lightweight aggregate	0.03-0.06	0.03-0.09 (-)
Concrete block, dense aggregate	0.02-0.04	0.02-0.06 (-)
Concrete block, lightweight agg.	0.03-0.06	0.02-0.06 (-)
Face brick, clay	0.02	0.02-0.07 (+)
Limestone	0.01	NA
Sandstone	0.07	NA

The use of steel reinforcement will usually lessen the above concrete values. (-) indicates a reduction, (+) an increase in dimension, and NA not available. In general, cement-based products shrink and fired clay products expand irreversibly as they equilibrate with the environment.

Live Load Movement

Deflection caused by structure or floor live loading should be considered for horizontal sealant joint openings, such as at floor joints in multi-story construction. The building structural engineer can supply live load deflection criteria for sealant joint design.

Actual live loads can be highly variable [8]. A multi-story building, with the same design live load for all floors, will have the actual live load vary from floor to floor and from one area of a floor to another. Very rarely will the live load be uniform everywhere. Where live load and thus deflection of a structure varies, the relative difference in live load deflection between floors should be considered in joint width design.

Most often live load deflection occurs after the joint has been sealed and, therefore, could be considered an irreversible narrowing of the joint opening, provided the loading conditions remain relatively static. If live loading will be highly variable, such as in a warehouse, then deflection could be treated as a reversible movement. The design professional should evaluate these situations and determine how the live load deflection is best accommodated.

Dead Load Movement

Deflection caused by structure or floor dead loading should also be considered for horizontal sealant joint openings. The building structural engineer can supply dead load deflection criteria for sealant joint design.

Dead load deflection of a structure usually occurs before a joint is sealed. There may be a portion that could occur after a joint has been sealed, for instance, when fixed equipment may be installed. Dead load deflection is an irreversible narrowing of the sealant joint opening width for most applications.

Wind Load Movement

Depending on building type, framing system, and the anticipated wind load, lateral sway or drift of the building and its affect on the sealant joint in the wall may have to be considered. The per story lateral sway or drift can be determined by the building structural engineer. Lateral sway or drift can occur both normal to and in the plane of the wall and both affects on the sealant joint will need to be considered.

Elastic Frame Shortening

Multi-story concrete structures, and to a lessor degree steel, shorten elastically almost immediately due to the application of loads [8] [9]. Frame shortening, the degree of which can be determined by a structural engineer, will cause an irreversible narrowing of horizontal sealant joint openings in multi-story construction. Frame shortening can be compensated for by building each floor level slightly higher, in effect negating most of the shortening, or the narrowing of joint width can become another performance factor considered in the design of the sealant joint.

Creep

The time dependent deformation of materials while loaded, in particular a concrete structure, should be included in sealant joint design. This deformation, which occurs at a decreasing rate as time progresses, can cause a continuing decrease in the width of horizontal joint openings in multi-story and other buildings. Creep, in contrast to elastic frame shortening, can occur over a long period of time [8] [9]. The building structural engineer can provide creep deflection criteria for sealant joint design.

Shrinkage

Concrete framed structures will undergo long-term shrinkage for a period of months [8] [9]. Shrinkage is due to the initial drying of the concrete mix water. The rate of shrinkage is dependent on the amount of water present, ambient temperatures, rate of air movement, relative humidity of the surrounding air, the shape and size of the concrete section, and the amount and type of aggregate in the concrete mix, among others. Table 2 and Reference [7] list guidelines for some shrinkage values for concrete and other materials. Shrinkage criteria can be provided by a structural engineer and included in sealant joint design or can be compensated for in the formwork. Shrinkage effects should be included in the design of horizontal joints in multi-story construction.

Construction Tolerances

ASTM and industry trade associations, among others, establish industry recognized standards for construction tolerances. For some materials or systems there are no industry recognized tolerances or the available tolerances are not directly applicable. In these instances, the design professional should evaluate the conditions and establish tolerances for the work. Tolerances should be indicated on the contract documents for the building, since they establish a level of quality and may affect the cost and performance of the work.

Material: Construction materials have a permissible variation for the exactness of their dimensions. For example, a face brick is nominally 57 mm (2-1/4 in) high by 203 mm (8 in) long by 89 mm (3-1/2 in) thick. Depending on the type of brick, the permissible manufacturing tolerances or variation could be as much as 4 to 6.4 mm (5/32 to 1/4 in) for a 203 mm (8 in) dimension, as established by Reference [10]. Variation of material dimension may have to be included as a performance factor in the design of a sealant joint. If material tolerance is not considered, a deficiently sized joint width could result.

Fabrication: Fabricated assemblies of materials also have dimensional tolerances. Usually factory fabrication will allow the use of smaller tolerances than job site fabrication. For example, unitized curtainwall frames fabricated in a factory may permit +/- 1.6 mm (1/16 in) tolerance on the length and width of the frames, while job site assembly of a face brick wall, may permit no better than +/- 6.4 mm (1/4 in) tolerance for the width of an expansion joint opening in the wall. Realistic fabrication tolerances should be established and enforced so that the designed joint opening width is attained and sealant performance isn't compromised, especially by a joint opening that becomes too narrow.

Erection: Frequently, wall materials or systems cannot be placed on a building exactly where called for by the contract documents. Some location tolerance for building components should be provided so that a deficient joint opening width does not occur. For example, a unitized curtainwall panel may be erected no closer then +/- 3.2 mm (1/8 in) to height or lateral locations shown by the contract documents. This erection variation will affect the width of sealant joints that occur between unitized frames. Erection tolerances must be intelligently developed so that they are realistic and also attainable at the building site.

BASIC JOINT MOVEMENTS

In addition to the effect of performance factors, there are four basic movements that butt type sealant joints experience. These movements are: compression, extension, longitudinal extension, and transverse extension. Longitudinal and transverse extension produce a shearing effect on the sealant joint. Thermal movement is usually the largest contributor; however, other performance factors can contribute to producing these movements. The following describes these movements.

Compression

A sealant joint that experiences predominantly compression (a narrowing of the opening width), is typically one where the sealant is installed during the cool or cold months of the year. Therefore, when the warm summer months occur the thermal growth of adjacent materials causes a narrowing of the joint opening which compresses the sealant.

Extension

A sealant joint that experiences predominantly extension (an increase in the opening width), is typically one where the sealant is installed during the warm months of the year. Therefore, when the cool or cold months occur the thermal contraction of adjacent materials causes a widening of the joint opening which stretches the sealant.

Extension and Compression

A sealant joint installed during the fall and spring months, or when temperatures are moderate, can experience both compression as well as extension since the sealant isn't installed at or near the hottest or coldest design temperatures. This results in compression during the summer months and extension during the winter months.

Longitudinal Extension

A sealant joint that experiences longitudinal extension (a vertical lengthwise displacement of one side of the joint opening relative to the other), is typically one that has different materials or systems forming the sides of the joint. For example, a brick masonry wall for one side and an aluminum curtainwall mullion for the other. These materials have different responses to temperature change (the brick will change dimension less than the aluminum), resulting in a diagonal lengthening of the sealant due to the differential movement between the materials in a vertical direction. This movement is dependent on each materials support locations and the unrestrained length of the respective materials, and usually reaches a peak only along part of the length of the sealant joint.

Transverse Extension

A sealant joint that experiences transverse extension (an out of plane movement, crosswise to the joint face, of one side of the joint opening relative to the other), is typically one that occurs at the juncture of walls that change plane, such as at a corner. As the materials forming the sides of the joint experience thermal movement, a diagonal lengthening of the sealant can occur crosswise to the plane of the sealant joint face.

Combinations of Movements

Frequently, joints must accommodate more than one of the above described movements. Examples include, the previously described extension with compression, as well as extension and/or compression combined with longitudinal or transverse shear. The design professional should evaluate the types of movement the joint will experience and design accordingly.

ILLUSTRATIVE EXAMPLES

For this paper, the following three examples of butt sealant joints will be illustrated with sample calculations. The examples are presented in both metric (on the left side) and inch-pound units (on the right side). Rounding-off has been performed so there will not be exact agreement.

Example One: Brick Masonry Expansion Joints

Brick masonry is used for the exterior walls of buildings because of its durability and aesthetic appearance. However, to maintain durability, the masonry surface must be divided into appropriately sized panels, separated by expansion joints, to control thermal movement and other performance factors. If properly designed expansion joints are not provided they will occur as uncontrolled cracking, with sometimes disastrous results.

For a building in Detroit, Michigan, the architect has decided that a red brick masonry wall will have expansion joints at 7.32 m (24 ft) maximum spacing. An extruded brick with little dimensional variation will be used which, in conjunction with the detailing of the wall, will permit a tolerance of +/- 3.2 mm (1/8 in) for the constructed joint opening width. No construction is expected to occur below 4°C (40°F). The above, as well as other performance factors that will effect joint design, are listed below as well as the source for the data.

Expansion joint spacing	7.32 m (24 ft)
Construction tolerance	+/-3.2 mm (1/8 in)
Coldest air temperature, Tw [3]	-16°C (3°F)
Hottest air temperature, Ta [3]	33°C (91°F)
Min. installation temperature, Ti	4°C (40°F)
Solar absorption coef., S (Table 1)	0.65-0.85
Moisture movement, Mr (Table 2)	0.02 percent
Moisture movement, Mi (Table 2)	0.02-0.07 percent
Thermal movement coef., A [2]	0.0000065 mm/mm/°C (0.0000036 in/in/°F)

The expected brick masonry thermal movement is determined as follows. The specific solar absorption coefficient for this brick is not known, so to be safe, the high end of the Table 1 range is used. The hottest summer wall surface temperature, Ts, is found by equation 1, using a constant for a high heat capacity material.

$$Ts = Ta + 42S = 33 + 42(0.85) = 69°C \quad \text{and} \quad Ts = Ta + 75S = 91 + 75(0.85) = 155°F$$

The coldest winter wall surface temperature, Tw, is -16°C (3°F). The maximum expected temperature difference, Td, is found using equation 2.

$$Td = Ts - Tw \qquad (2)$$

where
Td = temperature difference,
Ts = hottest summer wall surface temperature,
Tw = coldest winter wall surface temperature.

substituting

$$Td = 69 - (-16) = 85°C \quad \text{and} \quad Td = 155 - 3 = 152°F$$

If the installation temperature, Ti, is not known, the above values would be used. Since Ti is known, then Td is found using equations 3 and 4.

$$Td = Ts - Ti \quad (3) \quad \text{and} \quad Td = Ti - Tw \quad (4)$$

where
Td = temperature difference,
Ts = hottest summer wall surface temperature,
Ti = installation temperature,
Tw = coldest winter wall surface temperature.

substituting

summer temperature difference using equation 3

$$Td = 69 - 4 = 65°C \quad \text{and} \quad Td = 155 - 40 = 115°F$$

winter temperature difference using equation 4

$$Td = 4 - (-16) = 20°C \quad \text{and} \quad Td = 40 - 3 = 37°F$$

For joint design the largest difference would be used, which for this example is Td = 65°C (115°F). Thermal movement is determined using equation 5.

$$Lt = (L)(Td)(A) \qquad (5)$$

where
Lt = change in dimension due to thermal movement,
L = expansion joint spacing,
Td = temperature difference,
A = thermal movement coefficient.

substituting

$$Lt = (7.32)(1000)(65)(0.0000065)$$
$$= 3.0927 \text{ mm}$$

and

$$Lt = (24)(12)(115)(0.0000036)$$
$$= 0.1192 \text{ in}$$

The effect of moisture growth is determined as follows. The reversible moisture content value, Mr, is based on measurements made from an extreme wet (but not submerged) to an extreme dry external exposure [7]. A review of the technical data for this particular brick indicates a low rate of water absorption and neither an extreme wet or dry exposure is expected. The reversible moisture growth is expected to be negligible and will not be included. For irreversible moisture growth, Mi, a range of values is indicated. If specific data for a particular brick is unavailable it is advisable to use the Table 2 maximum value. Irreversible moisture movement is determined using equation 6.

$$Lm = (Mi/100)(L) \qquad (6)$$

where
- Lm = change in dimension due to moisture movement,
- Mi = irreversible moisture movement percent expressed as a decimal,
- L = expansion joint spacing.

substituting

$$Lm = (0.07/100)(7.32)(1000)$$
$$= 5.124 \text{ mm}$$

and

$$Lm = (0.07/100)(24)(12)$$
$$= 0.2016 \text{ in}$$

For this example, the expected reversible movement is caused by only thermal movement. There are two factors that contribute to a permanent narrowing of the joint: construction tolerance and irreversible moisture growth. The required joint width is determined using the movement capacity of the sealant and the thermal movement criteria to calculate a width to satisfy reversible movement. The permanent narrowing effects are then added to that width to arrive at the designed width of the joint opening.

A sealant with a +/- 50 percent movement capacity will be specified. The author does not think it is prudent to use a sealant at its rated movement capacity. Doing so does not provide any insurance against unknowns and any imprecision in establishing surface temperature and other performance factors. It also provides some extra capacity for unexpected conditions such as quality of workmanship. It should be noted that this approach, to the author's knowledge, is not universally followed by sealant manufacturers and other designers. Presently, the author uses a sealant at 80 percent of its rated movement capacity to provide capacity for unknowns or error. The amount of reduction should depend on the particular circumstances of a joint design. For this example, using the +/- 50 percent sealant at +/- 40 percent seems appropriate. Based on this, the required joint width, Wm, to satisfy the reversible movement criteria is found using equation 7.

$$Wm = Lt/B \qquad (7)$$

where

Wm = joint width required for movement,
Lt = change in dimension due to thermal movement,
B = sealant movement percentage expressed as a decimal.

substituting

$$Wm = (3.0927)/0.40 = 7.7318 \text{ mm} \quad \text{and} \quad Wm = (0.1192)/0.40 = 0.2980 \text{ in}$$

The final designed width, W, of the sealant joint is determined using equation 8.

$$W = Wm + Lm + C \qquad (8)$$

where

W = final designed joint width,
Wm = joint width required for movement,
Lm = change in dimension due to moisture movement,
C = construction tolerance.

substituting

$$W = 7.7318 + 5.124 + 3.2 = 16.0558 \text{ mm}$$

and

$$W = 0.2980 + 0.2016 + 0.125 = 0.6246 \text{ in}$$

and rounding off

$$W = 16 \text{ mm } (5/8 \text{ in})$$

The contract documents should indicate the final designed joint width with the permissible construction tolerance as follows.

16 mm (+/- 3.2 mm) and 5/8 in (+/- 1/8 in)

Reference [2] should be consulted for guidelines to establish the depth of sealant required for this joint width. If depth is not properly established, the ability of the sealant to accommodate movement can be seriously compromised.

For those who are more mathematically oriented, a detailed approach to establishing masonry surface temperatures and thermally induced movement of the brick is available in reference [12].

Example Two: Joints Between Different Materials or Systems

Frequently different materials or building systems abut on the face of buildings. The interface is usually a sealant joint which is a means of transition and a seal. Often, these sealant joints are subject to combinations of thermal movements that occur in more than one direction and perhaps differing tolerances of construction from the materials or systems on both sides of the joint. An often seen failure of these joints is longitudinal tearing of the sealant.

The masonry wall from Example One terminates at a clear anodized aluminum and glass curtainwall enclosed lobby. The curtainwall and masonry wall are both 7.32 m (24 ft) high and are separated by a sealant joint. Both wall systems are supported at the bottom on a foundation wall; therefore, all vertical thermal movement occurs at the top of the walls. Since the masonry wall will be erected before the curtainwall, the curtainwall could be erected at any temperature including the extremes. The expected construction tolerances for the sealant joint are +/- 3.2 mm (1/8 in) for the location of the end of the brick wall, Cb, and since the curtainwall will be erected as a stick system in the field, it is reasonable to expect +/- 3.2 mm (1/8 in) for exactness of location of the vertical mullions, Ca. The above, as well as other performance factors that will affect joint design, are listed below. The same data sources for the previous example were used.

Height of masonry wall	7.32 m (24 ft)
Height of curtainwall	7.32 m (24 ft)
Construction tolerance, Cb	+/-3.2 mm (1/8 in)
Construction tolerance, Ca	+/-3.2 mm (1/8 in)
Coldest air temperature, Tw	-16°C (3°F)
Hottest air temperature, Ta	33°C (91°F)
Solar absorption coef., brick, Sb	0.65-0.85
Solar absorption coef.,alum., Sa	0.60
Moisture movement, Mi	0.02-0.07 percent
Thermal movement coef., Ab brick	0.0000065 mm/mm/°C (0.0000036 in/in/°F)
Thermal movement coef., Aa alum.	0.0000238 mm/mm/°C (0.0000132 in/in/°F)

The expected thermal movement of the sealant joint separating the two systems has a vertical and horizontal component. The horizontal component causes a compression and extension of the joint while concurrently the vertical component causes a longitudinal shearing effect on the sealant. For this example, the horizontal component for the curtainwall will be accommodated within the curtainwall system, while the horizontal component for the brick masonry will be accommodated by the sealant joint. The vertical component is the differential movement that occurs between the increase or decrease in height of the two systems, due to their differing surface temperatures and coefficients of thermal movement, and the irreversible moisture growth of the brick masonry.

The expected thermal movement is determined as follows. The masonry wall surface temperatures were previously determined and are Tw = -16°C (3°F) and Ts = 69°C (155°F) with a maximum expected temperature difference of Td = 85°C (152°F). The masonry wall expansion joints are spaced at 7.32 m (24 ft) and one-half of that wall length should contribute to sealant joint movement. Some consultants indicate that the thermal expansion of extruded face brick in a vertical direction (the direction of the coring) is perhaps 22 percent greater than in the horizontal [11] [12]. Until independent testing provides values to be used, it would seem prudent to use the 22 percent increase. Thermal movement of the brick wall is determined using equation 5 as before.

horizontal thermal movement

$$Lt = 3.66(1000)(85)(0.0000065) = 2.0222 \text{ mm}$$

and

$$Lt = 12(12)(152)(0.0000036) = 0.0788 \text{ in}$$

vertical thermal movement

$$Lt = 7.32(1000)(85)(0.0000065)(1.22)$$
$$= 4.9340 \text{ mm}$$

and

$$Lt = 24(12)(152)(0.0000036)(1.22)$$
$$= 0.1923 \text{ in}$$

Thermal movement of the vertical aluminum mullion is as follows. The surface temperature extremes are Tw = -16°C (3°F) and by equation 1, using a constant for a low heat capacity material, Ts = 67°C (151°F) and by equation 2, Td = 83°C (148°F). Thermal movement is found using equation 5.

$$Lt = 7.32(1000)(83)(0.0000238)$$
$$= 14.4599 \text{ mm}$$

and

$$Lt = 24(12)(148)(0.0000132)$$
$$= 0.5626 \text{ in}$$

The effect of irreversible moisture movement of the face brick is found using equation 6 as follows.

horizontally

$$Lm = (0.07/100)(3.66)(1000)$$
$$= 2.562 \text{ mm}$$

and

$$Lm = (0.07/100)(12)(12)$$
$$= 0.1008 \text{ in}$$

vertically

$$Lm = (0.07/100)(7.32)(1000)$$
$$= 5.124 \text{ mm}$$

and

$$Lm = (0.07/100)(24)(12)$$
$$= 0.2016 \text{ in}$$

The sealant from Example One will be used at +/- 40 percent movement capacity. The joint width required to satisfy the horizontal thermal movement of the face brick is found by equation 7 as follows.

$$Wm = 2.0222/0.40 = 5.0555 \text{ mm} \quad \text{and} \quad Wm = 0.0788/0.40 = 0.1970 \text{ in}$$

The joint width required to satisfy the vertical component of thermal movement and the irreversible face brick growth, both of which cause a shearing effect on the sealant, is found as follows. The net vertical thermal movement, for this example, is the relative difference between the thermal movements of the two different materials that form the joint sides and is as follows.

14.4599 - 4.9340 = 9.5259 mm

and

0.5626 - 0.1923 = 0.3703 in

To this value is added the expected irreversible face brick growth. Although this is not a reversible movement it causes a diagonal lengthening of the sealant depending on the time of year that the curtainwall is installed. Since it could be installed at either extreme it must be included as follows to determine the total vertical movement.

9.5259 + 5.124 = 14.6499 mm

and

0.3703 + 0.2016 = 0.5719 in

The total vertical movement produces a diagonal lengthening of the sealant. To not exceed the +/- 40 percent capacity of the sealant, the diagonal length of the sealant after movement must not be greater than the joint width at rest plus 40 percent of that width. A simple trigonometric equation will provide the required joint width. If X represents the sealant joint width at rest, Y the total vertical movement, and 1.4X the diagonal sealant length after movement, the equation to find the joint width required to satisfy vertical movement is:

$$X^2 + Y^2 = (1.4X)^2 \qquad (9)$$

solving for X

$$X = \sqrt{Y^2/0.96}$$

where
X = joint width to accommodate vertical movement,
Y = total vertical movement.

substituting

$$X = \sqrt{(14.6499)^2/0.96}$$
$$= 14.9520 \text{ mm}$$

and

$$X = \sqrt{(0.5719)^2/0.96}$$
$$= 0.5837 \text{ in}$$

resulting in

joint width required for vertical movement, Wv, of 14.9520 mm (0.5837 in)

The designed width of the sealant joint is determined using equation 10 as follows.

$$W = Wv + Wh + Lm + Ca + Cb \quad (10)$$

where
W = final designed joint width,
Wv = joint width required for vertical movement,
Wh = joint width required for horizontal movement,
Lm = change in dimension due to moisture movement,
Ca = construction tolerance, aluminum,
Cb = construction tolerance, brick.

substituting

$$W = 14.9520 + 5.0555 + 2.562 + 3.2 + 3.2$$
$$= 28.9695 \text{ mm}$$

and

$$W = 0.5837 + 0.1970 + 0.1008 + 0.125 + 0.125$$
$$= 1.1315 \text{ in}$$

and rounding off

W = 29 mm (1-1/8 in)

The contract documents should indicate the final designed joint width with the permissible construction tolerance as follows.

29 mm (+/- 6.4 mm) and 1-1/8 in (+/- 1/4 in)

Reference [2] should be consulted for guidelines to establish the depth of sealant required for this joint width. If depth is not properly established the ability of the sealant to accommodate movement can be seriously compromised.

Example Three: Multi-Story Building Floor Line Joints

Curtainwalls on multi-story buildings are supported at each floor level with horizontal sealant joints also occurring at each floor level. These sealant joints must respond to many performance factors related to curtainwall design and materials, building floor loading and usage, characteristics of the building framing system, and construction tolerances among others.

For this last example a reinforced concrete framed multi-story building in Detroit, Michigan is clad with a dark anodized aluminum and glass curtainwall system. The curtainwall has a sawtooth profile so there will be reflected solar radiation to be included in the thermal movement calculations. The curtainwall is a unitized, shop-fabricated system, with the panels being one module wide by one floor high. At every floor level there is a horizontal sealant joint. The performance factors that will affect the joint design are listed below. The same data sources for Example One were used.

Spacing of sealant joints	3.962 m (13 ft)
Construction tolerance, Cf	+/-1.6 mm (1/16 in)
Construction tolerance, Ce	+/-3.2 mm (1/8 in)
Wind drift, normal and in plane, D	3.2 mm (1/8 in)
Coldest air temperature, Tw	-16°C (3°F)
Hottest air temperature, Ta	33°C (91°F)
Min. installation temperature, Ti	0°C (32°F)
Max. installation temperature, Ti	38°C (100°F)
Solar absorption coef., S, drk gry	0.80
Thermal movement coef., A, alum.	0.0000238 mm/mm/°C (0.0000132 in/in/°F)
Spandrel beam creep deflection, Cc	4.8 mm (3/16 in)
Spandrel beam live load defl., L	1.6 mm (1/16 in)

There are other performance factors related to the building framing system that must be evaluated.

The concrete columns will shorten due to shrinkage, creep, and elastic frame shortening. Some of these effects will occur before the curtainwall is erected and joints sealed. Theoretically, the total amount of each of these effects will not have to be compensated for in the joint design; however, determination of the partial values may be difficult. It has been reported that about 40 to 70 percent of the total axial shortening, including creep, takes place before a curtainwall system is erected. The author has seen total values for these effects range from 3.2 to 6.4 mm (1/8 to 1/4 in) for the indicated joint spacing. For this example, the formwork will be constructed slightly higher to compensate for most of these effects so they need not be considered directly in the sealant joint design. The use of the sealant at less than its rated movement capacity should provide enough extra capacity to compensate for the remaining indeterminate column shortening effects. The spandrel beam will also experience creep and shrinkage which is expressed as the single creep value above. The spandrel beam will be cambered to compensate for dead load deflection.

The expected thermal movement is determined as follows. The surface temperature extremes are Tw = -16°C (3°F) and by equation 1, using a constant for a low heat capacity material with reflected solar radiation, Ts = 91°C (195°F). Td is found using equations 3 and 4.

summer temperature difference using equation 3

Td = 91 - 0 and Td = 195 - 32
= 91°C = 163°F

winter temperature difference using equation 4

Td = 38 - (-16) and Td = 100 - 3
= 54°C = 97°F

For joint design, only the largest temperature difference will be used, which in this example is Td = 91°C (163°F). Vertical mullion thermal movement is found using equation 5.

Lt = 3.962(1000)(91)(0.0000238)
= 8.5809 mm

and

Lt = 13(12)(163)(0.0000132)
= 0.3356 in

The actual temperatures of the vertical mullions will be somewhere between the exterior winter or summer extreme and the interior ambient temperature. The actual temperature of the mullion will depend, among others, on degree of system insulation and the presence and quality of thermally broken mullions. However, use of the extreme surface temperatures will be conservative and provides some insurance if the building is unheated or uncooled for a period of time (i.e. power failure), and if the building thermostats may be set back to a minimum value over weekends or other extended periods.

Wind drift, D, the per story lateral drift or sway, can be compensated for with a trigonometric formula similar to as used in Example Two. By using equation 9 both the in-plane and normal sealant movement can be determined.

substituting

$$X = \sqrt{(3.2)^2/0.96}$$
$$= 3.2660 \text{ mm}$$

and

$$X = \sqrt{(0.1250)^2/0.96}$$
$$= 0.1276 \text{ in}$$

resulting in

joint width required for wind drift, D, of 3.2660 mm (0.1276 in)

The same sealant for the previous examples will be used at +/- 40 percent movement capacity. The joint width required to satisfy reversible movement is found using equation 11.

$$Wm = (Lt + D)/B \qquad (11)$$

where

Wm = joint width required for movement,
Lt = change in dimension due to thermal movement,
D = change in dimension due to wind drift,
B = sealant movement percentage expressed as a decimal.

substituting

$$Wm = (8.5809 + 3.2660)/0.40$$
$$= 29.6173 \text{ mm}$$

and

$$Wm = (0.3356 + 0.1276)/0.40$$
$$= 1.1580 \text{ in}$$

The final designed width of the sealant joint is determined using equation 12.

$$W = Wm + Cf + Ce + Cc + L \qquad (12)$$

where

W = final designed joint width,
Wm = joint width required for movement,
Cf = fabrication construction tolerance,
Ce = erection construction tolerance,
Cc = spandrel beam creep deflection,
L = spandrel beam live load deflection.

substituting

$$W = 29.6173 + 1.6 + 3.2 + 4.8 + 1.6$$
$$= 40.8173 \text{ mm}$$

and

$$W = 1.1580 + 0.063 + 0.125 + 0.188 + 0.063$$
$$= 1.5970 \text{ in}$$

and rounding off

$$W = 41 \text{ mm } (1\text{-}9/16 \text{ in})$$

The contract documents should indicate the final designed joint width with the permissible fabrication and erection construction tolerances as follows.

41 mm (+/- 4.8 mm)   and   1-9/16 in (+/- 3/16 in)

Reference [2] should be consulted for guidelines to establish the depth of sealant required for this joint width. If depth is not properly established the ability of the sealant to accommodate movement can be seriously compromised.

CONCLUSION

The design professional, as the examples illustrate, must evaluate each sealant joint from both a qualitative and quantitative position for the performance factors that will affect it, as well as the type(s) of movement the joint must accommodate. Also, it is advisable to not use the sealant at its rated capacity but at a lesser value. This provides some insurance since there is a degree of uncertainty in establishing the magnitude of many of the performance factors and tolerances. The design professional, using the simple mathematical approximations illustrated in this paper, will preclude many of the aforementioned failures which often lead to shortened building and building component durability.

REFERENCES

[1] Klosowski, J.M., Sealants in Construction, Marcel Dekker, Inc., New York, 1989.

[2] "C962-86, Standard Guide for Use of Elastomeric Joint Sealants", Annual Book of ASTM Standards, Volume 04.07, ASTM, Philadelphia, 1989.

[3] "Table 1, Climatic Conditions for the United States", Chapter 24, American Society of Heating, Refrigerating and Air Conditioning Engineers, Inc. (ASHRAE) Handbook of Fundamentals, ASHRAE, Atlanta, 1989.

[4] Handegord G.O., and Karpati, K.K., "Joint Movement and Sealant Selection", Canadian Building Digest 155, National Research Council of Canada, Ottawa, 1973.

[5] Latta, J.K., "Dimensional Changes Due to Temperature", Record of the DBR Building Science Seminar on Cracks, Movements and Joints in Buildings, NRCC 15477, National Research Council of Canada, Ottawa, 1972.

[6] "Estimation of Thermal and Moisture Movements and Stresses: Part 1", Building Research Establishment Digest 227, Building Research Station, Garston, Watford UK, 1979.

[7] "Estimation of Thermal and Moisture Movements and Stresses: Part 2", Building Research Establishment Digest 228, Building Research Station, Garston, Watford UK, 1979.

[8] "Building Movements and Joints", Portland Cement Association, Skokie, 1982.

[9] Rainger, P., Movement Control in the Fabric of Buildings, Nichols Publishing Company, New York, 1983.

[10] "C216-88a, Standard Specification for Facing Brick (Solid Masonry Units Made From Clay or Shale)", Annual Book of ASTM Standards, Volume 04.05, ASTM, Philadelphia, 1989.

[11] Grimm, C.T., "Designing Brick Masonry Walls to Avoid Structural Problems", Architectural Record, October, 1977, PP 125-128.

[12] Grimm, C.T., "Thermal Strain in Brick Masonry", Proceedings of the Second North American Masonry Conference, The Masonry Society, 1982.

Laboratory Investigation

Richard Wm. Tock

TEMPERATURE AND MOISTURE EFFECTS ON THE ENGINEERING PROPERTIES OF SILICONE SEALANTS

REFERENCE: Tock, R. Wm., "Temperature and Moisture Effects on the Engineering Properties of Silicone Sealants," Building Sealants: Materials, Properties, and Performance, ASTM STP 1069, Thomas F. O'Connor, Editor, American Society for Testing Materials, Philadelphia, 1990.

ABSTRACT: Physical property changes for two structural silicone rubber sealants were studied to determine the effects of expected variations in environmental temperature and moisture. Test coupons were subjected to uniaxial tensile forces at different isotherms and moisture levels. The elastic modulus at zero strain, stress at a function of extension, and relaxation times were determined for the two sealants.

For extensions of zero to thirty percent, and for temperatures over a range of 0°C to 30°C, both sealants were found to behave as ideal elastomers. The use of the model for an ideal rubber allowed the cross link density, tensile modulus, and stress to be quantified for a given isotherm. Normal, environmental temperature variations did not compromise sealant performance.

Moisture (immersion in water) appeared to have little effect on the measured mechanical properties. The pH of the water over a range of 5 to 12 also appeared not to affect sealant performance. However, strong sulfuric acid solutions (pH < 2.0) seriously compromised the integrity of both materials.

KEYWORDS: Silicone rubber sealants, viscoelastic, thermal changes, crosslinking, acid rain

Dr. Richard William Tock is a professor of Chemical Engineering at Texas Tech University in Lubbock, Texas, 79409.

INTRODUCTION

Structural silicone rubber sealants have permitted the rapid development of structural glazing technology. The use of structural adhesives by the construction industry to attach glass directly to a building frame has advanced rapidly, and most advances have occurred without benefit of standard design specifications. Generally manufacturers provide material specifications for their products and instruction on use, but there is seldom provision for an extensive tabulation of engineering performance characteristics. Data on sealant performance in actual structural applications have appeared in the literature, however, and more can be expected as their use continues to grow. [1][2][3][4].

This paper, however, attempts to describe the more fundamental properties and performance of two structural silicone sealants apart from their intended applications with structural glazing. Hence tensile coupons were cut from cured sheets of Dow Corning 983 (DC-983) and Dow Corning 795 (DC-795). The former is a high modulus sealant, and the latter is a medium modulus material. The test coupons were then subjected to uniaxial tensile testing in an Instron™. Various changes in experimental test conditions were imposed on the samples which included; temperature, strain rate, moisture, pH, dissolved oxidants and time of testing. In order to reduce the raw data, the concept of an ideal elastomer was utilized. Tock, et. al. [5] had earlier demonstrated that this concept adequately describes the stress-strain relationship of these silicone rubber sealants for extensions of less than fifty percent. Moreover, this model incorporates the effects of temperature and crosslink density into the basic equation. The viscoelastic characteristics of the sealants were addressed by relaxation data taken at a fixed extension and based on a Maxwell element.[5]

THEORETICAL APPROACH

Elastomers subjected to simple uniaxial tension are known to be a complex function of extension. They do not follow linear stress-strain (Hookean) relationships over the entire range of their rather large extensibilities. The simplest mathematical expression for an ideal elastomer can be written in the following manner for stress as a function of extension.

$$\sigma = NRT\ (\alpha - \frac{1}{\alpha^2}) \qquad (1)$$

where

σ = tensile stress, Pa (psi)
N = gmol/liter, a measure of crosslink density
R = gas constant
T = absolute temperature
α = extension (L/L_o)

The relationship between extension and engineering strain, along with the concept of modulus as the derivative of the stress-strain curve, gives the following relationship for the modulus of an ideal elastomer as a function of extension.

$$E = NRT\ (1 + \frac{2}{\alpha^3}) \tag{2}$$

The incorporation of Equation 1 into a Maxwell model for time dependent viscoelastic behavior yields yet another potential model for stress in a rubber sealant.

$$\sigma = K\eta(1 - e^{-\upsilon/\phi_r}) \tag{3}$$

where

K = strain rate
η = apparent viscosity
t = time
ϕ_r = E/η the characteristic relaxation time

The experimental stress-strain data were reduced by Equation (1) in order to obtain a value for N, the cross link density. This value of N was then used to estimate the elastic modulus at zero extension, i.e. α = 1 by Equation 2. A shear modulus could be predicted based on assumptions of isotropic material and Poisson's ratio of 0.5.

$$E_o = 3NRT$$
$$E = 2G(1 + \upsilon) \tag{4}$$

E_o = Young's modulus at zero extension
E = Young's or elastic modulus
υ = Poisson's ratio
G = shear modulus

Equation 3 was used to reduce the time dependent relaxation data in order to obtain values of relaxation time. Only a minimal amount of the latter data were collected due to experimental difficulties.

EXPERIMENTAL PROGRAM

Two large sheets of DC-983 and DC-795 were cast and cured according to the supplier's specifications. Test coupons were then cut from these sheets. Different orientation were selected (0° and 90°) in order to check for bias due to processing and curing. Isothermal conditions were maintained with an environmental chamber around each sample during the tensile tests. Relative humidity was not monitored or maintained, however. Temperatures from 0^oC to 90^oC were investigated. The tensile tests were also run with the samples immersed in different aqueous media. Ref. ASTM D412, D471, D1193.

The Instron testing equipment was capable of imposing constant strain rates of 0.254 centimeter/minute to 25.4 centimeter/minute (0.1 to 10 inches/min). However, because of the large extensions possible with elastomers, strain gauges were not used. Rather, the cross head movement was used to estimate extension. Based on the relaxation time experiments, it was noted that as much as one hour was needed between tests on the same sample in order to negate the memory of prior tests (viscoelastic behavior).

RESULTS AND DISCUSSION

N: Crosslink Density

As mentioned earlier, the crosslink densities for the two samples were estimated on a basis of ideal elastomeric behavior. These data are shown in Table 1 along with the initial moduli based on Equation 2 and the assumption that the extension, α, equals unity.

TABLE 1 -- Crosslink Density, N, and Initial Modulus of Elasticity, E_o, as a Function of Sealant Type and Environmental Variables; Time, Temperature, and Aqueous Immersion Medium.

Test Conditions	Sealant Type			
	DC-983		DC-795	
	E_o*MPa(psi)	**NX10^4	E_o*	**NX10^4
1. Initial Coupons				
t = 24°C	0.684(99.2)	2.77	0.309(44.8)	1.25
t = 49°C	0.621(90.1)	2.32	0.277(40.2)	1.12
2. Dry Storage For One Year				
t = 24°C	1.136(165)	4.60	0.259(37.6)	1.05
3. Coupons Stored Dry One Year: Then Immersed in Tap Water for;				
48 hrs @ 24°C	0.931(135)	3.77	0.380(55.1)	1.54
192 hrs @ 24°C			0.351(50.9)	1.42
4. Coupons Stored One Year: Then Immersed in Caustic & Mixed Oxidants For;				
pH = 12; $O_3 \approx$ 1200 ppm				
48 hrs @ 24°C	1.035(150)	4.19	0.343(49.7)	1.39
192 hrs @ 24°C	...	...	0.291(42.2)	1.18
5. Coupons Stored One Year: Then Immersed in H_2SO_4, pH<2.0				
(a) 5%H_2SO_4, 144 hrs @ 24°C	...	...	0.259(37.6)	1.05
(b) 20%H_2SO_4, 144 hrs @ 24°C	...	...	0.096(13.9)	0.39
(c) Concentrated H_2SO_4 72 hrs @ 24°C	1.558(226)	6.31	0.101(14.6)	0.41

E_o* = 3NRT; units, MPa (psi) 24°C = 75°F
**N = (mols of chins/cm^3); average error 8% 49°C = 120°F

The data shown in Table 1 reflect average values of N and E_o obtained for all strain rates. While the concept of an ideal elastomer considers only the existence of chemically bonded crosslinks, it is obvious that other parameters can mimic crosslinked structures. These factors are reflected in the differences in N for the two temperatures shown, i.e. a reduction of 10% to 15% as absolute temperature increased 7%. Particulate fillers can also effect N as will the distribution of chain lengths of polymer molecules between crosslinks [6].

The differences between a high modulus sealant (DC-983) and the medium modulus sealant (DC-795) were also highlighted by the magnitude of the crosslink density parameter. There are at least three times more crosslinks per unit volume for the high modulus material than for the medium modulus sealant. The modulus based on Equation 3 is a direct function of both N and T, where the temperature is absolute.

The effect of one year of aging on the samples appears to have either increased the effective crosslink density or left it unchanged. The change observed with the high modulus sealant after a year of storage was the more dramatic for the two silicones, and could be a signal of gradual embrittlement with age. Submerging the samples in various aqueous solutions appears to either enhance or diminish this aging behavior depending upon the solution. Water appears to slightly plasticize both sealants. An alkaline pH and mixed oxidants have no influence or only a slight effect, and again show a different response for the different sealants. This divergence in behavioral performance is best evidenced with the acid pH bath and immersion tests. The DC-983 showed a dramatic increase in N and evidenced surface aging. The medium modulus material on the otherhand reacted vigorously with the acid solution giving off gases and becoming very soft and compliant, i.e., N decreased. This was hypothesized to be due to the attack by acid on a carbonate filler in the medium modulus DC-795. The use of such fillers has not been confirmed with the supplier, however. It does suggest, however, that degradation and accelerated aging may occur with these structural sealants should they be exposed to acid rain conditions [7][8]. In most instances acid rains should not create conditions of bulk pH readings below 4.0. However collection or pooling of acid rain on exposed silicone seals, followed by evaporation of water with the concentration of acid, should results in lower pH levels. These low pH levels would tend to be localized on the exposed surfaces, however.

Relaxation Time ϕ

The relaxation time and viscosity parameter for the DC-983 is shown in Table 2 for different isotherms.

TABLE 2 -- Relaxation Time and Apparent Viscosity for DC-983 at Different Isotherms and Ten Percent Strain.

Temperature °C (°F)	Relaxation Time ϕ_r, min	Apparent Viscosity ηh Pa sec(psi min)
0 (32)	29	2.3×10^9 (5500)
24 (74)	8.3	7.4×10^8 (1800)
43 (109)	3.6	3.3×10^8 (800)

Table 2 indicates that temperature plays a more important role in the sealant's behavior than was indicated by the ideal elastomer model. While the direct dependence on absolute temperature appears to hold for lower tmperatures down to or near the glass transition temperature of the silicone sealant, it does not characterize the elevated temperature behavior. Here the shorter relaxation times are more characteristic of very viscous liquids as opposed to viscoelastic material solids. The data also suggest that a significant time period (minutes) is required for the state of stress in a sample to return to complete equilibrium conditions once the stress field is removed.

Tensile Stress Levels

Both modulus and tensile stress are a function of extension for rubber sealants. While the modulus is a material property, and in our study is directly related to the crosslinked density, N, the stress is not. Rather stress is a statistical parameter when it refers to failure; or it is an engineering term reflecting localized conditions in most other instances. Moreover stress at a given strain or displacement has been used to make engineering design judements. Hence it was considered desirable to report tensile stress in the samples at a common strain of 40% for our tests. These are shown in Table 3.

TABLE 3 -- Tensile Stress Level at Forty Percent Strain - MPa(psi)

Conditions of Test Strain Rate (2.54 cm/min) t = 24°C (75°F)	Sample DC-983 (psi) MPa	 DC-795 (psi) MPa
1. Sample after one year storage 24°C in air	(197) 1.36	(54.5) 0.38
2. Sample after one year storage then immersed in water (pH ≈ 7.0 24°C)	1.01 - 0.92 (146 - 134)	0.39 - 0.36 (57.5 - 53.3)
3. Sample after one year storage then immersed in caustic solution of water (pH > 11.0 24°C)	0.94 - 0.87 (137 - 126)	0.34 - 0.32 (49.8 - 46.1)
4. Sample after one year storage then immersed in H_2SO_4 acid solution (pH < 2.0 24°C)	1.33 (193)*	0.08 - 0.10 (11.7 - 14.7)*

*Failure occured at this level.

Table 3 clearly indicates the diminished stress levels created by the imposition of extreme environmental conditions. Clearly the reduced stress at a fixed extension indicates that water and the caustic-mixed oxidant solution plasticizes the sealants. They do not seriously compromise or restrict overall performance, however. The strong acid-aqueous solutions do, however, severaly reduce performance. Although the high modulus DC-983 has about the same stress level after exposure, its elongation at failure was now one-tenth that which the virgin material exhibited. The low modulus material (DC-795), on the other hand, became so weakened that serious questions concerning its ability to perform as a structural adhesive and sealant after exposure to strong acid environments need to be answered.

SUMMARY

Tensile material properties of two commercial silicone, structural sealants were exmained. These sealants were found to differ in their degree of crosslinking and also the magnitude of their elastic modulus by a factor of three. Temperature affected their performance most at elevated levels, but did not seriously compromise their properties up to a level of 100°C. Neither material was affected seriously by water immersion when pH levels were greater than 7.0. However, acid water solutions (pH < 2.0) degraded physical properties rather rapidly. The high modulus sealant was to found to be embrittled, while the low modulus material nearly dissolved to a viscous gum.

REFERENCES

[1] Bailey, J.R., "Experimental Behavior of Structurally Glazed Insulating Glass Units," PhD Dissertations, C.E., Texas Tech University, Lubbock, TX, 1989.

[2] Sandberg, L.B., and Ahlborn, T.M., "Combined Stress Behavior of Structural Glazing Joints," Journal of Structural Engineering, Vol. 115, No. 5, May, 1989, pp. 1212-1224.

[3] Sandberg, L.B., and Klosowski, J.M., "Structural Glazing: Behavior Details of Double-Bead Installations," Adhesives Age, 29(5), 1985, pp. 26-29.

[4] Vallabhan, C.V.G., et.al., "Thin Glass Plates on Elastic Supports," Journal of Structural Engineering, ASCE 3(11), 1985, pp. 2416-2426.

[5] Tock, R.W. et.al., "Viscoelastic Properties of Structural Silicone Rubber Sealants," Advances in Polymer Technology, Vol. 8, No. 3, 1988, pp. 317-324.

[6] Yanyo, L.C. and Kelley, F.N., "Effects of Chain Length Distribution on the Tearing Energy of Silicone Elastomers," Rubber Chemistry Technology, 60 Ma/Ap 1987, pp. 78-88.

[7] Melander, T., "Saving Lakes," Informator A.B. Gothenburg, Sweden, 1989, pp. 101.

[8] "Controversy Over Acid Rain," Chemistry and Engineering News, Vol. 67, #27, July 3, 1989, pp. 36-37.

W.S. Gutowski

ADHESIVE PROPERTIES OF SILICONE SEALANTS

REFERENCE: Gutowski, W. S., "**Adhesive Properties of Silicone Sealants,**" Building Sealants: Materials, Properties, and Performance, ASTM STP 1069, Thomas F. O'Connor, editor, American Society for Testing and Materials, Philadephia, 1990.

ABSTRACT: It is shown in this paper, that the strength of adhesion between the sealant and substrate depends upon surface properties of these materials expressed in terms of specific components of their surface energies and the acid-base interactions.

The predictive model regarding the relationship between the bond strength and the energy of acid-base interaction presented in this paper may be applicable in both dry and humid/wet environments.

KEYWORDS: silicone sealants, adhesion, acid-base interactions, shear strength, tensile strength, peel strength, fracture energy.

INTRODUCTION

The success of silicone sealants in modern technologies has its basis in their capacity to form strong chemical bonds with the surface of typical substrates used in curtain wall, e.g. aluminium, glass and granite. Resultant adhesive forces exerted across the interface exceed the cohesive forces between the sealant molecules and, thus, perfect initial adhesion is assured in the system. The emergence of various finishes applied to the surface of structural members, (e.g. a variety of polymeric coatings on aluminium, metallic and ceramic coatings on glass), however, results in significant variations of surface properties of the substrate which can lead to undesirable reduction of sealant adhesion.

Dr W.S. Gutowski is a Principle Research Scientist at the CSIRO Australia, Division of Building, Construction and Engineering, P.O. Box 56, Highett, Victoria 3190, Australia.

In order to provide long-term durability of the bond and to assure its sufficient strength, the silicone sealant must exhibit tenacious adhesion to the substrate's surface throughout the expected life of the structure. The adhesion forces must be strong enough to withstand all static and dynamic stresses applied to the sealant/substrate interface, e.g. due to the differential movements resulting from fluctuations of temperature or wind pressure, tectonic movements, dead-load weight of glass panels, and building frame deformation.

THEORIES OF ADHESION

Various theories on the mechanism of adhesion have been proposed in the literature [1-6], e.g.

(a) adsorption theory (see comments below),
(b) electronic theory – adhesion being achieved due to the electrostatic forces arising from the formation of the double layer of the electrical charge at the interface.
(c) diffusion theory – adhesion being achieved due to the mutual interdiffusion of polymer molecules across the interface, and
(d) mechanical adhesion – created due to the microscopic and macroscopic interlocking of the sealant or adhesive into irregularities of the substrate.

In certain systems any of the above mechanisms may be predominant, but in most cases the total adhesion involves several interactive mechanisms, e.g. adsorption and mechanical interlocking.

The adsorption theory has gained the widest acceptance. It proposes that materials adhere together as a result of strong interatomic and intermolecular forces exerted across the interface between the atoms and molecules of the substrate and sealant. These forces, in general, are referred to as primary and secondary bonds, whose typical energies are shown in Table 1. It can be seen from this table that even a small number of the ionic, covalent or donor-acceptor bonds should result in a significant increase of strength of any adhesive bond. The strength of hydrogen bonds is generally higher, according to Table 1, than the strength of typical van der Waals interactions.

DONOR-ACCEPTOR (ACID-BASE) INTERACTIONS IN ADHESION

All materials, including polymers (with the exception of saturated hydrocarbons), can be categorised as basic or acidic [7-9] according to whether they are proton donors or acceptors. Thus, the interactions of a polymeric sealant or adhesive with any organic or inorganic substrate can be considered in terms of acid-base interactions.

Table 1 – Typical bond types and their energies.

Type of forces	Bond energy (kJ/mole)
Primary forces	
ionic \	335 – 1050
hydrogen	80 – 42[a]
covalent /	63 – 920
Secondary (van der Waals) forces	
dipole – dipole	4 – 20
London (dispersion)	4 – 40
dipole – induced dipole	up to 20
Donor-acceptor bonds[b]	
Brønsted acid-base interactions	up to 1000
Lewis acid-base interactions	up to 80

[a] 42 kJ/mole in hydrogen bonds involving flourine.

[b] taken from Ref. [4].

These interactions are associated with the charge exchange between the materials in contact acting in the following alternative manner.

(a) An acid accepting an electron pair due to its incomplete electronic arrangements (Lewis theory) or proton donation according to the Brønsted theory, which leads to the formation of a strong covalent bond. Examples of materials acting in this manner include polymers such as PVC and PVDF, or inorganic materials like silica, Fe_20_3 or Fe_30_4.

(b) A base donating an electron pair (Lewis base) or accepting proton (Brønsted theory). Examples of relevant materials include polymethyl-methacrylate (PMMA), polycarbonate (PC), calcium carbonate (e.g. marble) and amorphous Al_20_3.

(c) Amphoteric, i.e., acting as both electron acceptor and electron donor. Examples of materials in this category include amides (e.g. polyamides), amines and alcohols (e.g. polyvinyl alcohol).

ENERGY OF INTERACTION AND FORCE OF INTERACTION

Any interfacial interactions can be conveniently analysed in terms of the energy (U) or enthalphy (ΔH) of interaction, since the interaction force and hence the strength of adhesive bond are defined as the first derivative of the interaction energy with relation to the separation distance 'r' between the atoms or molecules, i.e.

$$F = -\frac{dU}{dr}. \quad (1)$$

In adhesion science, the (negative) energy of interaction between materials 1 and 2 in immediate contact is known as the thermodynamic work of adhesion, W_A, which can be estimated using the following fundamental Dupré equation:

$$W_A = \gamma_1 + \gamma_2 - \gamma_{12}, \quad (2)$$

where γ_1 and γ_2 are the surface energies of materials 1 and 2 in contact, and γ_{12} is the interfacial energy.

An alternative expression for the estimation of work of adhesion was developed by Good and Girifalco [10], i.e.

$$W_A = 2\,\Phi(\gamma_1\,\gamma_2)^{1/2}, \quad (3)$$

where Φ is the interaction parameter given by

$$\Phi = (d_1 d_2)^{1/2} + (p_1 p_2)^{1/2}. \quad (4)$$

In Eq. (4), d and p are the dispersive and non-dispersive fractions of total surface energy of material 1 or 2, i.e.

$$d = \gamma^d/\gamma\text{, and} \quad (5)$$

$$p = \gamma^p/\gamma. \quad (6)$$

According to Fowkes [7], the total work of adhesion W_A comprises terms associated with the dispersive, i.e. W_A^d, and acid-base interactions, i.e. W_A^{ab}, components

$$W_A = W_A^d + W_A^{ab}. \quad (7)$$

The acid-base component of the thermodynamic work of adhesion can be calculated using the following expression:

$$W_A^{ab} = W_A - 2(\gamma_1^d / \gamma_2^d)^{1/2}. \quad (8)$$

Interactions related to hydrogen bonds are a sub-set of acid-base interactions.

ADHESIVE FRACTURE ENERGY

Structural bonds made with the use of sealants or adhesives frequently fail due to the initiation and propagation of flaws (cracks) which are developed either at the sealant/substrate interface or within the bulk sealant. The use of fracture mechanics to analyse the mechanism of failure and the expected service life of the bonded structure may thus be useful.

The adhesive fracture energy, G_c, required to propagate the crack is given by [4]

$$G_c = G_o + \psi , \quad (9)$$

where G_o is the energy required to propagate cracks in the absence of visco-elastic and plastic energy losses (this is a measure of adhesive forces exerted across the interface), and ψ is the energy dissipated within the adhesive or sealant at the propagating crack [11,12]

$$\psi = G_o \, f_v(\dot{a},T,\varepsilon). \quad (10)$$

the introduction of $\phi_v(\dot{a},T,\varepsilon) = 1 + f(\dot{a}, T, \varepsilon)$ into Eqs (9) and (10) leads to the following expression for the adhesive fracture energy [4]:

$$G_c = G_o \, \phi_v(\dot{a},T,\varepsilon). \quad (11)$$

Equation (11) explains that the total value of G_c for an elastic adhesive is highly dependent upon viscoelastic losses being a function of the crack propagation rate $\dot{a}$, temperature T, and the strain level ε. When viscoelastic losses are negligible, i.e. when $\phi_v(\dot{a},T,\varepsilon) \to 1$ and $f(\dot{a},T,\varepsilon) \to 0$, then the measured adhesive fracture G_c is equal to G_o. In this way a direct measure of the interatomic and intermolecular forces acting across the interface may be obtained.

For the joints which are bonded solely due to the physical, i.e., van der Waals interactions, and when the failure mode is purely interfacial, we have

$$G_c = W_A \,, \tag{12a}$$

where W_A, the thermodynamic work of adhesion, is given by Eqs (2) or (3). Thus, under these circumstances, the adhesive fracture energy equals the thermodynamic work of adhesion.

If, on the other hand, chemical bonds are acting across the interface, the fracture energy will include the chemical bond energy and

$$G_c >> W_A \,. \tag{12b}$$

EXPERIMENTAL

Materials

Glass, aluminium and steel, as well as a range of engineering plastics (such as polymethyl methacrylate (PMMA), acrylonitrile-butadiene-styrene terpolymer (ABS), nylon 6-6, high impact polystyrene (PS), vinyl copolymer, acetal, polypropylene – natural (PP) and filled (PP Filled), polyethylene – low density (LDPE), high density filled (HDPE), and ultra-high molecular weight (UHMW–PE)), were selected to cover a broad range of variability of surface properties of substrates expressed in terms of their surface energies γ_1, polar (γ_1^p) and dispersive (γ_1^d) components. Three silicone sealants, denoted A, B and C, were chosen for this work. Sealants A and C are qualified as structural, whilst B is a general purpose sealant.

Surface Preparation

All substrates were thoroughly cleaned by washing with ethyl alcohol. After overnight drying at room temperature they were wiped three times with ethyl alcohol (ABS, PS, vinyl copolymer, PMMA) or with MEK (all other substrates) and allowed to dry for one hour prior to bonding.

Cure of Specimens

All bonded assemblies were allowed to cure for four weeks prior to testing under the following conditions: two weeks at 50% RH, 23°C; one week at 98% RH, 38°C; and one week at 50% RH, 23°C.

Test Methods

Peel strength: This was determined using the specimens prepared in accordance with the ASTM C-794 standard modified slightly with regard to dimensions of substrates which were 25 mm wide and 100 mm long. Two specimens were tested for each experimental point at the strain rate equal to 50 mm/min.

Shear strength: This was determined by single lap-shear specimens, as illustrated in Fig. 1(a). Three specimens were used for each experimental point. The strain rate during testing was 10 mm/min.

Tensile strength: This was determined using specimens illustrated in Fig. 1(b). Three specimens were used for each experimental point. The strain rate during testing was 10 mm/min.

Fracture energy: The fracture energy of adhesive joints was determined using double cantilever beam (DCB) specimens. In order to minimise viscoelastic losses during testing, a very thin layer (0.05 mm) of sealant was used to bond substrates. The specimen geometry is shown in Fig. 1(c). The joint fracture energy was calculated using the following expression [4]:

$$G_{Ic} = \frac{F_c^2}{2B} \cdot \frac{\partial C}{\partial a} , \qquad (13)$$

where F_c is the force at the onset of crack propagation, B is the specimen width, a is the crack length, and C is the specimen compliance, i.e. displacement/force.

Surface energy of substrates and sealants: This was determined from the wettability studies using the following Eqs [4,13]:

$$0.5\ \gamma_2^{(1)}\ (1 + \cos\theta^{(1)}) = \left(\gamma_1^d \gamma_2^{d(1)}\right)^{1/2} + \left(\gamma_1^p \gamma_2^{p(1)}\right)^{1/2} \qquad (14)$$

$$0.5\ \gamma_2^{(2)}\ (1 + \cos\theta^{(2)}) = \left(\gamma_1^d \gamma_2^{d(2)}\right)^{1/2} + \left(\gamma_1^p \gamma_2^{p(2)}\right)^{1/2} \qquad (15)$$

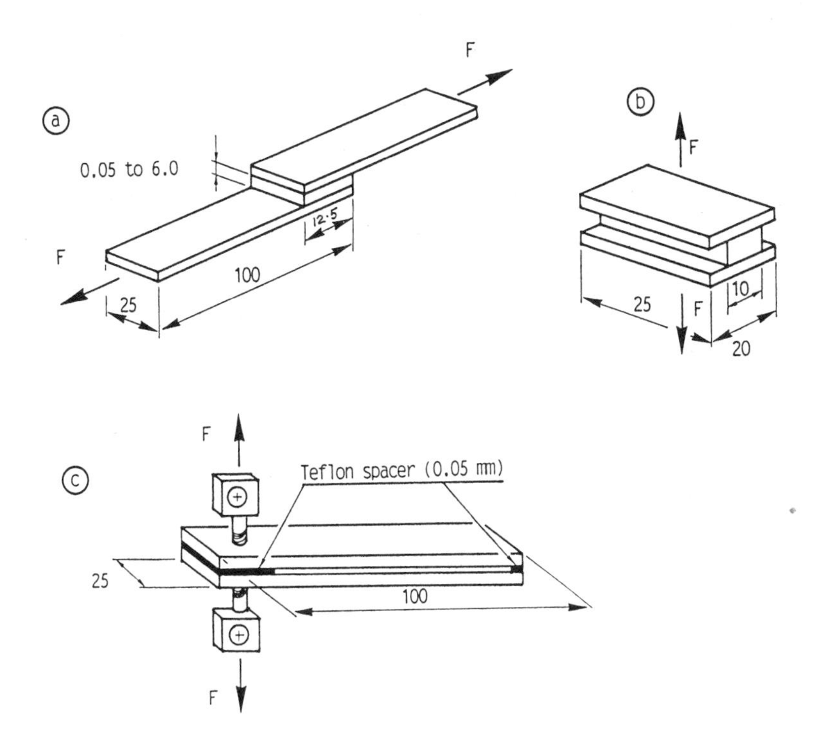

FIG. 1. Geometry of specimens used throughout experiments: (a) single lap-shear specimen; (b) tensile specimen; (c) double cantilever beam specimen (DCB) for determining the adhesive fracture energy, G_{Ic}.

which are solved for unknown parameters γ_1^d and γ_1^p ; the dispersive and non-dispersive components of the total surface energy of the substrate

$$\gamma_1 = \gamma_1^d + \gamma_1^p . \tag{16}$$

The parameters θ_1 and θ_2 in Eqs (14) and (15) are the equilibrium contact angles exhibited by the test liquids 1 and 2 deposited on the substrate's surface. The test liquids of known surface properties, i.e., γ_2, γ_2^d and γ_2^p, used during the experiments were water, formamide and glycerol. For more particulars regarding the determination of a solid's surface energy, see Refs [4,5,13].

RESULTS AND DISCUSSION

Sealant's surface energy with reference to acid-base interactions: It is shown in the literature [14] that the surface energy of a cured silicone sealant (polydimethylsiloxane) determined at the sealant/air interface is 21.7 mJ/m^2 and its polarity (at this particular surface) is 0.05. Similar results are obtained by this author, as shown in Table 2 below, for all sealants investigated.

Table 2 – Surface properties of silicone sealants.

Interface	γ_2 (mJ/m^2)	γ_2^p (mJ/m^2)	Polarity 'p'
Sealant 'A'			
uncured sealant/N_2	14.21	5.80	0.41
cured sealant/air	18.09	0.40	0.02
cured sealant/LDPE[a]	18.39	2.13	0.11
cured sealant/PS[b]	18.10	7.25	0.41
cured sealant/ABS[b]	21.12	6.92	0.33
Sealant 'B'			
uncured sealant/N_2	15.79	12.24	0.77
cured sealant/air	19.32	0.00	0.00
cured sealant/LDPE[a]	18.00	1.48	0.08
cured sealant/PP Filled[a]	19.63	3.97	0.20
Sealant 'C'			
uncured sealant/N_2	18.10	7.42	0.41
cured sealant/air	17.75	0.50	0.03
cured sealant/LDPE[a]	24.50	7.55	0.31
cured sealant/Acetal[a]	23.85	7.63	0.32
cured sealant/PS[b]	20.30	8.12	0.40

[a] Sealant peeled off from the substrate.

[b] Substrate dissolved in MEK.

It is suggested in this paper, that the above properties of the 'cured sealant/air' interface, and particularly γ_2^p (non-dispersive component of sealant's surface energy), are not relevant from the viewpoint of adhesive bond formation between the sealant and the substrate. The process of the formation of adhesive bond between the silicone sealant and substrate takes considerable time whilst the sealant is still in its uncured state. During this process, the mobility of molecules and flexibility of chains allows them to obtain preferential orientation with regard to the substrate's reactive sites taking part in acid-base interactions. It is assumed in this work that the polarity of the uncured sealant should be the factor contributing to the process of strength development.

In order to investigate the above assumption, experiments were carried out to determine surface energies of the sealants A, B and C at the sealant/air and sealant/substrate interface.

In these experiments, sealants were cast on a range of substrates, e.g. LDPE, PP, Acetal, PS and ABS. After a one month cure the sealant sample was gently peeled off the LDPE, PP, HDPE, Acetal and UHMW-PE, whilst in the case of ABS and PS the substrates were dissolved in MEK and thoroughly washed in a substantial quantity of MEK to remove any traces of the substrate from the interface exposed. Surface energy of uncured sealants was determined using fresh sealant films made and handled under dry N_2.

It is apparent from the data presented in Table 2 that all uncured sealants investigated exhibit high polarity ranging from 0.41 for sealants A and C to 0.77 for sealant B. This is indicative of the sealant's capacity to form bonds attributed to the acid-base interactions with appropriate substrates. High polarity of the uncured sealant is shown to be in sharp contrast with the essentially non-polar character of the cured sealant at the surface exposed to the air.

Analysis of the acid-base character of sealants at the sealant/substrate interface established during curing process, reveals that, in agreement with the earlier suggestion, the sealant molecules show a significant orientation effect with the quantity of the polar groups oriented towards the substrate dependent upon the polarity (or acid-base character) of the substrate in contact.

Taking the above into account, it was decided that the surface properties of the uncured silicone sealant are more relevant in estimating the energy of acid-base interactions (see Eq. (8)) with a given substrate than those of the cured sealant at the sealant/air interface.

Relationship Between the Bond Strength and Surface Properties of Bond Components

Surface energies of all substrates used in experiments, and energy of interaction (W_A and W_A^{ab}) between these substrates and sealant C are given in Table 3.

Table 3 – Surface properties of substrates used in bonding with sealants A and C and relevant thermodynamic work of adhesion W_A and energy of acid-base interactions W_A^{ab} (values of W_A and W_A^{ab} relevant to bonds with sealant C).

Material	γ_1 (mJ/m^2)	γ_1^p (mJ/m^2)	γ_1^d (mJ/m^2)	p	d	Φ	W_A (mJ/m^2)	W_A^{ab} (mJ/m^2)
Glass	59.0	33.6	25.4	0.57	0.43	0.9871	64.51	31.57
Aluminium	38.5	15.4	23.1	0.40	0.60	1.000	52.80	21.38
Nylon 6-6	39.44	16.11	23.33	0.41	0.59	1.000	53.44	21.86
ABS	39.97	12.68	27.29	0.32	0.68	0.996	53.56	19.21
Vinyl copolymer	43.36	13.99	29.37	0.32	0.68	0.995	55.78	20.36
PS	41.00	12.44	28.56	0.34	0.66	0.9974	54.34	19.41
PMMA	37.84	11.15	26.69	0.295	0.705	0.9927	52.27	18.25
HDPE	37.26	9.58	27.68	0.25	0.75	0.985	51.17	16.79
UHMU–PE	28.47	4.33	24.14	0.15	0.85	0.9511	43.41	11.29
Acetal	29.68	4.12	26.56	0.11	0.89	0.9370	43.43	11.06
PP	24.36	3.82	20.54	0.16	0.84	0.9601	40.32	10.70
PP filled	24.16	3.73	20.43	0.15	0.85	0.9597	40.14	10.59
LDPE	26.34	2.65	23.69	0.10	0.90	0.9312	40.66	8.85

Φ, W_A and W_A^{ab} calculated using the following sealant properties: γ_2 = 18.09 mJ/m^2; γ_2^p = 7.42 mJ/m^2; γ_2^d = 10.68 mJ/m^2 (all sealant properties refer to its uncured state).

Figure 2 illustrates the relationship between the shear and peel strength and the energy of acid-base interactions (W_A^{ab}) for the range of substrates bonded with the silicone sealant A. The pattern of this relationship is identical for both testing modes. It is noticeable that the bond strength increases monotonically with the increase of the energy of acid-base interactions as calculated by Eq. (8).

Interesting observations are made with regard to the failure mode of the systems investigated. The lowest strength and 100% adhesive failure (delamination at the sealant/substrate interface) occur with the substrates

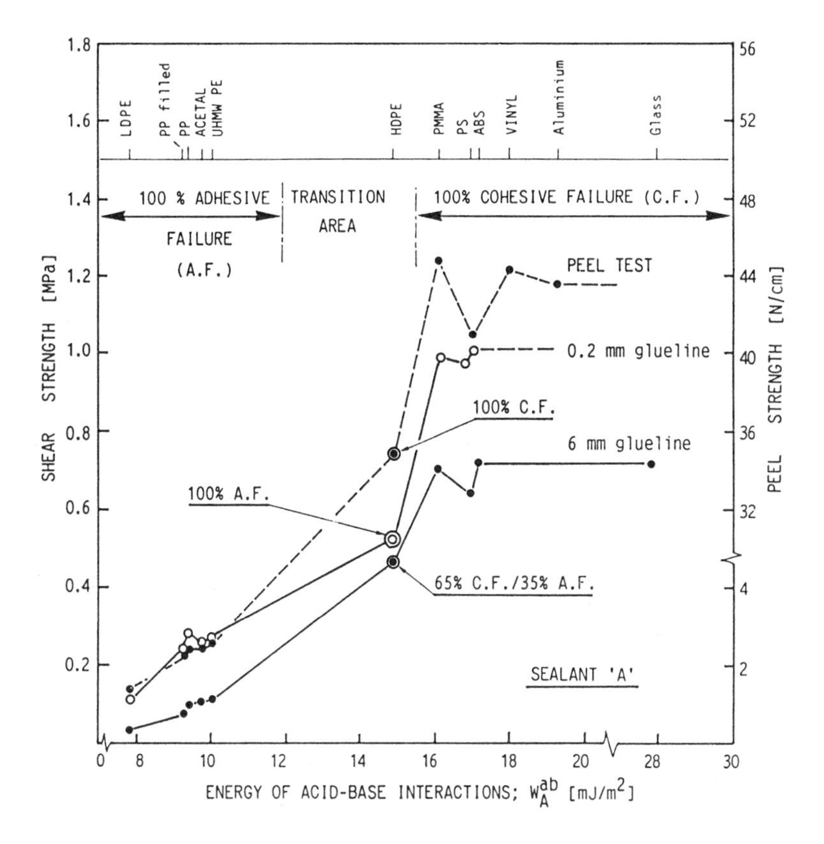

FIG. 2. The relationship between the strength (shear and peel) and energy of acid-base interactions, W_A^{ab}, for the bonds made with the sealant 'A' and variety of organic and inorganic substrates.

exhibiting the lowest polar component of their total surface energy, e.g. LDPE with $\gamma_1^p = 2.65$ mJ/m^2 (see Table 3). As the value of γ_2^p for subsequent materials increases, the resultant bond strength also increases monotonically for such plastics as PP, Acetal and UHMW PE. An interesting transition point has been observed in bonds made with HDPE which exhibited 100% adhesive failure at 0.2 mm sealant bead thickness, and the mixed failure mode (i.e. 35% adhesive/65% cohesive failure) in the case of 6.0 mm thick sealant bead.

The same failure mode occurs in shear and peel tests for all systems investigated, with the exception, again, of HDPE which gives 100% cohesive failure in peel tests. This phenomenom will require further investigations, since it

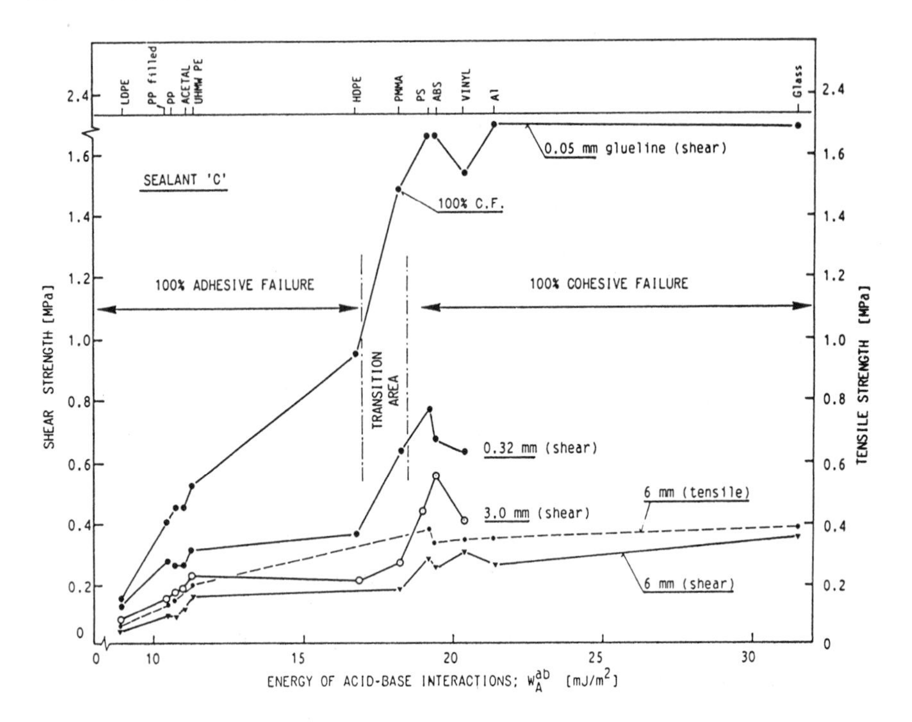

FIG. 3. The relationship between the strength (shear and tensile) and energy of acid-base interactions, W_A^{ab}, for the bonds made with the sealant 'C' and variety of organic and inorganic substrates.

may indicate that the peel test is less severe under certain conditions than either the tensile or shear test. The observed transition between the adhesive and cohesive modes of failure occurs at the point where the energy of acid-base interactions, W_A^{ab}, equals about 14 to 15 mJ/m^2. All other systems with the energy of acid-base interactions greater than 16 mJ/m^2 exhibit 100% cohesive failure within the sealant, whether the substrate is of an organic nature (plastics such as PS, PMMA, ABS, Nylon 6-6) or an inorganic nature (aluminium, glass), within the scope of the experiment.

The relationship between the strength of bonds made with sealant C and relevant energies of acid-base interactions is shown in Fig. 3. The pattern of this relationship is similar to that obtained with sealant A, as illustrated earlier in Fig. 2. All substrates exhibiting low energy of acid-base interactions with the sealant show 100% adhesive failure. Bonds with HDPE which produced mixed (cohesive/adhesive) failure with sealant A in this case, all failed 100% adhesively. The transition in this case occurred at W_A^{ab} equal to about 19 mJ/m^2 for bonds with ABS. All bonds made for the 'ABS/sealant C' system failed 100% cohesively at any thickness of the sealant bead within the range 0.05 to 3.0 mm. When the bead thickness approached 6 mm, however, the failure mode changed to the mixed one exhibiting the average 20% adhesive/80% cohesive mode. Peel test specimens for this system broke 100% cohesively. It can be summarised for this sealant that, again, there is a monotonic strength increase with the increase of the energy of acid-base interactions across the sealant/substrate interface. All systems exhibiting W_A^{ab} up to about 19 mJ/m^2 show 100% adhesive failure, whilst above 20 mJ/m^2 100% the cohesive mode occurs with the transition (mixed: cohesive/adhesive mode) observed for the system that exhibits W_A^{ab} = 19.4 mJ/m^2.

Fracture Energy of Bonds

Fracture energy of the sealant/substrate bonds was determined using DCB specimens. Numerical values found for G_{Ic} are listed in Table 4, which indicates also the failure mode. As explained earlier, the presence of viscoelastic losses during sealant deformation may result in an increased value of G_{Ic} . It is apparent from Table 4, however, that for those sealant/substrate systems which

Table 4 – Values of G_{Ic} and W_A for bonds with sealant C and A.

Substrate	Sealant C			Sealant A		
	W_A (mJ/m^2)	G_{Ic} (mJ/m^2)	Failure mode[a]	W_A (mJ/m^2)	G_{Ic} (mJ/m^2)	Failure mode[a]
LDPE	40.66	21.08	100%A	36.03		
Acetal	43.43	36.34	100%A	38.48	50.48	100%A
PP Filled	40.14	33.00	100%A	35.56		
PP	40.32	38.04	100%A	35.80		
UHMW–PE	43.41	50.65	100%A	38.26	56.36	100%A
HDPE	51.17	68.00	100%A	45.32	220.00	100%A
PMMA	52.27	320.00	100%A	46.03		
PS	54.34	1854.00	100%C	48.14		
ABS	53.56	2561.60	100%C	47.47		

[a]A – adhesive failure at the interface; C – cohesive failure within the sealant.

exhibit 100% adhesive failure (i.e. those where the crack propagates along the sealant/substrate interface), the experimental values of the adhesive fracture energy G_{Ic} are in reasonable agreement with the theoretical value of the thermodynamic work of adhesion W_A calculated using Eq. (3).

Interesting comments on the bonding mechanism can now be made considering the energies of typical primary and secondary bonds listed in Table 1. It is apparent from this table that the energy of covalent, ionic and donor-acceptor related bonds may be about 20 to 50 times greater than those attributed to physical interactions (i.e. van der Waals interactions) which, for purely interfacial failure, should result in the interaction energy as calculated by Eq. (3).

If the primary (chemical) bonds are predominant at the interface of some sealant/substrate systems, their energy of interaction should be 20 to 50 times greater (see Eq. (12b)) than the maximum value of W_A in Table 3 or 4 which still results in the 100% adhesive failure mode. This energy should thus be within the range of at least 1000 to 2500 mJ/m^2 for the systems exhibiting 100% cohesive failure, i.e. for the sealant or adhesive for which the adhesive forces exerted across the interface exceed cohesive strength of the sealant or adhesive.

As can be seen from Table 4, this is the case applicable to the bonds made with ABS and PS. It is also seen that the bond of sealant C with PMMA and the bond of sealant A with HDPE may show a presence of some chemical bonds since the adhesive fracture energy G_{Ic} = 320 mJ/m^2 and 220 mJ/m^2 is significantly greater than the theoretical work of adhesion for these systems, i.e. W_A = 52.27 mJ/m^2 and 45.32 mJ/m^2 respectively. Thus based on the above, the adhesive fracture energy provides good insight as to the nature of forces acting across the sealant/substrate interface.

Influence of the Sealant Bead Thickness on Bond Strength

Figure 4 illustrates the influence of the sealant bead thickness on the shear strength of bonds made with a variety of substrates used in this work. The two upper curves with the margin of standard deviation indicated are obtained from the results regarding strength of the bonds that failed 100% cohesively at a given glueline thickness, i.e., those made with PS, Vinyl, ABS, Glass, Aluminium, etc. for the sealant A (upper curve) and sealant C (lower curve).

The two lower characteristics refer to the bond strength *v.* glueline thickness for two different plastics (PP and UHMEPE) which produced entirely adhesive failure of bonds made with sealant C.

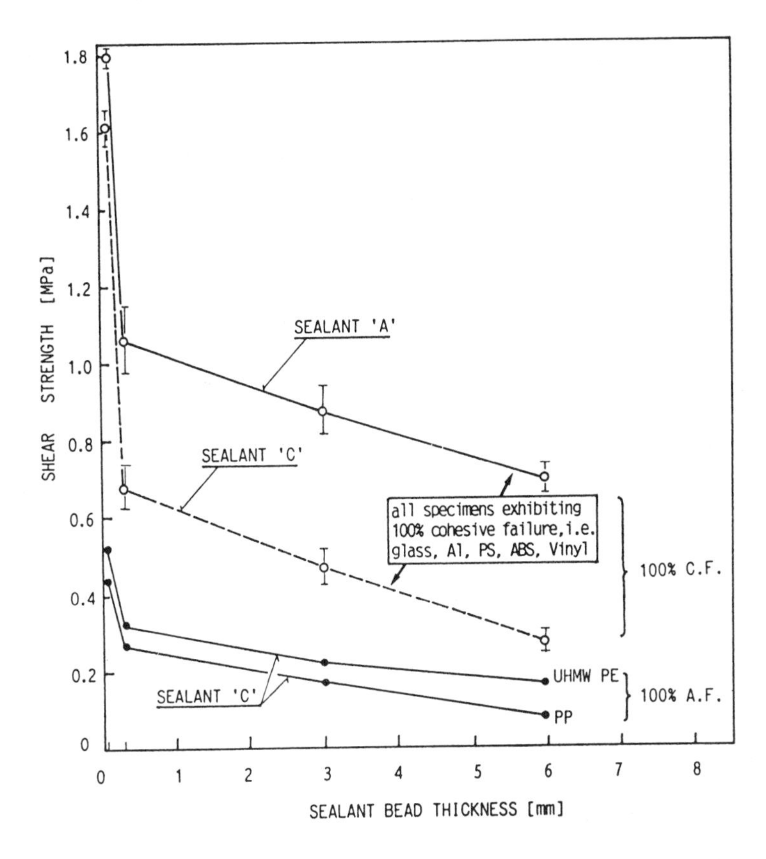

FIG. 4. The relationship between shear strength and the thickness of a sealant bead for the systems exhibiting 100% cohesive failure (C.F.) and 100% adhesive failure (A.F.).

The highest strength was obtained with a very thin sealant bead due to the fact that the stress between the substrates is transferred predominantly in shear with a presence of only small peel stress at the tips of the bond. Increasing the sealant bead thickness from 0.05 to 0.3 mm results in a dramatic drop of the bond strength (of about 60%). A further increase of the bead thickness from 0.3 to 0.6 mm results in a much less significant strength decrease.

It is apparent from Fig. 4 and from the earlier data that the general pattern of the 'strength *v.* bondline thickness' relationship is similar for both types of the failure mode – whether 100% adhesive (see graphs for PP, LDPE) or 100% cohesive (e.g. for glass, aluminium, ABS, PS, Nylon 6-6).

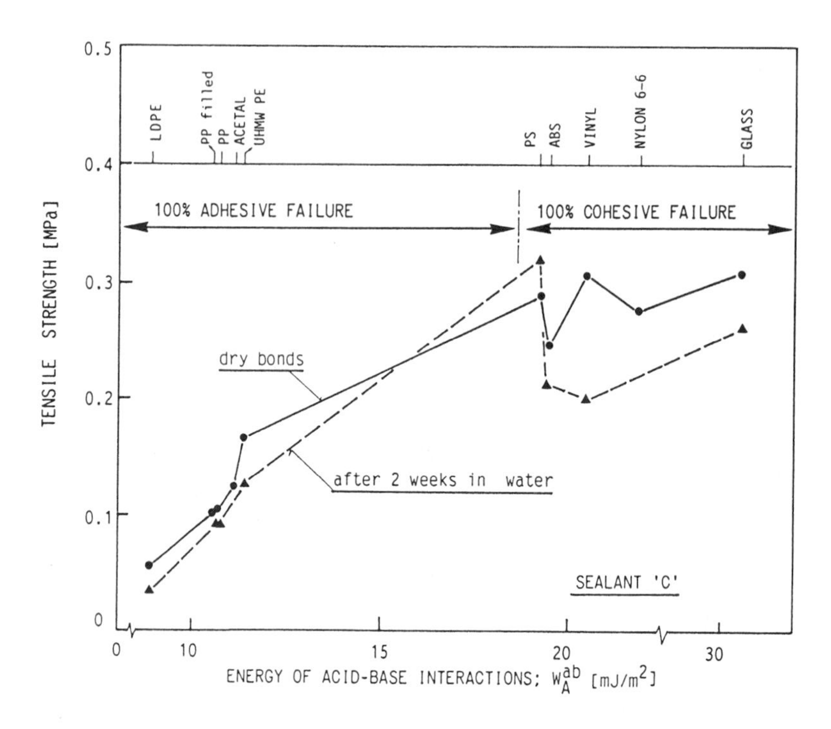

FIG. 5. Influence of water immersion on the 'strength *v.* W_A^{ab} characteristic for a range of substrates bonded with sealant 'C'.

For sealant C, both the shear and tensile strength was determined at the bead thickness equal to 6.0 mm, using identical joint geometry. As shown in Fig. 3, the tensile strength was about 20% greater than shear strength.

Influence of Water on Bond Strength

Tensile specimens, as illustrated in Fig. 1 were prepared using sealant C and the following substrates: LDPE, PP, Acetal, UHMWPE, PS, ABS, Vinyl, Nylon 6-6, and Glass. After one month cure, 50% of the specimens were immersed in H_2O for two weeks, whilst the remainder were left to be tested in dry conditions.

Figure 5 illustrates the resultant relationship between the strength of 'wet' and 'dry' bonds and W_A^{ab}, the energy of acid-base interactions calculated using Eq. (8). It is apparent from this Figure that the strength of 'wet' bonds is slightly less (except for PS) than those tested in dry conditions, and that the general relationship 'strength *v.* W_A^{ab}' is similar for either – the bonds maintained in the dry or wet environment prior to testing. The influence of water on strength of the bonds made with a variety of sealants and substrates is a subject of further research.

CONCLUSIONS

1. It is shown in this work that the strength of adhesion between silicone sealants and a range of organic and inorganic substrates investigated can be attributed to the acid-base interactions between the substrate and sealant.

2. The strength of bonds (whether tested in shear, tensile or peel) increases monotonically with the increase of the energy of acid-base interactions, W_A^{ab}, whose value can be quantified by Eq. (8).

3. In practical terms, the value of the energy of acid-base interactions is directly related to the value of the polar component of total surface energy of the substrate (γ_1^p) and the sealant in its uncured state (γ_2^p). This indicates that, in order to increase the strength of adhesion of a given sealant to the substrate of interest, it is necessary to increase the polarity of the substrate. The data show, that the 100% cohesive failure within the sealant can be obtained if the value of γ_1^p is greater than about 12-13 mJ/m^2.

4. Within the scope of the experiments there are three areas regarding failure mode in the relationship 'strength *v.* W_A^{ab}':

   (a) 100% adhesive failure for W_A^{ab} = 0 to 14 mJ/m^2,
   (b) mixed mode (adhesive/cohesive failure) for W_A^{ab} = 14 to 19 mJ/m^2,
   (c) 100% cohesive failure within sealant for $W_A^{ab} \geq$ 19 mJ/m^2.

5. Adhesive fracture energy studies indicate that the systems exhibiting $W_A^{ab} \geq$ 19 mJ/m^2 have fracture energies G_{Ic} in excess of 1800 mJ/m^2, which is about 40 to 50 times greater than the theoretical value of the thermodynamic work of adhesion, W_A, relevant to the secondary interaction associated with 100% interfacial bond failure. This is indicative of the presence of strong chemical bonds whose energy is typically 20 to 50 times greater than those attributed to the secondary (physical) interactions.

6. Sealant molecules at the sealant/substrate interface exhibit significant orientation effect due to the acid-base interactions between nondispersive groups present at the substrate's surface and those of the sealant during formation of adhesive bond. The quantity of these groups exposed at the sealant/substrate interface is dependent upon the polarity of the substrate surface.

ACKNOWLEDGMENTS

The author wishes to acknowledge with appreciation the help of Mrs Magda Morehouse and Lee Russell in determining surface energies of materials investigated, Mr Dick Pankevicius in adhesive fracture energy studies and Mr Anthony Cerra for his help in samples testing and the preparation of this paper. The financial support by research grant from BRDAC (Building Research and Development Advisory Committee), Australia, is gratefully acknowledged.

REFERENCES

[1] Fowkes, F.M., 'Attractive Forces at Interfaces', Industrial and Engineering Chemistry, Vol. 56, 1964, pp. 40–52.

[2] Good, R.J., 'Intermolecular and Interatomic Forces', in Treatise on Adhesion and Adhesive, Vol. 1: Theory, Marcell Dekker Inc., New York, 1967.

[3] Israelachvilli, J.N., Intermolecular and Surface Forces, Academic Press, London, 1985.

[4] Kinloch, A.J., Adhesion and Adhesives, Chapman and Hall, London, New York, 1987.

[5] Wu, S., Polymer Interface and Adhesion, Marcell Dekker, New York, Basel, 1982.

[6] Dzialoshinskii, I.E., Lifshitz, E.M. and Pitaevskii, L.P., Advances in Physics, Vol. X, 1961, pp. 165–209.

[7] Fowkes, F.M., 'Acid-Base Interactions in Polymer Adhesion', in Physico-Chemical Aspects of Polymer Surfaces, Vol. 2 (Ed. K.L. Mittal), Plenum Press, New York and London, 1978.

[8] Gutman, V., Donor-Acceptor Approach to Molecular Interaction, Plenum Press, New York and London, 1983, pp. 583–603.

[9] Schultz, J., Lavielle, L. and Martin, C., 'The Role of Interface in Carbon Fibre-Epoxy Composites', Journal of Adhesion, Vol. 23, 1987, p. 45.

[10] Good, R.J. and Girifalco, L.A., Journal of Physical Chemistry, Vol. 64, 1987, p. 561.

[11] Gent, A.N. and Schultz, J., Journal of Adhesion, Vol. 3, 1972, p. 281.

[12] Gent, A.N. and Kinloch, A.J., Journal of Polymer Science, Vol. A2, No. 9, 1971, p. 659.

[13] Kaelble, D.H. and Uy, K.C., Journal of Adhesion, Vol. 2, 1970, p. 51.

[14] Klosowski, J.M., 'Durability of Silicone Sealants' in Adhesives, Sealants and Coatings For Space and Harsh Environments, (Ed. L.H. Lee), Plenum Press, New York and London, 1988.

Herbert Stoegbauer and Andreas T. Wolf

THE INFLUENCE OF HEAT AGEING ON ONE-PART CONSTRUCTION SILICONE SEALANTS

REFERENCE: Stoegbauer, H. and Wolf, A. T., **"The Influence of Heat Ageing on One-Part Construction Silicone Sealants,"** Building Sealants: Materials, Properties, and Performance, ASTM STP 1069, Thomas F. O'Connor, editor, American Society for Testing and Materials, Philadephia, 1990.

ABSTRACT: The paper studies the effects of heat ageing on the elongation at break and tensile strength of 10 commercially available, standard construction silicone sealants, based on acetoxy, alkoxy, aminoxy and oxime cure. Mechanical properties are measured on ISO 8339 tensile test joints after exposure to 100, 150, 180 and 200 °C as well as room temperature (as a control) over a period of 1, 3 and 6 months. The study shows that the heat stability of those one-part silicone sealants investigated is influenced mainly by their cure chemistry and only to a lesser extent by their formulation. A relative ranking of the heat stability of the sealants is derived based on the percentage change of their mechanical properties.

KEY WORDS: silicone sealant, heat stability, elongation at break, tensile strength

In many building applications, the proper functioning of a joint seal is determined by the heat (high-temperature) resistance of the sealant. The specifier will, however, find very few test methods and requirement criteria for this property. Although most sealant manufacturers provide information on the maximum permissible service temperature for their sealants, few give details of the test methods employed. This study will examine the influence of heat ageing on one-part silicone construction sealants. The results obtained will permit classification of the sealants according to their heat resistance.

Ing. Stoegbauer is Technical Service Engineer at Dow Corning GmbH, Rheingaustraße 53, 6200 Wiesbaden 13, West-Germany. Dr. Wolf is Technical Service & Development Section Manager at Dow Corning S.A., Parc Industriel, 6198 Seneffe, Belgium

CAUSES OF THE THERMAL STABILITY OF SILICONE ELASTOMERS

In contrast to purely organic sealants, silicone sealants exhibit excellent thermal stability. Resistances of up to 200 °C and, for special formulations, even 250 °C are quoted in the relevant literature. This high resistance to thermal or thermal-oxidative degradation is a characteristic of silicon-organic polymers. It is primarily determined by the chemical nature of the polymer backbone. The energies of the Si-C and Si-O bonds are considerably higher than that of a C-C bond. Consequently, thermal-oxidative degradation of silicone polymers starts at much higher temperatures than is the case with organic polymers. Significant splitting of the Si-C bond only occurs above 200 °C, while the Si-O bond is stable to above 300 °C.

Besides the silicone polymer, silicone sealants also contain other components, e.g. catalysts, fillers and other additives which also have some influence on the heat resistance. The resistance of silicone sealants at higher temperatures is also determined by environmental conditions. High humidity, for example, causes much faster degradation of the polymer properties at higher temperatures than occurs at the same temperatures in dry air. A further example is the higher resistance of silicon-organic polymers in a vacuum or inert atmosphere than in the presence of oxygen. The thermal resistance of a sealant is thus determined by the temperature, the duration of the exposure and by the environmental conditions.

In silicone elastomers, both oxidative cleavage of the Si-C bond as well as depolymerisation of the siloxane chain occur at high temperatures. The rate of oxidative Si-C cleavage is determined by the oxygen permeability of the sealant. The chemical nature of the side group also has a major influence on the thermal stability. Aromatic side groups result in a greater stability than alkyl groups; the longer an alkyl group is, the more quickly oxidative Si-C cleavage occurs. The thermal stability of the silicone elastomer increases with the following side groups at the polysiloxane chain: $-C_2H_5 < -CH_3 < -C_6H_5$. The oxidative cleavage of the Si-C bond leads to further bond formation which, in turn, increases the crosslinking density. The sealant embrittles, this being demonstrated by increases in the indentation hardness and the modulus. At extremely high temperatures, this process can lead to complete degradation of the silicone elastomer into silica. In contrast, thermal depolymerisation causes a reduction in cure density due to the cleavage of bonds. The sealant becomes softer, this being revealed in a reduction in modulus and indentation hardness. At higher temperatures, the sealant surface becomes sticky. Depolymerisation is accelerated by the presence of metal salt catalysts in the cured silicone sealant. Steam combined with higher temperatures leads to hydrolytic degradation of the polymer chains.

Depolymerisation and oxidative degradation can be slowed down by adding metal oxides to the sealant formulation. This permits the formulation of silicone sealants which are stable up to 250 °C and can thus be used in applications subject to extremely high temperatures.

EXPERIMENTAL WORK

10 commercially available, one-part construction silicone sealants were used for this study. All sealants listed in Table 1 are standard formulations, i.e. without any special heat-stabilising additives. All sealant samples were within first 3 months of their shelf-life. Tensile test joints as per ISO 8339 [1] were prepared with sealant dimensions of 12x12x50 mm^3 and float glass as substrate. They were then conditioned for 28 days in the standard climate (23 °C and 50% rel. humidity). Three specimens were each exposed to temperatures of 100, 150, 180 and 200 °C and room temperature (as a control) over a period of 1, 3 and 6 months. After completion of the heat ageing, the specimens were conditioned for one hour at room temperature and then extended using a tensile test equipment at a rate of 6 mm/min. Elongation at break and tensile strength were evaluated and averaged over each set of 3 specimens.

TABLE 1 -- Initial tensile properties according to ISO 8339

Product	Cure System	Tensile Strength, MPa	Elongation at break, %
Si1	Acetoxy	0.53	30
Si2	Acetoxy	0.49	30
Si3	Acetoxy	0.35	60
Si4	Acetoxy	0.49	70
Si5	Alkoxy	0.62	160
Si6	Alkoxy	0.58	95
Si7	Alkoxy	0.48	160
Si8	Alkoxy	0.79	80
Si9	Oxime	0.52	150
Si10	Aminoxy	0.41	520

RESULTS AND DISCUSSION

All silicone sealants tested suffered adhesive failure in the tensile test. The elongation at break and the tensile strength are illustrated graphically below as a function of temperature for the different conditioning periods.

Acetoxy Silicones

The acetoxy silicones tested include widely differing formulations. Si1 and Si2 are relatively high-modulus silicones, formulated without the addition of silicone fluid as plasticiser. In contrast, Si3 contains larger quantities of silicone fluid, while a compatible, high-boiling-point mineral oil was added to Si4 as a non-silicone extender.

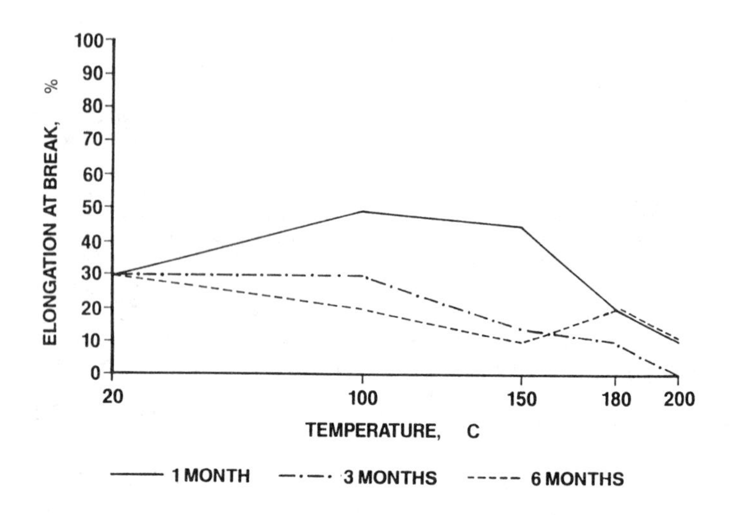

FIG. 1 -- Elongation behaviour of sealant Si1

Fig. 1 to 4 illustrate the behaviour of elongation at break and tensile strength of sealants Si1 and Si2 according to the various conditioning periods and temperatures. As was expected with these formulations, both sealants achieve only low initial elongation at break. The elasticity improves during the one-month heat ageing, i.e. elongation at break and tensile strength increase. In the case of Si1, this effect occurs at temperatures of up to 150 °C; at higher temperatures and/or

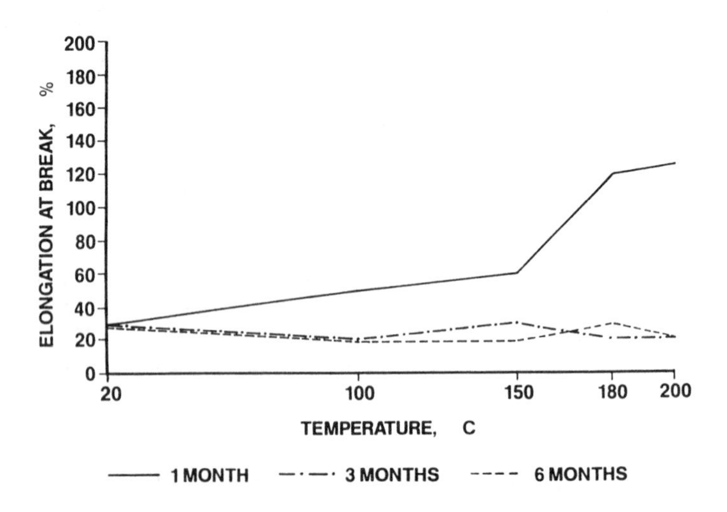

FIG. 2 -- Elongation behaviour of sealant Si2

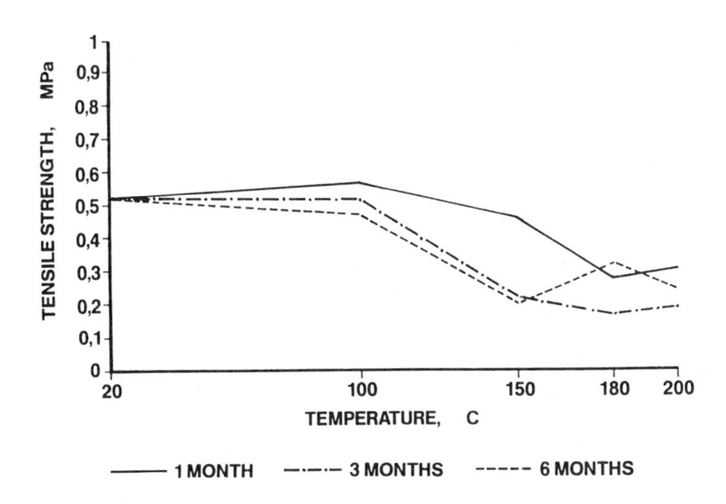

FIG. 3 -- Tensile behaviour of sealant Si1

longer conditioning periods the elasticity starts to degrade again. In the case of Si2, this effect is even more pronounced: after one month conditioning at temperatures of up to 200 °C, elongation at break rises constantly, while tensile strength starts to fall off again above 150 °C, but remains above the initial value. This phenomenon is well known for many types of rubber and is the result of the additional formation of crosslinks. The mechanical properties are thus improved in the ini-

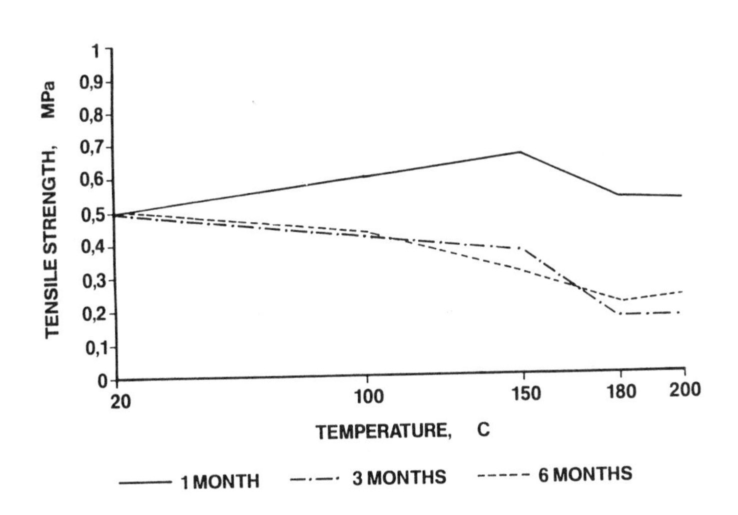

FIG. 4 -- Tensile behaviour of sealant Si2

tial stage of heat ageing. These values worsen, however, after longer periods of heat ageing, especially at higher temperatures. In the case of acetoxy silicone Si2, the degradation of the mechanical properties is particularly low.

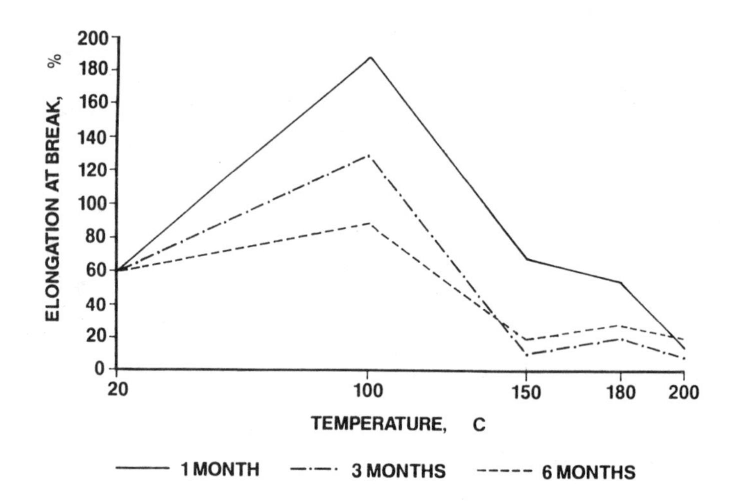

FIG. 5 -- Elongation behaviour of sealant Si3

As illustrated in Figs. 5 and 6, the mechanical values of sealant Si3, which is plasticised with silicone fluid, increase for all conditioning periods at 100 °C. Above this temperature, elongation at break declines rapidly if the conditioning period exceeds one month. At 200

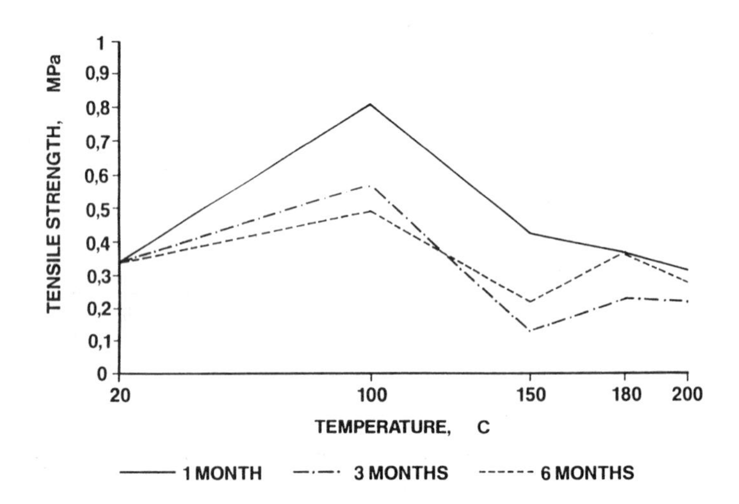

FIG. 6 -- Tensile behaviour of sealant Si3

°C, elongation falls to a half or third of the original value, irrespective of the conditioning period.

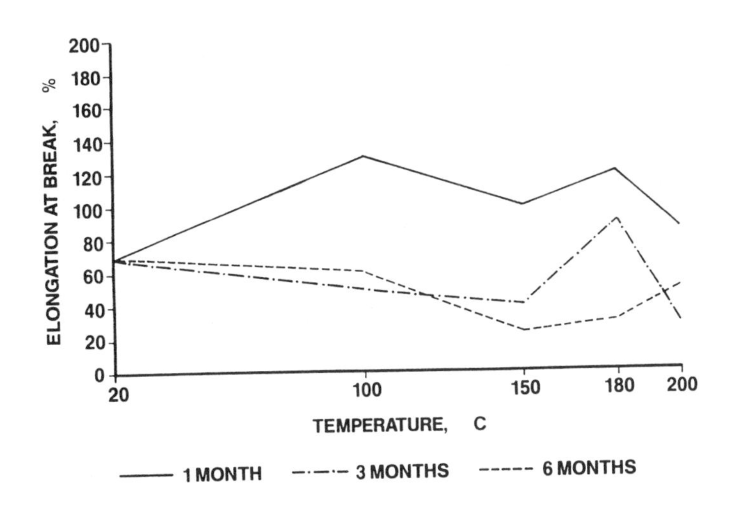

FIG. 7 -- Elongation behaviour of sealant Si4

The behaviour of sealant Si4 (see Fig. 7 and 8), plasticised with a non-silicone extender, is similar to that of the non-plasticised sealant Si1. During one-month heat ageing at 100 °C, the mechanical values improve; both elongation at break and tensile strength increase. Higher temperatures do not cause any significant worsening of the mechanical

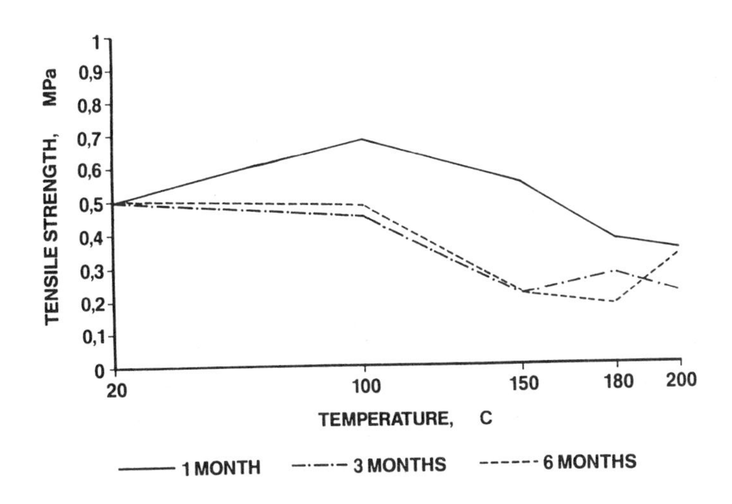

FIG. 8 -- Tensile behaviour of sealant Si4

values in the case of one-month heat-ageing either: while elongation at break always remains above its initial level, tensile strength falls slightly. The sealant thus becomes softer. If the conditioning period exceeds one month, elongation at break and tensile strength fall sharply at 150 ^{o}C, while slightly better results are obtained at higher temperatures. This may be the result of the evaporation of the non-silicone plasticiser. As the tensile strength and elongation at break change at an approximately uniform rate, the sealant modulus is not increased significantly by the heat ageing. Contrary to expectations, the migration of the non-silicone plasticiser does not simply lead to hardening of the sealant. A possible explanation may be that this organic extender catalyses depolymerisation of the polysiloxane at higher temperatures which, in turn, overcompensates the hardening resulting from its evaporation.

Alkoxy Silicones

The behaviour of the alkoxy silicones upon heat ageing is illustrated in Fig. 9 to 16. A common feature of all alkoxy silicones is the fact that the mechanical values improve with heat ageing up to one month duration and at temperatures up to 150 ^{o}C. This is caused by the post cure already described under acetoxy silicones. If the conditioning period exceeds one month, the mechanical properties worsen markedly at or above 150 ^{o}C. After six months conditioning, only minimal elongations

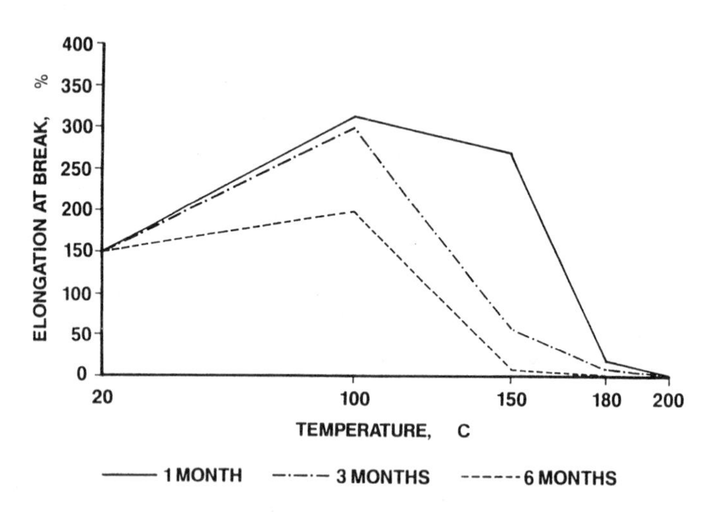

FIG. 9 -- Elongation behaviour of sealant Si5

at break are obtained. The tensile strength only falls slightly over the same period, i.e. the sealants undergo extreme hardening. Above 150 ^{o}C, degradation of the mechanical properties proceeds so rapidly that elongation at break virtually falls to zero after 3 months conditioning.

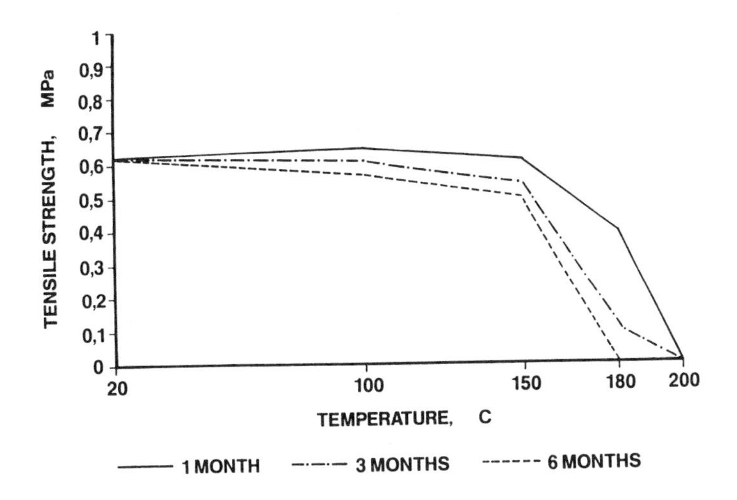

FIG. 10 -- Tensile behaviour of sealant Si5

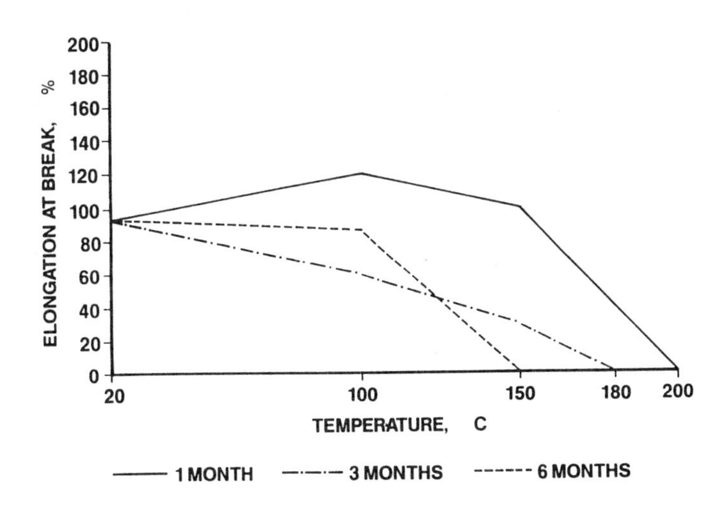

FIG. 11 - Elongation behaviour of sealant Si6

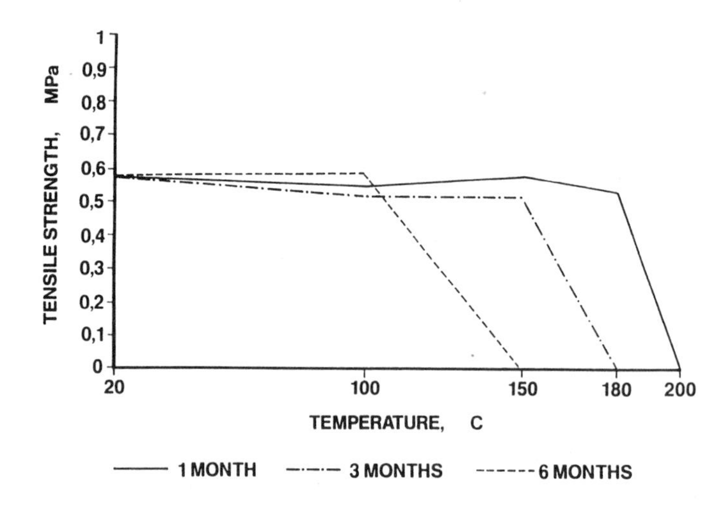

FIG. 12 -- Tensile behaviour of sealant Si6

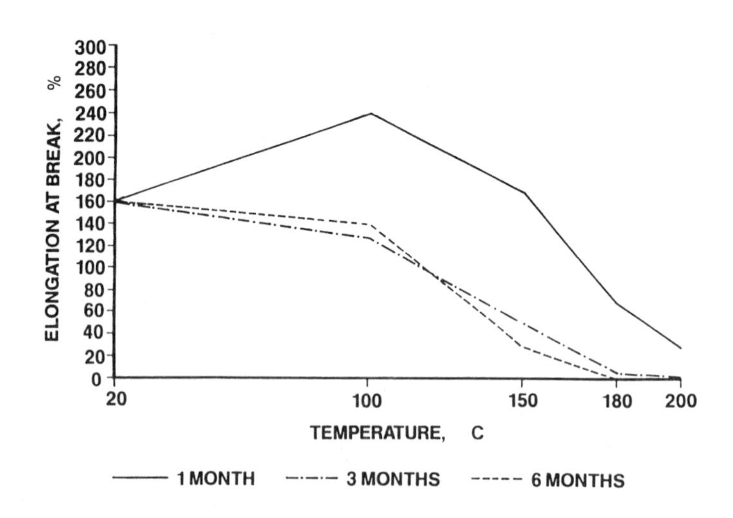

FIG. 13 -- Elongation behaviour of sealant Si7

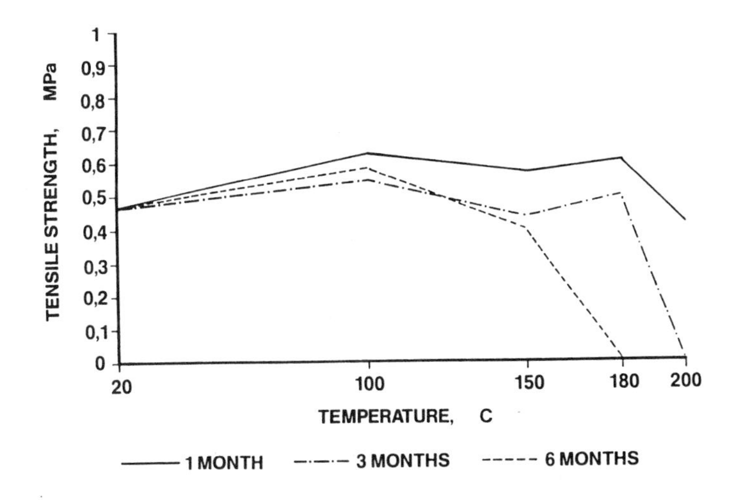

FIG. 14 -- Tensile behaviour of sealant Si7

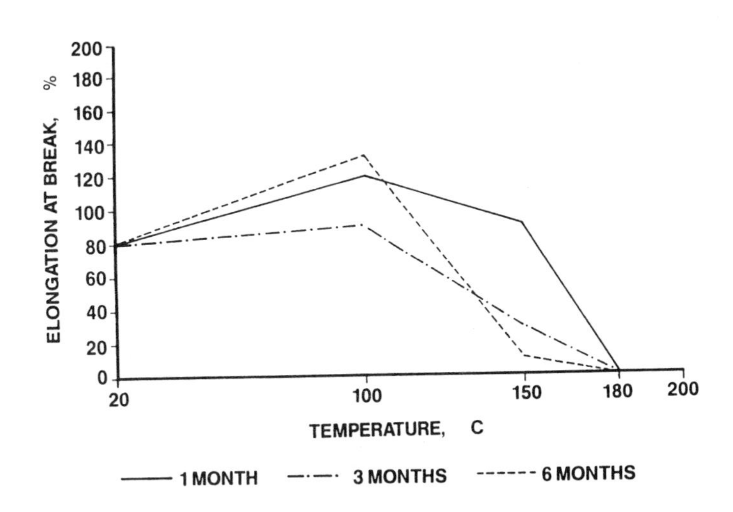

FIG. 15 -- Elongation behaviour of sealant Si8

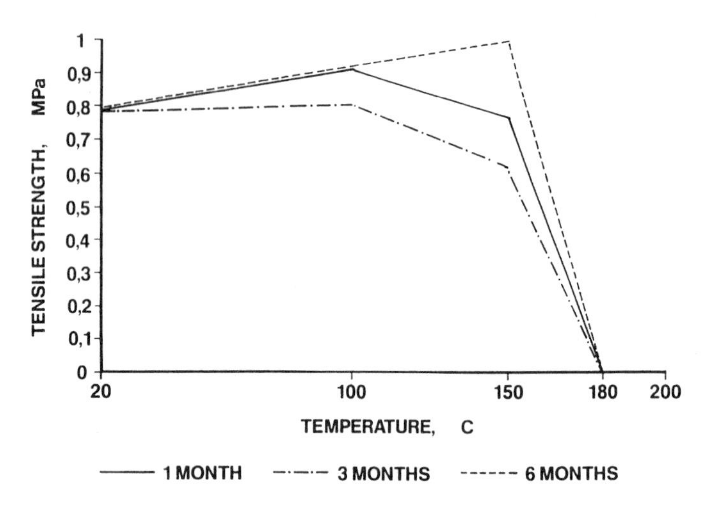

FIG. 16 -- Tensile behaviour of sealant Si8

Oxime Silicone

As illustrated in Fig. 17 and 18, the tensile strength of the tested oxime silicone is only slightly influenced by the temperature and period of conditioning. In contrast, the elongation at break falls by half upon 100 °C conditioning and by two-thirds with 150 °C condition-

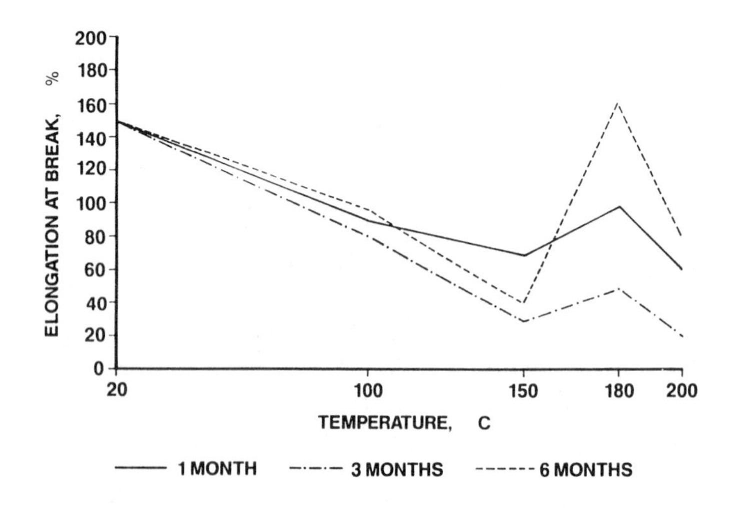

FIG. 17 -- Elongation behaviour of sealant Si9

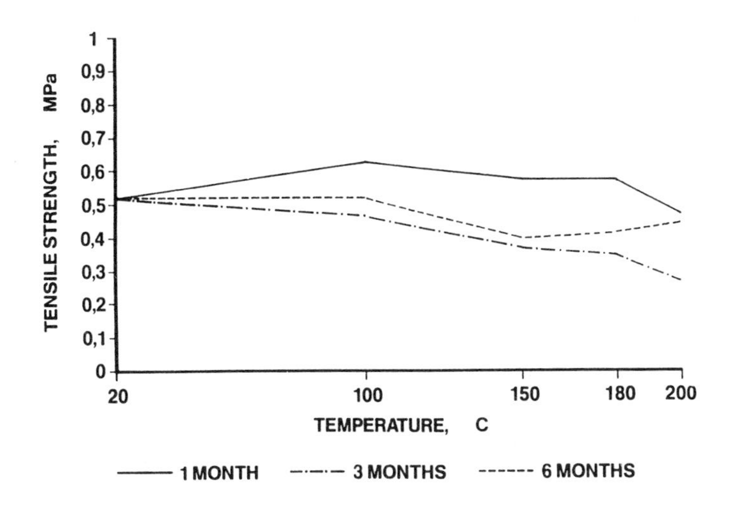

FIG. 18 -- Tensile behaviour of sealant Si9

ing. Compared to 150 °C conditioning, higher elongation at break is achieved upon 180 °C conditioning, but the value obtained at 200 °C is lower than that at 180 °C. Obviously, at least two effects are involved in this complex behaviour: at temperatures of up to 150 °C, post cure occurs due to the formation of additional crosslinks. This results in a sharp fall in the elongation at break, while the tensile strength remains virtually constant due to the increasing crosslinking. Depolymerisation takes effect at 180 °C, resulting in a slight softening of the silicone sealant.

Aminoxy Silicone

The aminoxy silicone sealant tested is an extremely soft-elastic sealant with high movement capability. As illustrated in Fig. 19 and 20, conditioning at high temperatures causes a reduction in elongation at break, although this only falls by a third in the case of six months conditioning at 150 °C. In contrast, the tensile strength increases slightly, leading to a slight increase in the modulus of the sealant. At 180 °C, elongation at break falls to half the original value. With heat ageing up to 180 °C, only a slight difference can be determined between the mechanical values after one month and the mechanical values after six months ageing. Thus, heat ageing only causes a very slow change in the sealant properties. At 200 °C, in contrast, acceptable elongation at break values are only obtained after one month's heat ageing.

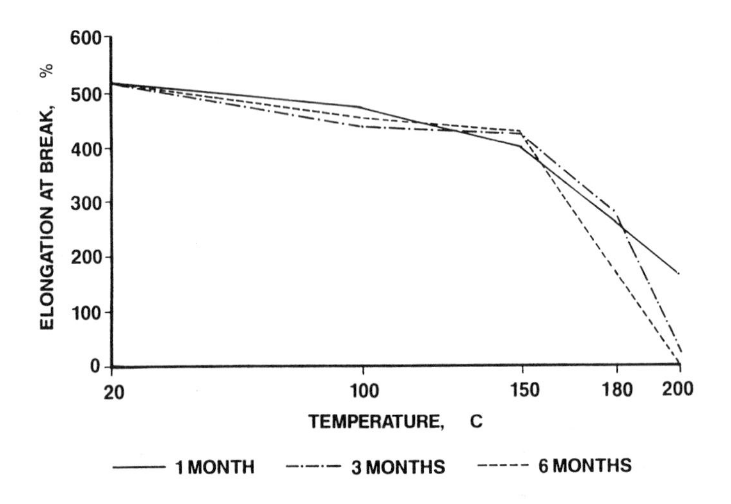

FIG. 19 -- Elongation behaviour of sealant Si10

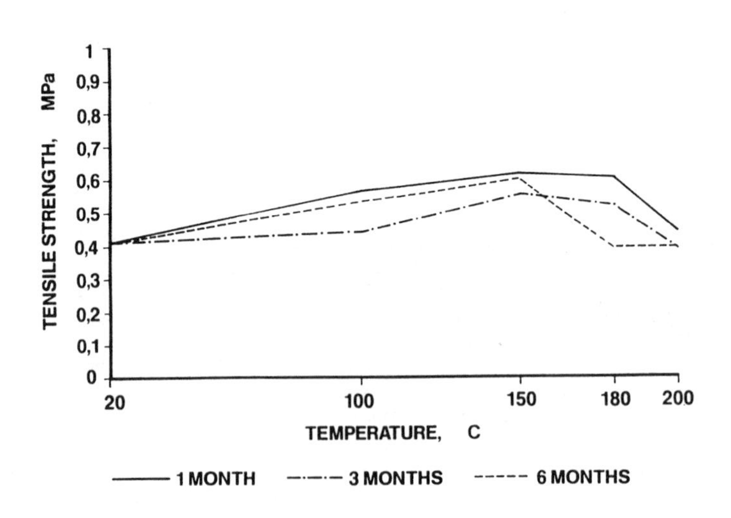

FIG. 20 -- Tensile behaviour of sealant Si10

CLASSIFICATION OF THE THERMAL STABILITY OF SEALANTS

Since no standard procedure exists for the classification of sealants according to their thermal stability, the data provided by the manufacturers are based on various processes and assessment criteria. The test methods differ in terms of the duration of heat ageing; the assessment is based on various material characteristics, e.g. Shore A hardness, modulus, elongation at break, tensile strength and, finally, varying degrees of change are tolerated for these mechanical values.

The sealants examined in this study are classified in Table 2 according to their thermal stability. A maximum increase or decrease in elongation at break or tensile strength by 50% of the initial value was tolerated. As can be seen, the changes in tensile strength result in different maximum service temperatures than those obtained from changes in elongation at break. Although variations in the maximum service temperatures occur within a group of silicone sealants sharing the same cure chemistry - which can be attributed to differences in the formulations - the heat stability is mainly influence by the cure chemistry itself. Following the results obtained in this study, the heat stability of the different sealants can be ranked in the following way:

$$\text{Acetoxy} \sim \text{Aminoxy} > \text{Oxime} \gg \text{Alkoxy}$$

In some applications, the absolute values for tensile strength and elongation at break which apply at the service temperature can be even more important than the percentage changes of these values. If a silicone sealant is used in an adhesive joint, the tensile strength remaining at high temperatures will certainly be of interest. If, however, movements of a particular scale are to be absorbed, a certain minimum elongation at break must be ensured.

TABLE 2 -- Maximum service temperatures based on minimum retention of elongation at break (EB) and tensile strength (TS)

Sealant	max. T (EB), °C	max. T (TS), °C
Si1	180	150
Si2	200	200
Si3	150	200
Si4	150	150
Si5	120	150
Si6	120	100
Si7	120	150
Si8	120	150
Si9	120	200
Si10	150	200

RECOMMENDATIONS FOR FUTURE STANDARDISATION WORK

The sealants studied here demonstrate widely varying behaviour under thermal stress, even though they all belong to one material class. The behaviour patterns would undoubtedly differ even more if sealants from other material classes, e.g. polysulphides or polyurethanes, were included in such a study.

In order to ensure durability of sealants at high temperatures, application-specific test and requirement standards for the thermal stability of sealants must be established as a matter of urgency.

ACKNOWLEDGEMENTS

The authors would like to express particular thanks to Helge Hanke for carrying out all the tensile adhesion tests.

REFERENCES

[1] ISO 8339 "Building Construction - Jointing Products - Sealants - Determination of Tensile Properties", International Standardisation Organisation, 1984

John Beech and Christopher Mansfield

THE WATER RESISTANCE OF SEALANTS FOR CONSTRUCTION

REFERENCE: Beech, J.C., and Mansfield, C., "The Water Resistance of Sealants for Construction", Building Sealants: Materials, Properties and Performance ASTM STP 1069, Thomas F. O'Connor, ed., American Society for Testing and Materials, Philadelphia, 1990.

ABSTRACT:
When sealants are exposed to water in the service environment a number of distinct phenomena may occur which can influence their performance, and may contribute to the failure of the sealant to perform the intended functions. The primary effects with which we are concerned are first, the absorption of water by the sealant, which may cause either softening or enhanced cure, depending upon the chemical nature of the sealant, its cure mechanism and state of cure at the time it is first exposed to moisture. Secondly, the diffusion of liquid water to the sealant/primer/substrate interfaces may impair the adhesive bond of the sealant to the joint surfaces. The laboratory evaluation of these effects is based on tensile testing in which the modulus and the extensibility of sealant test joints after varying durations of immersion in water are used as indices of performance. A system of performance evaluation based upon changes in these indices with water immersion is proposed, and discussed in relation to the results of BRE research results, and those of other workers in the UK.

KEYWORDS: test methods; wet environments; water absorption; adhesive bond; performance rating.

Sealants used in joints in building and construction may be exposed to a wide range of service conditions in which water may have a significant effect upon performance. These range from joints in the external walls of buildings, where exposure to water is usually intermittant and of short duration; to fully immersed joints in water retaining structures, where the sealant must withstand prolonged exposure to moisture.

The retention of water by porous materials such as concrete may present problems in securing adequate adhesion. At the time of application, the presence of water at the joint surface, whether liquid or solid, will militate against securing an adequate bond between the surface and the sealant, and the use of a suitable primer will usually be essential. The continuous transport of liquid water through a porous substrate and its absorption by the sealant may also influence the subsequent performance of the sealants in many service situations.

In recent years considerable research effort has been devoted to the question of water resistance, in both the USA and the UK, and in particular to the development of appropriate test methods for its evaluation. This work has found a focus in the activities of the International Standards Organisation technical sub-committee (TC59/SC8) dealing with jointing products. Earlier work done in the UK, by both the Building Research Establishment (BRE) and the Water Research Centre (WRC), was reported at a R.I.L.E.M. Symposium in Boras, Sweden in 1988 [1], [2]. At present the British Standards Institution is considering proposals for water resistance tests, appropriate to different degrees of exposure to water in service, but is likely to await the outcome of discussions within ISO/TC59/SC8 before making any decisions. At the 1989 meeting of the ISO Committee in Düsseldorf, it was agreed that a Working Group should be established to consider the question of water resistance: it appears that this will comprise UK and US representation only, since there is no significant research activity in any of the other countries represented.

This paper proposes some principles for the laboratory testing of sealants used in wet conditions and for the evaluation of test data. Some recent BRE data are used to illustrate the application of these proposals, and a rating system is described and illustrated by reference to these and other UK data for the effects of water on sealant properties.

Principles & Procedures for laboratory assessment

At least three phenomena may occur when sealant samples are immersed in water [1]. These will affect certain properties measured in the laboratory, referred to as "performance indices".

First, an incompletely-cured material may continue to change chemically (e.g. by cross-linking), which may be evident in the measurement modulus (or stiffness) of water-immersed samples after short periods of immersion. This effect has been found by Gill [3] with one-part moisture-curing materials, but is likely in practice to be of small extent compared to the other two phenomena. Of these, the absorption of water by the sealant leads to softening and plasticisation, and is manifested by a reduction of the measured modulus. (Two distinct phenomena, involving absorption by, respectively, the sealant polymer and by the filler, were identified by Aubrey and Beech [4]).

Thirdly, the diffusion of water through the porous substrate, and through the sealant itself, to the sealant/primer/substrate interfaces is likely eventually to cause a degree of impairment of the sealant bond to the joint surface. It would be possible to assess the magnitude of this effect, after (or during) prescribed conditions of exposure to water, by subjecting appropriate sealant/substrate specimens to repeated cyclic movement at appropriate amplitudes. Performance in such laboratory tests might be quantified by reference to the amplitude of movement and to the number of cycles before failure is observed. Alternatively, the extent of failure after a prescribed total number of movement cycles may be used.

However this procedure has two major practical disadvantages: it requires long test durations to obtain results, and the test data obtained do not lend themselves readily to unambiguous expression to allow a simple performance index to be derived. In practice, therefore a simple tensile test to destruction is preferred, after immersion of the samples in water for a prescribed period (or periods) related to the expected service conditions.

From such tests, using at least 3 replicate specimens, two performance indices may be readily derived:

Modulus (M): this may be taken as the force to extend a standard specimen by a given amount: 25 per cent is used in this work (F_{25}) since this extension is not greatly in excess of the maximum strain values to which sealants are commonly subjected in correctly-designed joints. A change in modulus (ΔM) is indicative of hardening or softening of the sealant.

Extensibility (E) is measured as the percentage elongation at which the maximum force is observed during the tensile test. E may be regarded as an index of bond integrity.

A further essential observation to be made in these tests is the nature of the failure, whether adhesive or cohesive in character; and whether there is any change in this when water-immersed specimens are compared with controls maintained in air at standard conditions.

Datum Conditions

In assessing changes in the performance indices after exposure to water it is necessary to compare the relevant mean values from the laboratory tests with those obtained from control specimens. Although control samples of the same age as those immersed in water may be used for this purpose, this does not always provide true stable datum values of the relevant measured properties as a basis for comparison.

In the laboratory it is not uncommon to find progressively smaller though significant changes in the measured values of sealant modulus occurring for periods of up to 12 months after preparation of the specimens.

In general, therefore, it is recommended that the datum condition for measurement of stable "fully-cured" values of performance indices should be storage in air at 23°C and 50% r.h. for the longest period of water-immersion used in the absorption tests (in this case, 84 days).

Comparison of mean values for immersed and control specimens allows the calculation of the modulus difference (ΔM) and of the extensibility difference (ΔE), these being the putative effects of exposure to water on these key performance factors for the sealant.

Evaluation: the Rating of Water Resistance

These considerations lead us to a simple diagrammatic representation of the effects of water on sealant properties, which allows the changes with increasing duration of immersion to be charted, and forms the basis of a simple rating system for water resistance at a number of levels (or degrees) of exposure to water in service. Some tentative proposals for the latter will be advanced later in the paper.

Figure 1 indicates the basis for the proposed model in which data are plotted on $\Delta E/\Delta M$ co-ordinates. Experimental data, including those reported herein, had indicated that the majority of results fall within the ranges of modulus and extensibility differences shown: i.e. ΔE from total loss of bond to 100% increase; ΔM showing reductions of up to 60%.

A statistical examination of the experimental data for sealants which had modulus values indicating relative completeness of cure at the time of immersion had shown (on the basis of replicate specimens) that at the 1% significance level, differences in E were not significant for values within ± 37%; while ΔM values were significant when outside the range ± 22%. This suggested the basis for dividing the area of the $\Delta E/\Delta M$ diagram into nine equal areas, in which the data falling could be rated in performance terms relative to the significance of the measured differences in both modulus and extensibility.

The cardinal principles of the rating system based on fig 1 may be expressed thus:

- bond integrity being regarded as the primary performance index, a significant increase in extensibility is given a positive rating, and vice versa.

- positive and negative changes in the extensibility are then ranked taking account of any softening of the sealant caused by water absorption. If the sealant is softened, this will mask any loss of bond integrity which is caused by immersion, since lower stresses will occur at the sealant/surface interface for a given extension of the wet sealant.

Thus data points falling in areas to the right of area "0" represent improved extension values and these areas have been denoted with positive rating values. But whereas data in the area denoted "+1" show apparent increases in mean E values for significantly softened sealant, the area "+3" represents sealant specimens having improved bond integrity without significant softening and plasticisation. A similar argument applies to the negative ΔE values

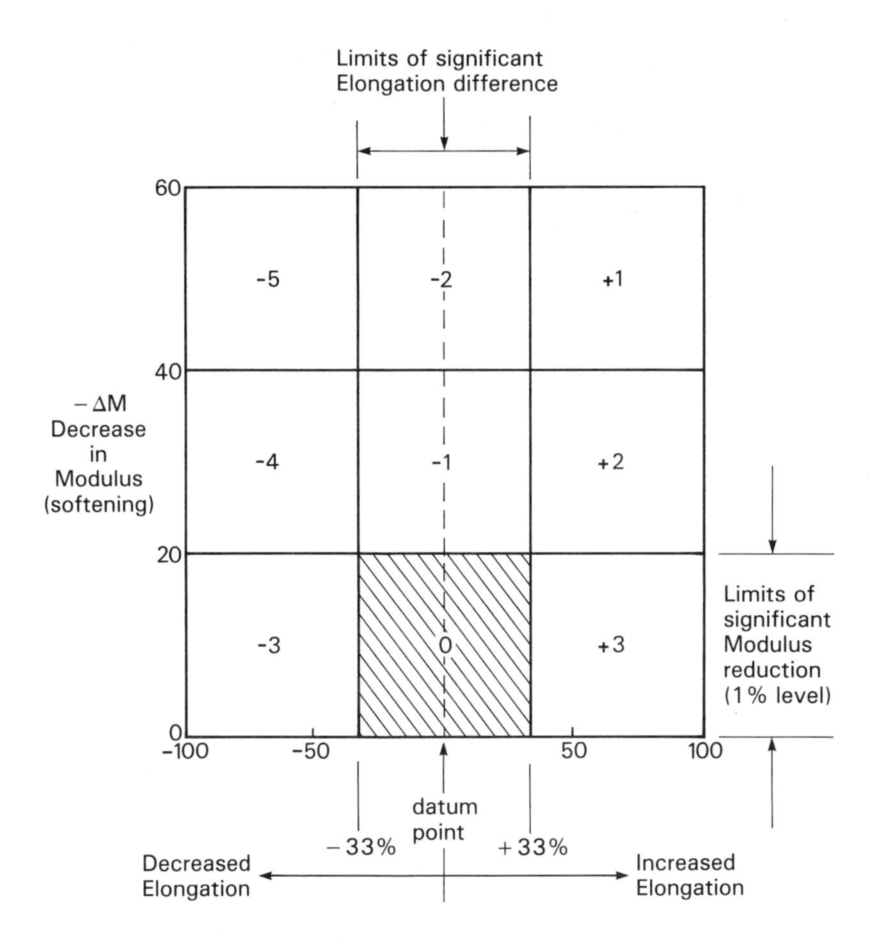

FIG. 1 -- Extensibility/modulus effects with water immersion

e.g. rating "-5" is given to sealants which, despite a considerable degree of softening, are found to have lost bond integrity markedly when subjected to tensile strain after water immersion.

The few data points which fall below the section of the $\Delta E/\Delta M$ diagram in figure 1 (i.e. showing a small degree of hardening) can be included in the rating by arbitrarily allocating them to the immediately adjacent area on the diagram.

Criteria for Testing and Performance Requirements

It will be necessary to specify the duration of water immersion for laboratory tests for water resistance in relation to the intended conditions of service. Work by Gill [2], as well as data to be discussed in this paper [5] suggest that for sealants to be used in water retaining structures, samples may be required to be immersed in water for 6, or even 12 months to allow a realistic assessment of suitability for use in these conditions.

A tentative schedule for testing is given in Table 1.

TABLE 1 -- Water Immersion Schedule

Exposure to water in service conditions	Period of water immersion			
	7 days	28 days	16 weeks	48 weeks
Intermittant, short duration e.g. vertical surfaces on exterior of buildings	✓	...	...	...
Moderate: e.g. horizontal surfaces which may become flooded	✓	✓	✓	...
Severe: continuously immersed [a]	...	✓	✓	✓

[a]This schedule would also be appropriate for sealants required to withstand contact with potentially aggressive aqueous media such as sea water, river water, sewage.

The schedules are designed to ensure that sealants for use in each of the identified service conditions will demonstrate a minimum level of performance in respect of water resistance. In addition, by securing data at intermediate periods of immersion, information may be obtained about the rate of change of sealant properties during immersion, as an aid to prediction of service lifetimes.

Regarding the performance requirements for sealants tested according to such a schedule, a number of possible limits may be envisaged. First, and most simply, a minimum rating in the model put forward earlier (figure 1) may be stipulated. For example, a requirement of "-1 rating or better" would ensure the sealant exhibited no loss in mean extensibility of more than 33%, and a mean reduction of modulus not greater than 40% after the period of water immersion appropriate to its intended service environment. (It will be recalled that these differences relate to the appropriate datum condition of the sealant).

Alternatively, the ΔE and ΔM values which are acceptable after appropriate water immersion may be quoted without reference to the rating system. A third option is to indicate the areas of the ΔE/ΔM diagram in which the data points for the relevant mean values for the water-immersed samples should fall.

EXPERIMENTAL

Materials & Experimental Procedure

Nine commercial sealants, supplied for use in construction where wet conditions may occur, were used in the test programme. These comprised four two-part polysulphide-based sealants, three two-part polyurethanes and two silicone sealants.

For each sealant 10 replicate specimens were prepared for each period of water immersion or control storage. Specimens comprised a 12 x 12 x 50 mm bead of sealant between substrate blocks of either aluminium alloy (AL) or cement mortar (CM), and were prepared according to British Standard Specifications [6]. Primers recommended by the manufacturers were applied to the substrate surfaces as instructed. The specimens were preconditioned initially for 56 days at 23°C and 50% rh before storage in either tap water or in air at 23°C/50% rh for periods of 7, 14, 28 or 84 days.

Five each of water-immersed and control specimens were extended to failure at 5 mm/min, and force/elongation data recorded. The remaining 5 specimens of each were extended to 200% of their initial widths, and held at this extension using suitable spacers for 24 hours. Any failures in cohesion or adhesion were then noted; the nature of then failures was also observed for the samples tested to destruction.

From the experimental data, two performance indices were derived: an estimate of the modulus was obtained from the mean force (for 10 specimens) recorded at 25% extension; while the extensibility was calculated from the mean value of the elongation at which maximum force was recorded for those specimens tested to failure (normally 5 in number).

The water immersion test gives acceptable reproducibility. The measurement of extensibility is inherently more variable than that of modulus: for well-cured sealants, coefficients of variation less than 20% are typical. For slow-curing materials such as Si1, however, values as high as 60% were noted for individual elongation data.

Results

The complete results of this study are to be published elsewhere [5]. Selected data are given in figures 2, 3 and 4 to illustrate changes in the two performance indices for three of the materials tested: one of each of the generic sealant types included in the study. Presentation of the data in this form allows any changes due, respectively, to softening of the sealant, and to bond impairment, to be readily charted with increasing durations of immersion in water. The data points relate to differences between water-immersed and control samples at 7, 14, 28 and 84 days, for aluminium (AL) and cement mortar (CM) substrates.

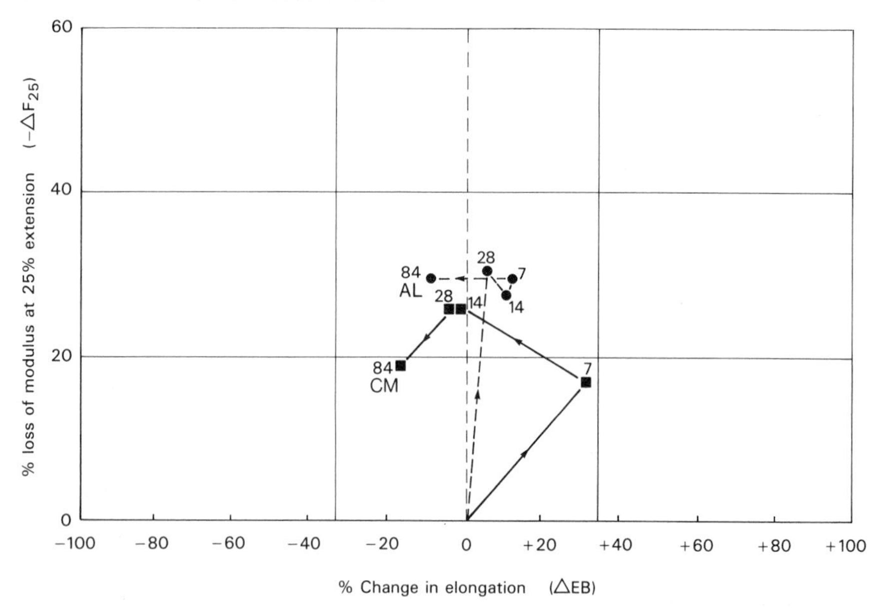

FIG. 2 -- Effect of water immersion on sealant PS1

DISCUSSION

The primary aim of the study was to evaluate the proposals outlined earlier for test methods for the water resistance of sealants and for the evaluation of the test data.

The results provide, incidentally, information regarding the suitability and relative performance of the sealants tested for use in wet environments, and illustrate some striking differences in performance between the two substrates used. In general performance is better with cement mortar, since the sealants are intended for use in joints in concrete and other porous materials.

Figures 2 and 4 exhibit a typical feature of the $\Delta M/\Delta E$ changes for these sealants with water immersion: at 7 days extensibility increases as a result of softening of the sealant but there is rarely any evidence of bond loss. With further exposure to water, modulus changes are generally small but there is often evidence that the bond has been impaired. In some cases the resulting loss of extensibility is severe: see for example the effect of water on the silicone sealant (Si.1) for both substrates in figure 4. This sealant is seen

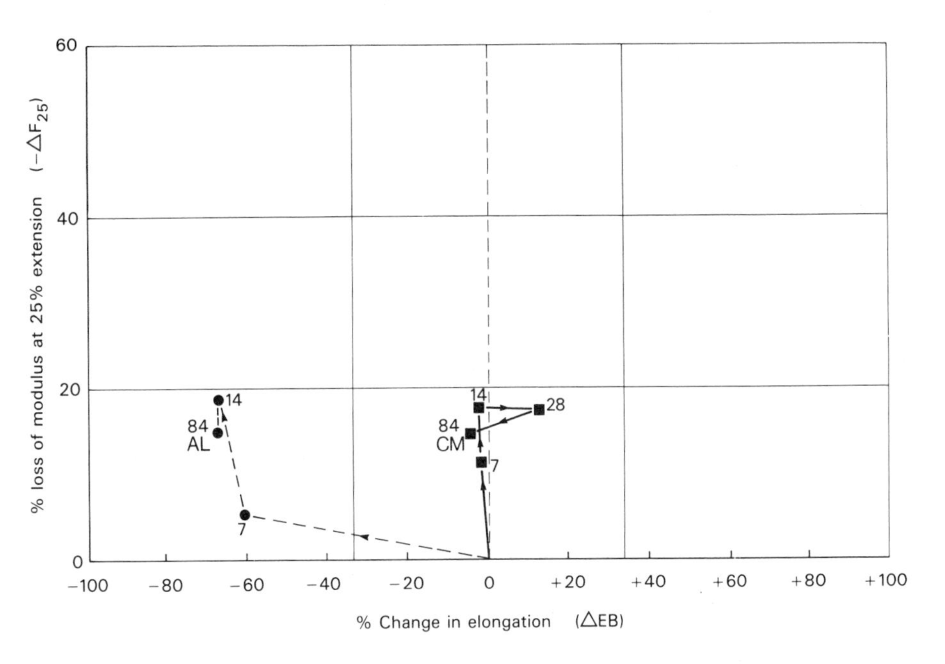

FIG. 3 -- Effect of water immersion on sealant PU3

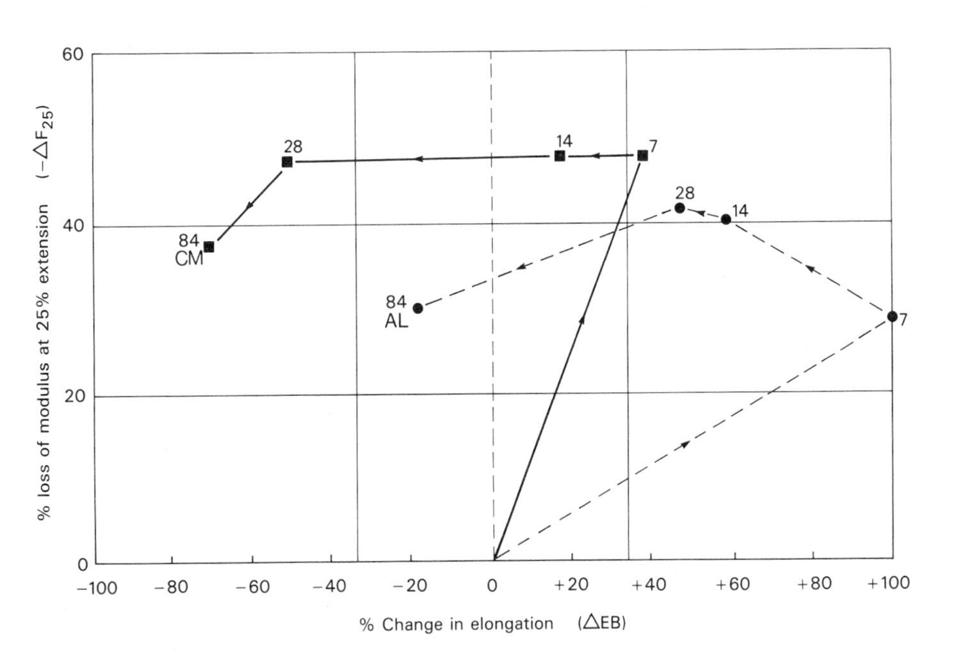

FIG. 4 -- Effect of water immersion on sealant Si.1.

to be poorly suited for joints in wet environments, and untypically, its performance is worse with the CM than with the aluminium substrate. In this respect the performance of sealant PU3 (figure 3) is more typical, and its performance with the CM substrate indicates excellent water resistance even with prolonged water immersion.

The results of the maintained extension tests did not give much additional information on the water resistance of the materials tested. It was found that when failures did occur, they were observed during the initial extension of the specimens. It is inferred that during the 24 hour period of maintained extension, some degree of stress relaxation may occur to reduce the effective stress within the sealant. In view of the additional time required to carry out this test it is not considered that it is justified to include maintained extension as part of a standard water resistance test.

Failures of water-immersed samples were invariably adhesive in character, whether in maintained extension or elongation to failure.

Performance Rating of Tested Sealants

In figures 2, 3 and 4 the ΔM and ΔE data plotted are based on datum values measured with initial control specimens. As noted previously the datum used for measuring the percentage changes in E and M can strongly affect the performance rating of sealants which are not well-cured when water immersion commences. This can be demonstrated for the three sealants used in the illustrations, restricting the comparison to data for CM substrates only.

Table 2 gives some mean data for percentage changes in M and E of the three sealants illustrated previously, using three different datum conditions: initial control values (t = 0), well-cured controls, (t = 84) and using controls of the same age as the water-immersed samples.

Sealant PU3 is typical of the best performance measured with CM substrates showing relatively stable properties: water immersion appeared to cause little softening and no significant loss of bonding. This sealant and PSI were well-cured before immersion so that the ratings of both are not much affected by the choice of the datum condition.

In contrast the silicone sealant cured slowly, so that choice of the initial control values as datum may not fully reflect in the performance ratings the effects of water on modulus and extensibility. As can be seen from figure 5, the state of cure before immersion can influence the effect of water on sealant modulus with increasing duration of exposure.

Water Immersion Requirements

It will be evident from figures 2 to 4 that data obtained after a short period of immersion (7d) may be a poor guide to performance after prolonged immersion (this was also found with the other sealants

TABLE 2 -- Effect of datum condition on ratings after water immersion

Sealant Code	Immersion Period days	Datum t = 0			Datum t = 84d			Datum: same age controls		
		ΔM%	ΔE%	Rating	ΔM%	ΔE%	Rating	ΔM%	ΔE%	Rating
PS.1	7	-25	+18	-1	-21	+31	-1	-17	+30	0
	14	-26	+6	-1	-23	+18	-1	-26	-1	-1
	28	-23	-13	-1	-20	-4	-1	-26	-3	-1
	84	-23	-26	-1	-19	-17	0	-19	-17	0
PU3	7	-21	+11	-1	-16	0	0	-11	-2	0
	14	-27	+16	-1	-23	+4	-1	-18	-2	0
	28	-26	+18	-1	-21	+6	-1	-18	+12	0
	84	-20	+8	0	-15	-3	0	-15	-3	0
Si1	7	-48	+37	+1	-42	-13	-2	-46	+15	-2
	14	-48	+17	-2	-43	-26	-2	-47	-8	-2
	28	-47	-51	-2	-42	-69	-5	-47	-62	-5
	84	-37	-63	-4	-30	-77	-4	-30	-77	-4

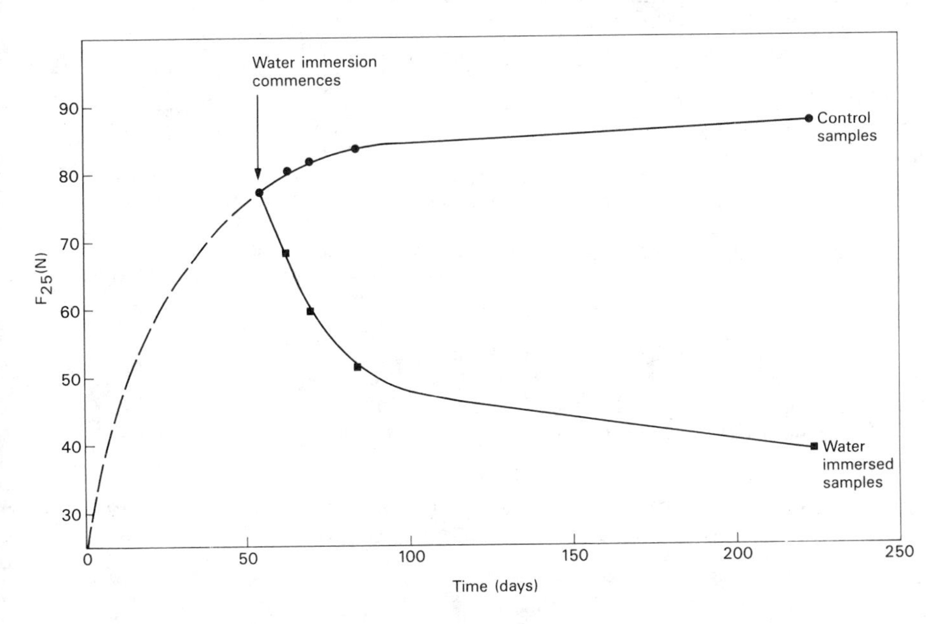

FIG. 5 -- Modulus changes of silicone sealant in air and in water

for which data are not reported here). The use of immersion periods as short as 4 days, as for example in French standard test methods, is not sufficient as an assessment of "water resistance", for which the period of immersion must be related to the degree of exposure to water expected in service.

Regarding sealants used in joints which are subjected to prolonged water immersion, there was evidence from data for a number of the sealants tested that further changes in extensibility were likely with periods of immersion longer than 84 days, though modulus values had generally stabilised at this time. This finding confirms the results obtained by Gill [2], and it is considered therefore that laboratory testing should include water immersion for at least 6 months when it is required to simulate the effects of service environment of joints in water retaining structures.

CONCLUSIONS

The laboratory assessment of the water resistance of construction sealants requires periods of immersion sufficient to simulate the changes in performance-related properties which will affect performance in service. It is proposed that these changes should be estimated by measurement of two performance indices from tensile test data: the extensibility and modulus. By the choice of an appropriate datum condition, the differences in these two performance indices form the basis of a rating system, which reflects the changes in properties which most strongly influence sealant performance in joints in wet environments: softening of the material due to absorption of water, and impairment of the bond to the joint surfaces.

REFERENCES

[1] Beech, J.C., "The Performance of Sealants under Wet Conditions", presented at R.I.L.E.M. Symposium on Building Joint Sealants, Boras, Sweden, May 1988, available as reprint from BRE.

[2] Gill, B.W., "Performance Requirements for Building and Civil Engineering Sealants used in Wet Conditions". ibid, Statens Provningsanstalt, Boras, Sweden, 1988.

[3] Gill, B.W: personal communication.

[4] Aubrey, D.W., and Beech, J.C., "The Influence of moisture on Building Joint Sealants", Building and Environment, Vol. 24, No. 2, 1989, pp. 179-180.

[5] Mansfield, C: "Tests for the Water Resistance of Construction Sealants". To be published in Construction and Building Materials, March 1990.

[6] British Standard: Building and construction sealants. BS 3712: Part 4: 1985. Method of test for tensile extension and recovery. British Standards Institution.

I. Luis Gomez[1] and M. R. Colella[2]

USE OF Q-U-V TO PREDICT THE EDGE COMPATIBILITY OF COMMERCIAL SEALANTS WITH LAMINATED GLASS

REFERENCE: Gomez, I. L. and Colella, M. R., **"Use of Q-U-V To Predict The Edge Compatibility of Commercial Sealants With Laminated Glass"**, Building Sealants: Materials, Properties, and Performance, *ASTM STP 1069*, Thomas F. O'Connor, Editor, American Society for Testing and Materials, Philadelphia, 1990.

ABSTRACT: An important factor to be considered when commercial sealants contact the edges of laminated glass is that of the compatibility of the sealants with the polyvinyl butyral (PVB) interlayer. Lack of compatibility with sealants, which is manifested by visible narrow edge-band defects in the laminates, is more detrimental to the aesthetic than to any of the other functional properties of these laminates. For the most part, however, the functionality of sealants is unaffected by these compatibility problems since they are created by the by-products of the normal curing reaction of the sealants. The information presented here is aimed at helping laminated glass users make more informed judgments regarding the selection of sealants. It uses mainly the Q-U-V laboratory accelerated weathering tester to predict these compatibility problems.

KEY WORDS: sealants, Q-U-V laboratory accelerated weathering tester, compatibility, fluorescent UV lamps, condensation, high-modulus silicones, low-modulus silicones, butyl/polyisobutylene tapes, edge defect, hardness, dirt pick-up, barrier properties, polyvinyl butyral, PVB laminated glass

INTRODUCTION

This paper deals with the use of Q-U-V, a laboratory accelerated weathering tester, to predict the edge compatibility of commercial sealants with laminated glass.

Up to now, Monsanto has been reporting on the compatibility of commercial sealants with Saflex PVB interlayers in laminated architectural glass (Reference 1). The first bulletin on sealants' compatibility, for example, describes the results of (1) 12 months of Florida and Arizona, and (2) 2400 hours of laboratory accelerated, Q-U-V, exposures. The second bulletin on this issue, recently published, updates these exposures, discussing the results of (1) 30 months of Florida, (2) 24 months of Arizona, and (3) 3200 hours of Q-U-V exposures. Throughout these bulletins, frequent references have been made of (1) the Q-U-V accelerated weathering tester and its alternating capability of providing ultraviolet and condensation exposures; (2) sample preparation techniques; and (3) edge defect measurements. This edge defect is mostly manifested as a delamination band. Other important test features are described such as:

[1]Science Fellow, Monsanto Chemical Company, Springfield, MA 01151.
[2]Technical Service Specialist, Monsanto Chemical Company, Springfield, MA 01151.

1. The length of the UV cycle and the type of UV lamps selected to accelerate weathering;
2. The temperatures and conditions of the Q-U-V chamber during the condensation cycle, which alternates with the UV cycle;
3. The two UV/condensation alternating cycles used in these studies.

   a. 16-8 cycle:16 hours UV/no condensation, followed by 8 hours condensation/no UV;
   b. 8-4 cycle:8 hours UV/no condensation, followed by 4 hours condensation/no UV; and
   c. The effect of condensation alone and no UV light (this exposure was conducted in the Q-C-T), however, have not been reported before.

   For more details on temperatures and conditions of the UV/condensation alternating cycles, see Table I.

In this paper we will describe the three test features and conditions listed above and:

1. The close correlation found between the 16-8 cycle and the prolonged outdoor exposures;
2. A parallel evaluation in the Q-U-V unit with the 16-8 and 8-4 cycles;
3. The weatherability of the sealants, per se (as measured by changes in sealants' hardness); and
4. Some tentative conclusions derived from the analyses of the available data.

TEST CONDITIONS

Fluorescent UV Lamps:

Only UVB-313 lamps with peak emission at 313 nm have been used for this evaluation. Due to their shortwave UV emission, and because shorter wavelengths have higher energy, these lamps act very fast in inducing sealant curing reaction, aging, and other changes in the sealant which could promote edge defects in the area of the laminates contacting the sealant. Furthermore, since the effect of lamp aging on spectrum emission for this particular wavelength is minimal, reproducibility of their accelerating effect is very good. Moreover, since these lamps are used at temperatures >60°C (>140°F), these rather severe exposure conditions also accelerate changes in the sealant which could propitiate migration of volatiles and/or plasticizers into the PVB interlayer. In addition to acceleration, one can depend on the reproducibility of the accelerating effect of these lamps. Although not yet tested, it is expected that the longer wavelength spectrums emitted by the UVA-340 and UVA-351 will produce the same results but with longer exposure. For more information on operating procedures for using fluorescent ultraviolet (UV) and condensation apparatus, see the ASTM G53 Method.

Condensation Cycle:

The condensation cycle which alternates with the UV cycle has several functions; first, since condensation is conducted at rather high temperatures, it contributes to acceleration of the exposure conditions and hence, aging of the sealant. Second, it helps to detect if sealant-to-glass edge adhesion is adequate to prevent laminate damage. Third, it also helps to assess if sealant water vapor transmission is low enough to provide the required protection.

During the condensation cycle, the Q-U-V simulates rain and dew by condensing hot vapor on the surface of the laminates, thus causing some wetness during a rather high percentage of the exposure time. More on this later.

Table I -- Temperatures and Conditions of the UV/Condensation Alternating Cycles

Cycle	Conditions
16-8	16 hours UV @ 66°C (150°F), no condensation 8 hours condensation @ 60°C (140°F), no UV light
8-4	8 hours UV @ 60°C (140°F), no condensation 4 hours condensation @ 45°C (113°F), no UV light
24	24 hours condensation only at 60°C (140°F), no UV light

Outdoor Exposure

Samples were mounted according to the standard practice for atmospheric environmental exposure testing of non-metallic materials, ASTM G7 Method.

TEST RESULTS

Outdoor Exposure vs the 16-8 Cycle and the 16-8 vs the 8-4 Cycles

Figures 1 through 9 illustrate the average general performance of five high-modulus, six low-modulus silicones, and six butyl/polyisobutylene tapes in Florida, Arizona, and Q-U-V 16-8 cycle exposures. See Tables 2 and 3 for a list of the sealants initially evaluated. As shown for each of the sealants tested, the correlation between the three types of exposure conditions is quite good. In most cases, not only the shape of the edge defect curve but also the magnitude of the edge defect is closely matched.

Figures 10, 11 and 12 illustrate the general performance of the same sealants evaluated in the above figures. This time, however, we used the Q-U-V 8-4 hour cycle. As seen, the 8-4 hour cycle edge defect plots do not closely approximate those of the 16-8 hour cycle (Figures 1, 4 and 7), and one can make the following observations:

1. High Modulus Silicone Sealant - Average Edge Defect:

Magnitude of the edge defect of the cut edge[a] is significantly lower than that of the autoclaved edge[b] which is not what we have seen upon outdoor exposure. Also, it tends to plateau at about 1.5 mm (0.06 inch), while with the 16-8 hour cycle, it plateaus at about 3.5 mm (0.14 inch).

2. Low Modulus Silicone Sealant - Average Edge Defect:

Autoclaved edge shows about the same value measured when a similar set of samples were exposed using the 16-8 cycle. The cut edge, however, never plateaus and the values are significantly higher than that of the 16-8 cycle.

3. Butyl/Polyisobutylene - Average Edge Defect:

Overall values are higher than those measured when using the 16-8 cycle. Furthermore, the average cut edge initially shows lower edge defect values than the autoclaved edge and then after about 2500 hours of Q-U-V exposure, there is a sudden increase in edge defect growth.

[a]A laminate edge created by cutting an original laminate into smaller sections.
[b]An original laminate edge exposed to the autoclaving environment (oil or air) during manufacture of the laminate.

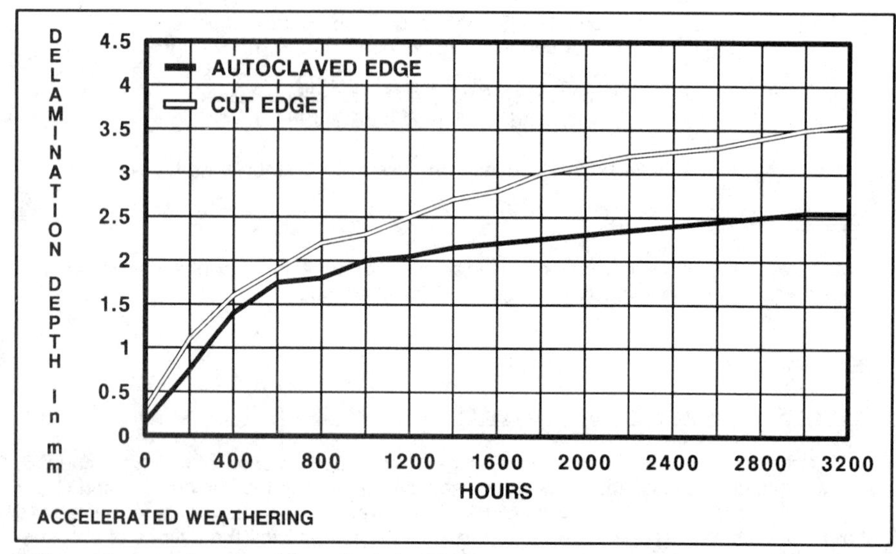

Figure 1 -- Average defect depth in Q-U-V exposure, 16-8 cycle high-modulus silicone sealant.

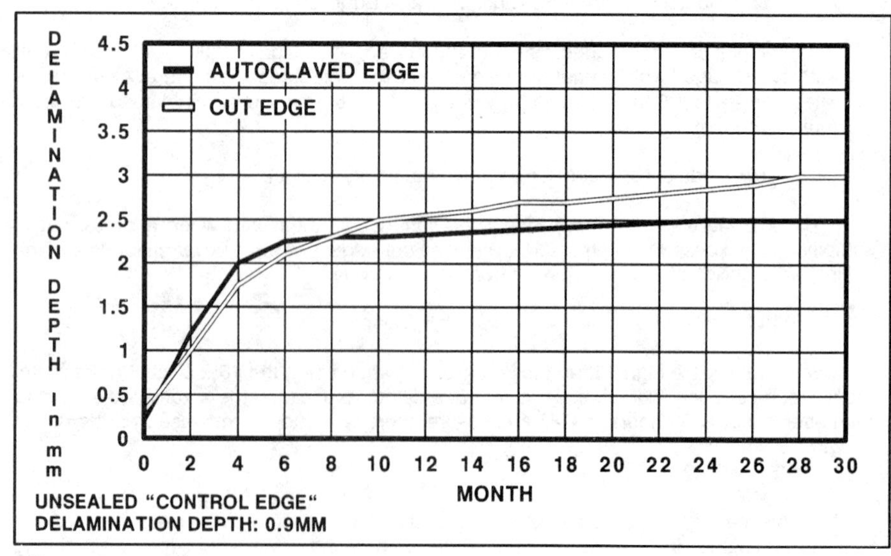

Figure 2 -- Average defect depth in Florida exposure high-modulus silicone sealant.

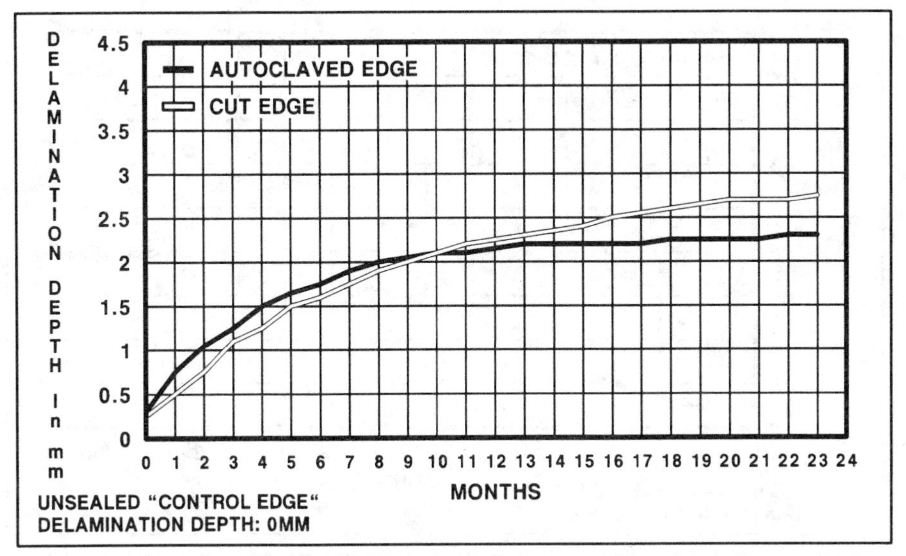

Figure 3 -- Average defect depth in Arizona exposure high-modulus silicone sealant.

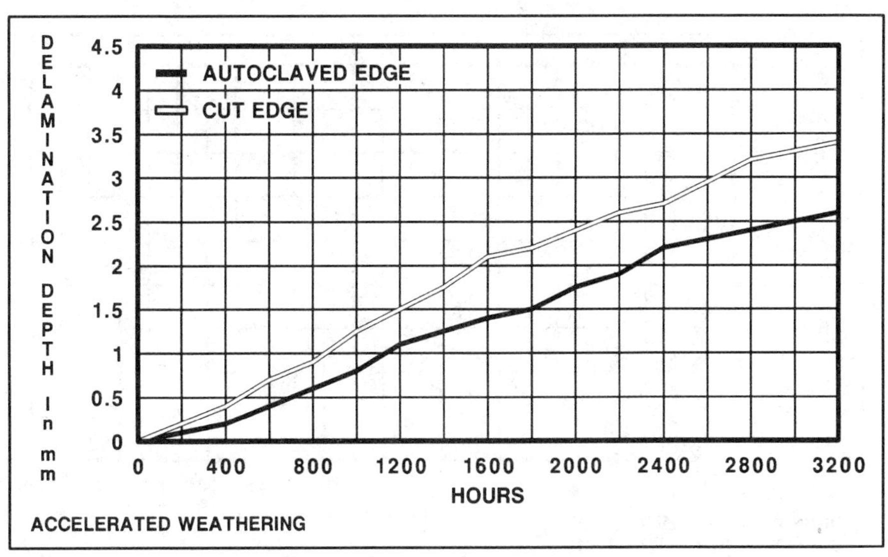

Figure 4 -- Average defect depth in Q-U-V exposure, 16-8 cycle low-modulus silicone sealant.

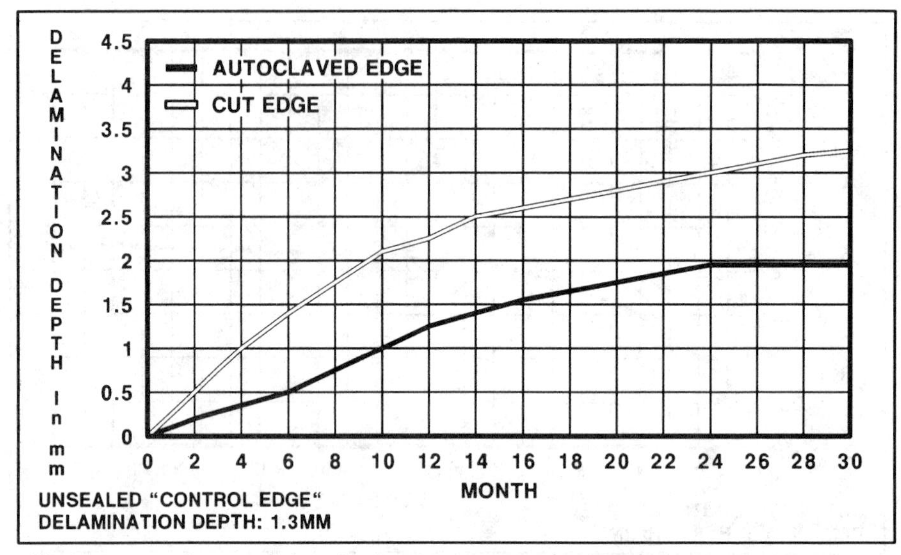

Figure 5 -- Average defect depth in Florida exposure low-modulus silicone sealant.

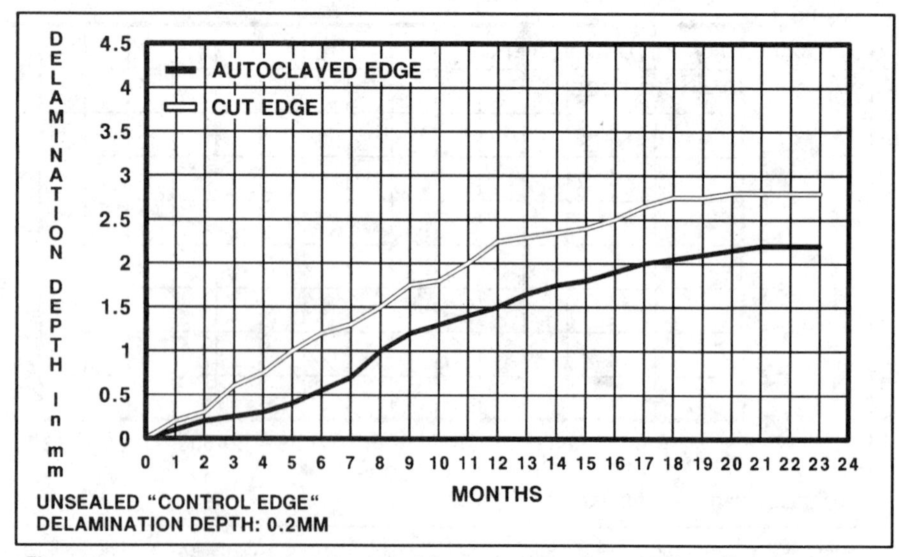

Figure 6 -- Average defect depth in Arizona exposure low-modulus silicone sealant.

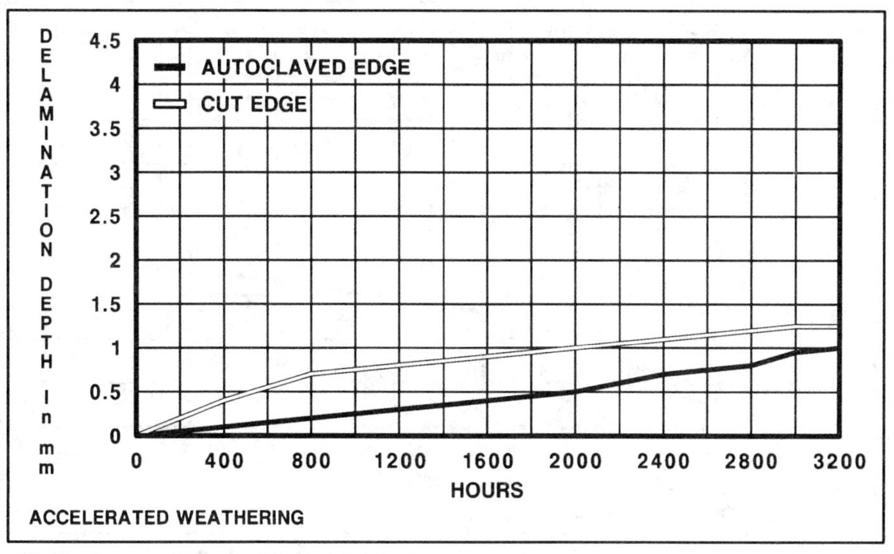

Figure 7 -- Average defect depth in Q-U-V exposure, 16-8 cycle butyl/polyisobutylene.

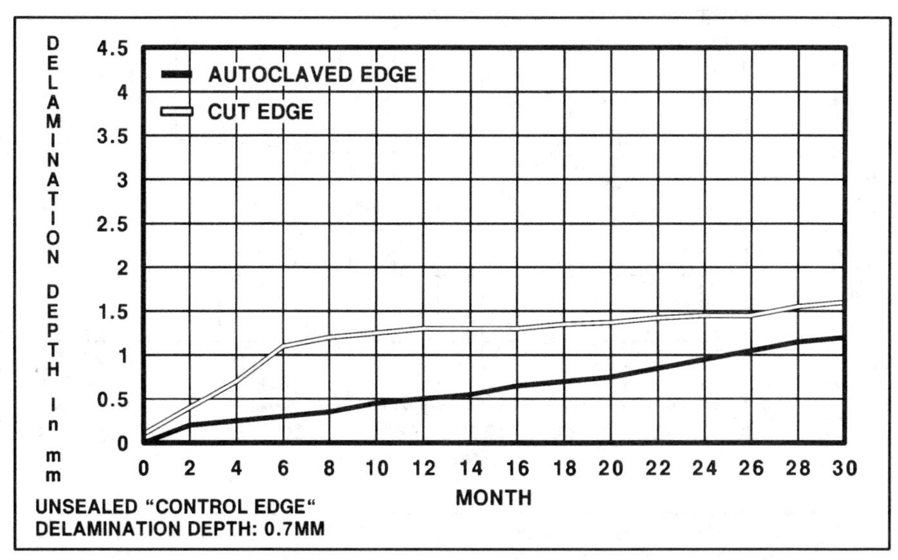

Figure 8 -- Average defect depth in Florida exposure butyl/polyisobutylene.

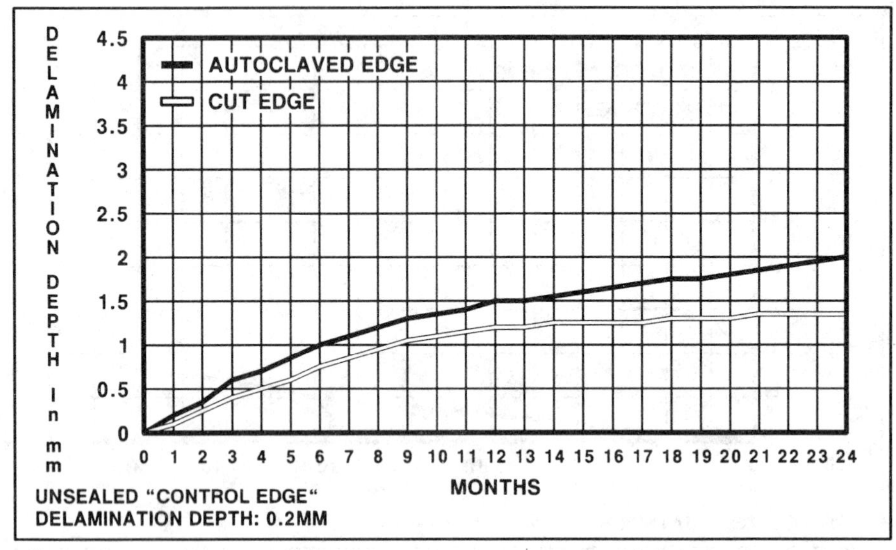

Figure 9 -- Average defect depth in Arizona exposure butyl/polyisobutylene.

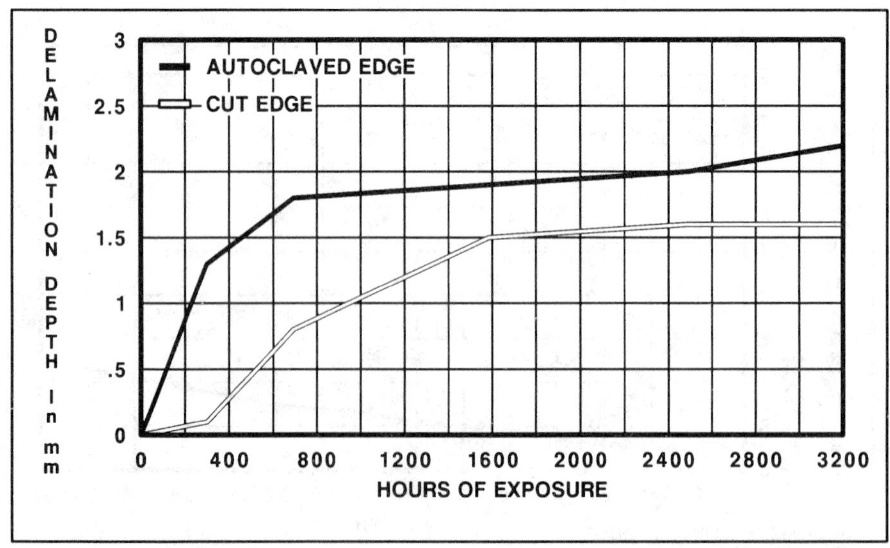

Figure 10 -- Average defect depth in Q-U-V exposure, 8-4 cycle high-modulus silicone sealant.

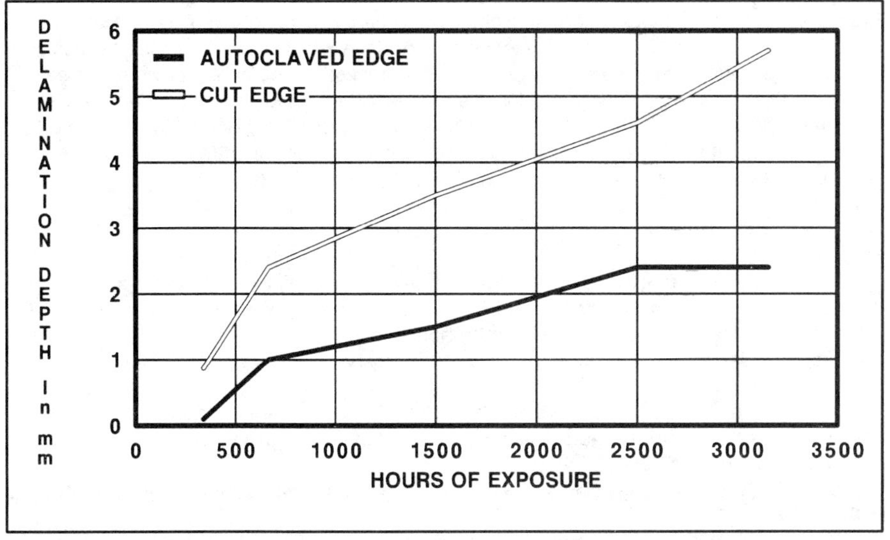

Figure 11 -- Average defect depth in Q-U-V exposure, 8-4 cycle low-modulus silicone sealant.

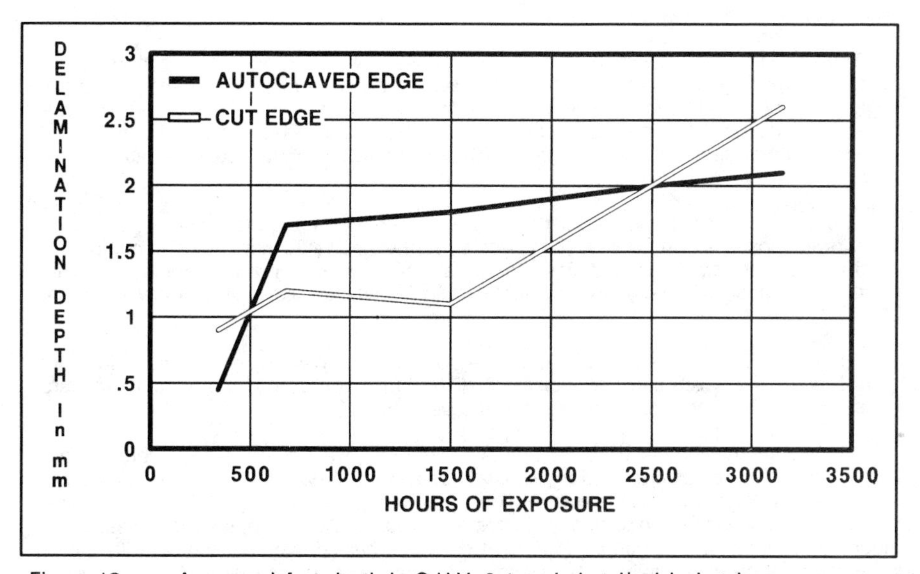

Figure 12 -- Average defect depth in Q-U-V, 8-4 cycle butyl/polyisobutylene.

Obviously, the correlation between the more frequent cycle changes of the 8-4 cycle and outdoors is not as good as that of the 16-8 cycle and outdoors. The reason why the 16-8 hour cycle reproduces more closely the Florida exposure is not fully known. One might expect just the opposite results since the 8-4 hour cycle seems to reproduce closer the south Florida weathering changes where sunlight can be immediately followed by rain and vice versa. One explanation is that the changes taking place in the sealant and at the sealed edge of the laminate are governed by slow moving processes, i.e., cross-linking of the sealant as part of the sealant curing reaction, permeability throughout the sealant and/or interlayer, migration of plasticizers from sealants to interlayer, or vice versa, do not respond to the quicker changes of the 8-4 hour cycle (for sealants' composition, mechanism of sealant curing reaction, by-product of sealants curing reaction, see Reference 2).

The Effect of Condensation Alone and No UV Light

As part of the edge compatibility studies of sealants with laminated architectural glass (LAG), an identical set to the ones used in the previous studies were exposed in the Q-C-T Condensation Tester at 60°C (140°F) for 24 hours (no UV light). Direct inspection of the exposed samples at about 1500 hours of Q-C-T exposure suggests the following:

1. Unlike the Q-U-V exposure with the 16-8 cycle, the magnitude of edge defect is about the same for both the autoclaved and cut edges and significantly greater than the ones observed in the Q-U-V.
2. Some of the well-known silicone sealants showed unexpected surface crumbling in the Q-C-T.
3. Butyl rubber sealants and tapes which are known and used for their good moisture barrier properties do not protect the edge of the laminates as expected.

It seems that condensation alone at 60°C (140°F) is too severe to assess edge compatibility of laminated glass with sealants as compared to outdoor exposure. It also suggests that moisture barrier properties of the sealants tested is not high enough to use these sealants in direct contact with water at 60°C (140°F) and consequently prolonged exposure of the sealed laminated glass to water and/or complete immersion should be avoided.

Weatherability of the Sealants

Tables 2 and 3 represent the "C" Shore hardness measured on some of the sealants tested in this program before they were exposed outdoors and after the sealants in the sealed laminates had been exposed in Florida for 12, 24, and 30 months and in Arizona for 12 and 24 months. Although not included in these figures, the changes in "C" hardness of the samples exposed in the Q-U-V with the 16-8 cycle for 2500 hours show the same overall values.

Although changes in color and dirt pick-up were also monitored, the changes that should be followed closely are those in "C" hardness since they may influence the edge compatibility of the sealants with laminated glass. Increase in "C" hardness indicates some cross-linking of the polymeric component of the sealant and/or some losses of plasticizers, tackifiers, etc.

As shown, the increases in hardness are, for the most part, about the same for both outdoor locations which seem to indicate that the humidity factor of the Florida exposure does not appear to accelerate changes in the sealants.

With the information discussed above as background, we are now ready to present details of the Florida exposure program, which is a very active part of the ongoing Monsanto sealant evaluation effort. For example, Figure 13 was taken at the Sub-Tropical Testing Service in Florida where the sealed laminates are exposed. As shown, sealed laminates are mounted on racks at a 45° angle. Figure 14 depicts a sealed laminate sample with a very well-defined edge defect frame. This overall performance suggests that the edge defect

TABLE 2

SHORE HARDNESS "C"
FLORIDA

SEALANT	CONTROL	12 MOS	24 MOS	30 MOS	△"C" H
DOW999	7	7	7	7	0
GE1200	9	10	10	10	+1
RHOD3B	4	5	6	7	+3
TREMCO PR	4	6	8	9	+5
DOW790	3	3	3	3	0
DOW795	7	7	10	11	+4
GESIL N	7	13	20	20	+13
SILPRUF	4	5	6	6	+2
SILGLAZE	4	5	6	7	+3
999NMK	8	8	8	8	0
795NMK	8	7	10	11	+3
790NMK	2	3	3	3	+1
ADCOBOND R-900	3	13	25	30	+27
SONOLASTIC NP1	10	11	12	12	+2
TAPE 440	0	5	5	5	+5
POLYSHIM	9	8	10	10	+1
PRC 438	8	11	16	16	+8
SST-800	2	6	6	6	+4

TABLE 3

SHORE HARDNESS "C"
ARIZONA

SEALANT	CONTROL	12 MOS	24 MOS	△ "C" H
DOW999	7	6	7	0
GE1200	9	8	9	0
RHOD3B	4	5	6	+2
TREMCO PR	4	7	9	+5
DOW790	3	2	3	0
DOW795	7	7	10	+3
GESIL N	7	13	18	+11
SILPRUF	4	5	5	+2
SILGLAZE	4	7	8	+1
999NMK	8	8	9	+1
795NMK	8	8	8	0
790NMK	2	2	3	+1
ADCOBOND R-900	3	5	15	+12
SONOLASTIC NP1	10	11	12	+2
TAPE 440	0	0	1	+1
POLY SHIM	9	10	15	+4
PRC 438	8	10	15	+7
SST-800	2	1	2	0

Figure 13 -- Exposure of sealants at the Sub-Tropical Testing Service site.

Figure 14 -- Sealed laminate sample with a well-defined edge defect frame.

has reached a plateau. Figure 15 represents another typical edge defect. The edge defect, however, is somewhat more irregular. Figure 16 shows a sealed laminate sample practically free from visible edge defects. Figure 17 represents a comparison between a sealant causing an edge defect and a sealant that is very compatible with the PVB interlayer.

CONCLUSIONS

Throughout this presentation we have discussed the usefulness of the Q-U-V apparatus (16-8 cycle) to replicate several field performance factors which architects, engineers, and glaziers must consider in selecting sealants to best satisfy building project requirements when installing laminated glass (see Reference 3 for sealant technology in glazing systems). These performance factors are:

1. Compatibility of the sealants with the laminated glass, especially if the sealant or sealant by-products are likely to contact the interlayer, i.e., butt laminated glass installation.
2. Sealant weatherability. Does it deteriorate or change physically over time?

Also, we have shown that more frequent cycle changes in the Q-U-V, the 8-4 cycle, which was intended to provide for more acceleration, does not closely replicate outdoor exposures. Furthermore, and as detected by the Q-C-T unit, the moisture barrier properties of the tested sealants is not high enough to use these sealants in prolonged exposure to water vapor. Moreover, these accelerated techniques may be used for optimizing sealants selection for LAG.

Understanding the compatibility issues of sealants with the PVB interlayer of LAG is a critical requirement in making the proper choice of sealants.

REFERENCES

1. Monsanto Chemical Company, "A Study of Sealant Compatibility with Saflex® PVB Interlayer in Laminated Architectural Glass", Technical Bulletin No. 1512, Parts I & II, St. Louis, MO.

2. Skeist Laboratories, Inc., Sealants IV, Livingston, New Jersey.

3. American Society for Testing and Materials, Sealant Technology in Glazing Systems, Special Technical Publication 638, Philadelphia, PA.

Figure 15 -- Laminate sample with an irregular shape edge defect.

Figure 16 -- Laminate sample with very low level of edge defect.

Figure 17 -- Comparison between a sealant causing edge defect and a sealant rather compatible with the PVB interlayer.

Laura E. Gish

CRITERIA FOR DESIGN AND TESTING OF SEALANTS USED IN SINGLE-PLY ROOFING AND WATERPROOFING SYSTEMS

REFERENCE: Gish, L. E., "Criteria for Design and Testing of Sealants Used in Single-Ply Roofing and Waterproofing Systems," Building Sealants: Materials, Properties andPerformance, STP 1069, Thomas F. O'Connor, Ed., American Society of Testing and Materials, Philadelphia, 1990.

ABSTRACT: Roofing membranes are subjected to large temperature gradients, ultraviolet radiation from the sun, building movement, constant and intermittent water immersion, and microbial attack. Waterproofing membranes are subjected to many of the same stresses. In both cases sealants are generally used as the first line of defense for the seams in the membrane system. Each of the single ply roofing and waterproofing systems in use today utilizes at least one if not several sealants whose application is unique to that system.

Sealants used with each of the generic single ply types are described in terms of their function and required performance.

Criteria for testing such sealant products are discussed with emphasis on the design and interpretation of meaningful laboratory test methods, those that will rapidly predict the performance of a candidate sealant in conditions of actual use.

Systems discussed include EPDM, Neoprene, Butyl, and Hypalon elastomers, CPE and PVC thermoplastics, and modified bitumen.

KEYWORDS: single-ply, roofing, waterproofing, sealant, weathering

Within the last 10 years single-ply roofing of non residential buildings has gained rapidly in popularity, supplanting hot

Laura E. Gish is a Senior Adhesives Chemist at Carlisle SynTec Systems, P. O. Box 7000, Carlisle, PA 17013

applied bituminous roofing as the most popular roofing method. Single-ply membranes of EPDM are recognized as the most popular of the single-ply membranes on the market, and reportedly outsell all other types of non-residential roofing. Each of the other single-ply systems enjoy popularity in niche markets, where special characteristics of a particular membrane or system make it preferable. For example, CPE, Hypalon, and to some extent Neoprene membranes offer oil resistance, Butyl offers superior resistance to air and water vapor transmission, the thermoplastics can be used where a light color and the ease and perceived security of a heat welded seam is desired, and modified bitumen offers a blend of single-ply and built up system technology.

Most of the single-ply systems are used both in roofing and waterproofing, with several notable exceptions. Because of its sensitivity to attack from ozone and other airborne degradents, Butyl membrane is not specified for roofing or exposed waterproofing applications. Conversely, non-reinforced reinforced CPE, and Hypalon, are not specified for waterproofing applications because of their sensitivity to constant water immersion. Reinforced Hypalon products are used in waterproofing and pond lining. Each manufacturer or marketer imposes its own use restrictions.

Each of the single-ply systems utilizes adhesives and sealants to join the pieces of membrane and to seal the termination points of the system. This paper will discuss the various sealants used with these systems and attempt to quantify the properties of a particular sealant which make it appropriate for use in a given application.

FUNCTION OF SEALANTS USED IN SINGLE-PLY ROOFING AND WATERPROOFING SYSTEMS

In single-ply roofing and waterproofing, as with many sealant applications, the sealant often serves as the first line of defense for the seams and terminations in the system. As such it is subjected to, and must resist, the same stresses as the membrane material itself, but must be easily applied in the field.

There are four basic types of sealant product used commonly with single-ply membranes, although individual systems may differ in their requirements. LAP SEALANT or SEAM CAULK is used at the edge of the membrane field splices to provide a smooth transition between sheets, and to serve as the primary sealant for the splice by resisting moisture and physical damage by traffic over the splice edge. (See Figure 1, Membrane Splice)

WATER CUT-OFF MASTIC or WATER STOP SEALANT is used in termination details where it will be under compression. This material is designed to remain soft and permanently tacky. Its function is to resist the intrusion of liquid water at the top of pipe seals and behind the termination bar when the membrane is turned up a wall. It is also used to seal the compression ring in roof drains where it keeps water from flowing under the edge of the membrane. (See Figure 2, Pre-molded Vent Pipe and Extension, and Figure 3, Roof Drain)

POURABLE SEALER is used in "pitch pockets" which are boxes built around irregular roof penetrations that cannot be sealed using normal flashing details. Pourable sealer materials generally cure quickly to form a durable, rubbery film which can resist movement and the elements while maintaining adhesion to both the membrane and substrate penetration, effectively sealing the break in the membrane. (See Figure 4, New Pourable Sealer Pocket)

NIGHT SEALANT is a material used to form a temporary seal for the membrane when work must be stopped at night or in the face of approaching inclement weather. The sealant must adhere well enough to the membrane, and form a strong enough film to resist a potential hydrostatic pressure head of several inches. It must also be easily removed from the membrane and substrate so that its use does not interfere with installation of the next sheet of membrane(see Fig. 5).

One other sealant is used in certain roofing systems. IN-SEAM sealant is used within the lap joint itself to seal the bottom edge of the membrane and around fastening plates which penetrate the lower of the two sheets to be bonded in certain mechanically fastened systems. The use of this sealant which develops very high adhesion once cured provides a redundant seal, extra insurance against the penetration of water into the seam, and a final barrier in case the adhesive seam would be compromised. (Refer to Figure 1)

Although the types of sealants used for single-ply systems are basically similar, each generic class of membrane requires sealants that are specific for that material.

EPDM systems, both for roofing and waterproofing, use a caulk grade EPDM based sealant for sealing field seams bonded with liquid applied adhesives. This material generally does not vulcanize, but cures (dries) by means of solvent release. The water cut off mastic used in EPDM systems is generally a butyl or polyisobutylene based product, but materials based on a host of polymers are specified by the various systems manufacturers. The pourable sealer is most often a two-part polyurethane sealant which sets within an hour or so to provide structural integrity to the "pitch pocket". One part sealants are also specified by some manufacturers. The night sealant can be anything from a two-part solvent free polyurethane to asphalt based roof mastic. A night sealant specifically recommended for a particular system is usually available from the manufacturer of the roofing or waterproofing system.

Butyl waterproofing systems employ the same sealant products as EPDM systems.

In Neoprene roofing and waterproofing, the sealants are often the same as those used in EPDM and Butyl systems. Some manufacturers specify a lap sealant made of Neoprene rather than EPDM to give the system more oil resistance. In special applications, a Nitrile

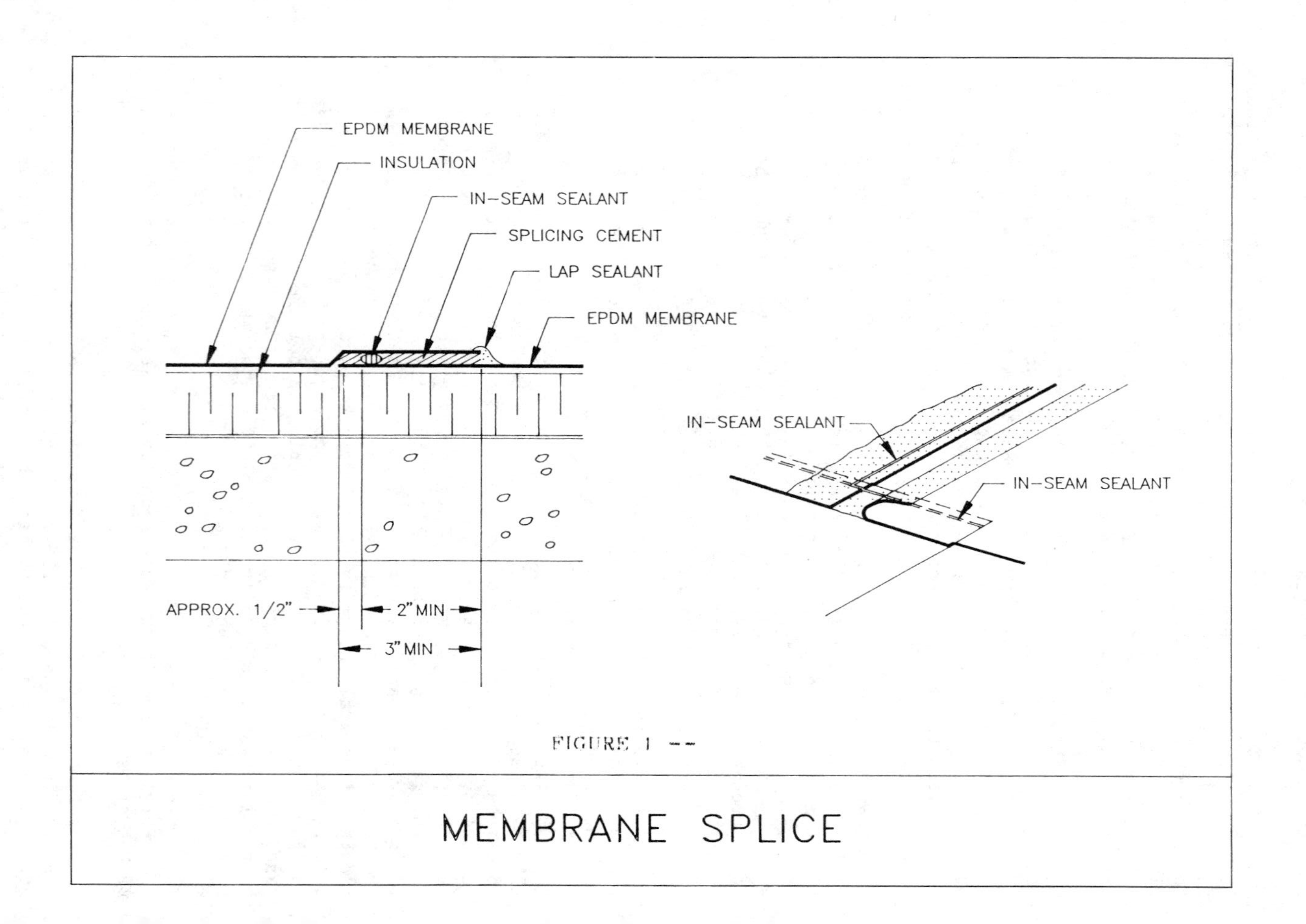
EPDM MEMBRANE
INSULATION
IN-SEAM SEALANT
SPLICING CEMENT
LAP SEALANT
EPDM MEMBRANE
APPROX. 1/2"
2" MIN
3" MIN
IN-SEAM SEALANT
IN-SEAM SEALANT
MEMBRANE SPLICE

FIGURE 1 —

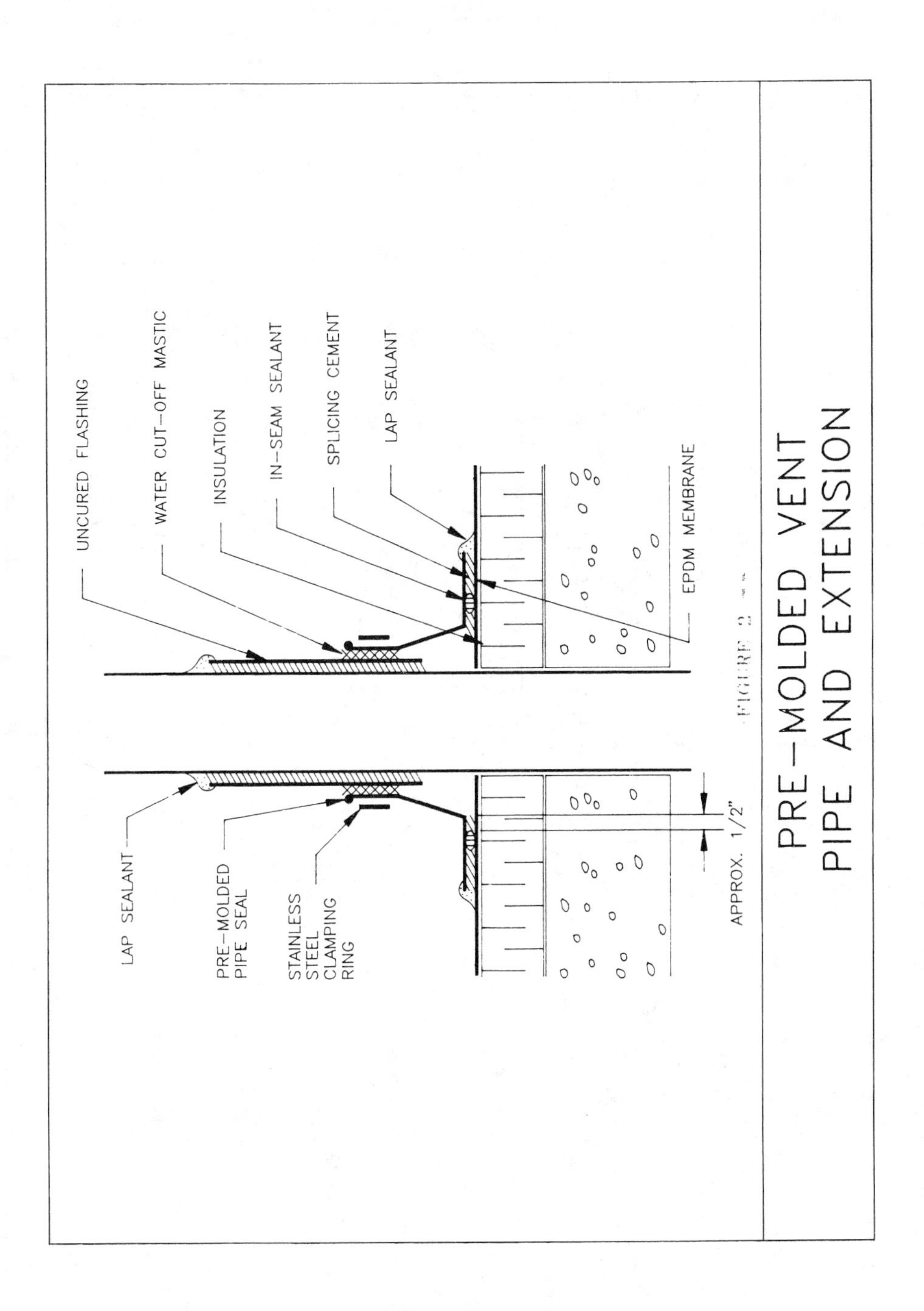
UNCURED FLASHING
WATER CUT-OFF MASTIC
INSULATION
IN-SEAM SEALANT
SPLICING CEMENT
LAP SEALANT
EPDM MEMBRANE
LAP SEALANT
PRE-MOLDED PIPE SEAL
STAINLESS STEEL CLAMPING RING
APPROX. 1/2"
PRE-MOLDED VENT PIPE AND EXTENSION

FIGURE 2

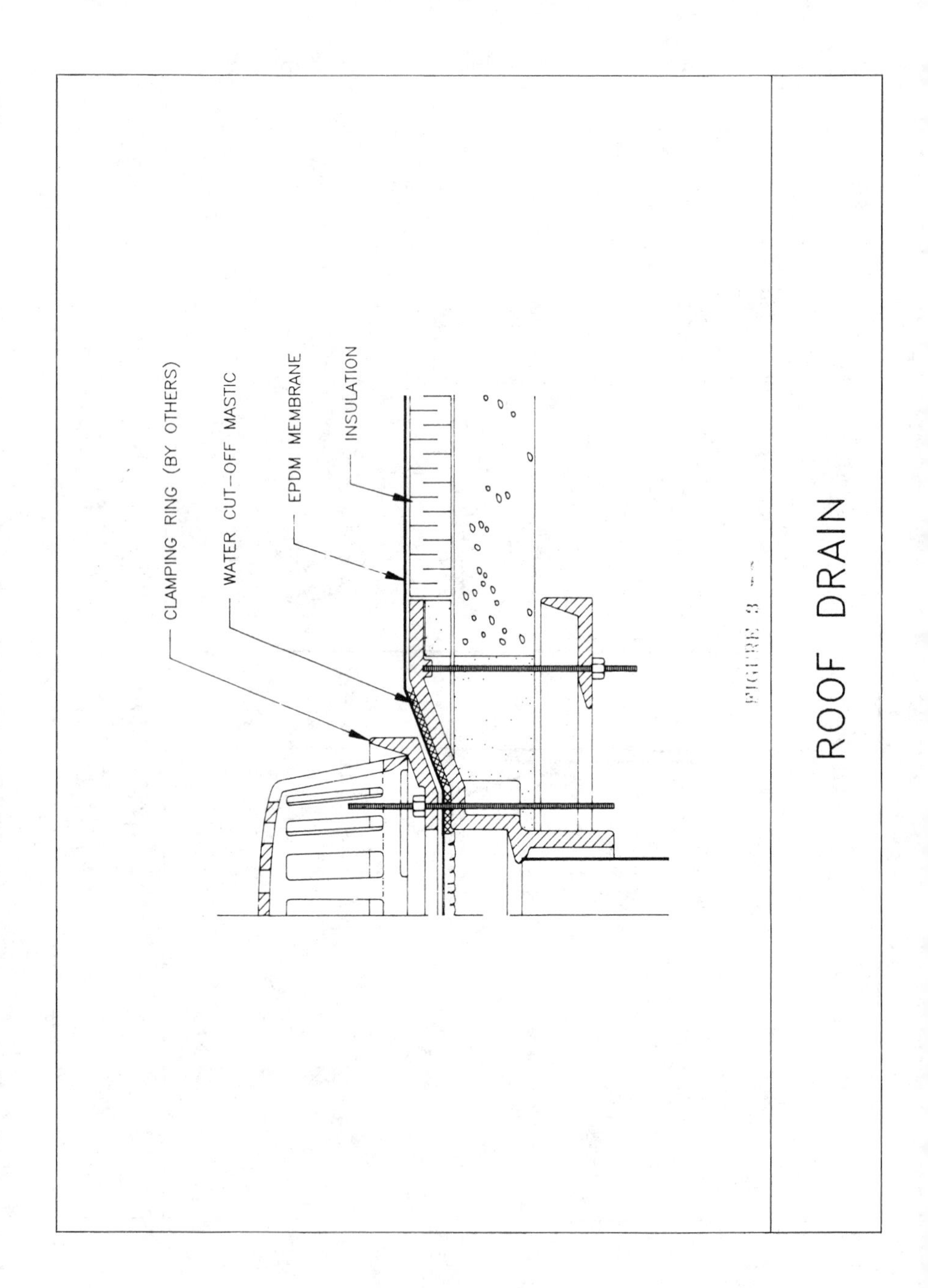
CLAMPING RING (BY OTHERS)
WATER CUT-OFF MASTIC
EPDM MEMBRANE
INSULATION
FIGURE 3
ROOF DRAIN

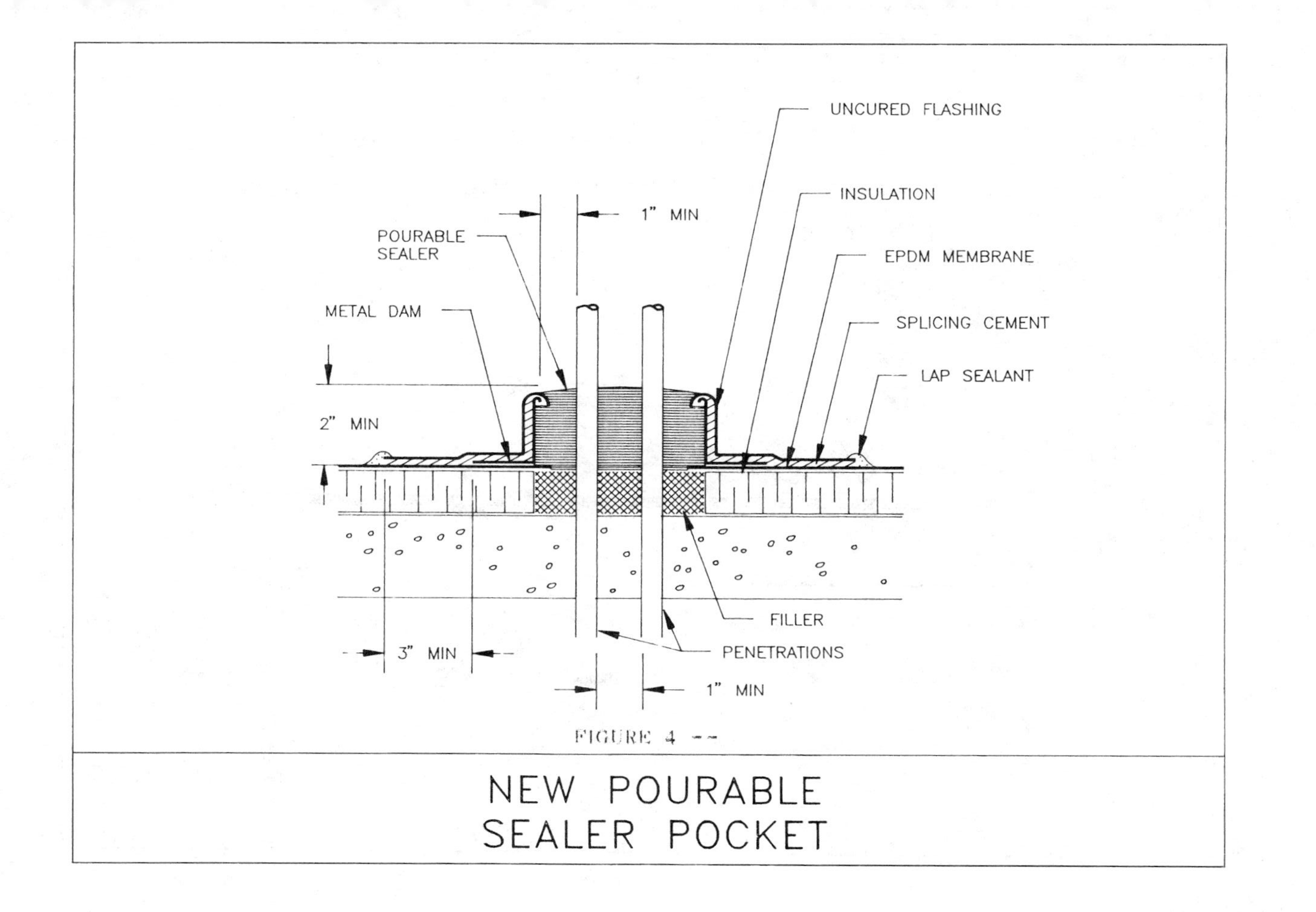

FIGURE 4 --

NEW POURABLE SEALER POCKET

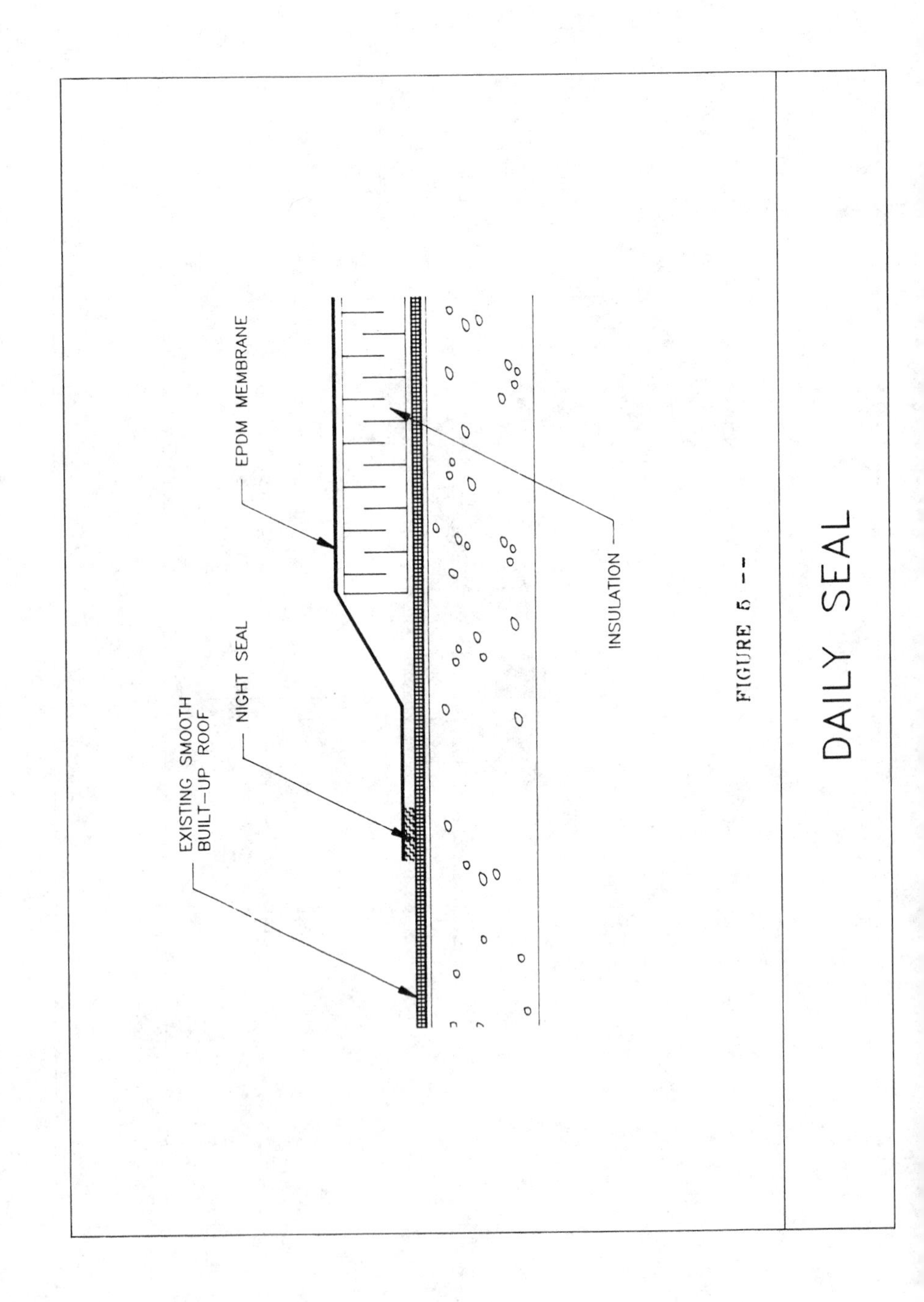

FIGURE 5 -- DAILY SEAL

rubber based lap caulk has been specified as an extra measure of protection where oil resistance is required.

The heat welded seams of Hypalon systems are not generally sealed with a lap sealant, unless the membrane must be cut to fit or patched. In this case, the exposed reinforcing cords are sealed with a thin mastic applied from a squeeze bottle. The purpose of the sealant is to protect the membrane reinforcement from exposure to water, either in a ponding situation or as humidity. Water cut off mastic, pourable sealer, and night sealant are used in the same way as for EPDM systems, but often one sealant product serves several of these functions.

In PVC roofing systems, where both heat welding and solvent welding are commonly employed to bond the field seams, a PVC based seam caulk is sometimes specified to seal all field seams. Sometimes it is used only around penetrations through the roof and to seal cut edges of reinforcing fabric, depending on the manufacturer's specification. A water cut off mastic is used behind termination bars and in compression seals. As in the systems described previously, pourable sealer and night sealant are also used.

There are two major marketers of CPE systems for roofing. Their details are very dissimilar as is their approach to the use of sealants. One company specifies sealants marketed by other companies for use in sealing pitch pockets, flashing and terminations of their CPE roofing system. The other company markets its own brand of pourable sealer, water cut off mastic, and night sealant to be used with the system. Both firms specify their own seam sealer to be used on all seam edges.

The seams of modified bitumen single-ply systems are either sealed with a layer of hot asphalt, in hot mop application systems, or sealed in place with the flame from the torch which is used to bond the seam. Most other detail work where pourable sealer, night sealant, or water cut off mastic would be used in elastomeric or thermoplastic systems is accomplished with asphalt, more like traditional built up roofing.

Nearly all roofing systems also employ a caulking or sealing compound to be used on vertical counterflashing and on the top edge of termination bar which is attached to a vertical wall. These sealants may be the seam sealer, if one is used with the system, or a polyurethane or silicone caulk not marketed by the membrane manufacturer. Each system and each manufacturer has a slightly different specification for these details.

No matter what the chemical make-up of the roofing or waterproofing membrane, the criteria used for selecting a particular sealant to serve a particular purpose in the roofing or waterproofing system will be similar. These criteria are outlined here.

A. HEAT EXPOSURE If the membrane is to be used for roofing in a non-ballasted system, surface temperatures of 180 F or higher can be expected daily in the summer months where a black membrane is used. If the membrane is white, 150 F on the membrane surface is commonly measured in full sunshine. The sealants used in the system will be exposed to the same temperature as the membrane and must be able to function after long exposure at the upper service temperature, and be resistant to temperature cycling. In below grade waterproofing, high temperature resistance is not as important, as long as the membrane is completely covered by the backfill.

Temperatures at the membrane level in ballasted roofing systems are generally not as high, but the sealants are often not covered by the ballast and therefore must be as temperature resistant as those used in non-ballasted systems.

B. **HOT WATER EXPOSURE** Many non residential roofs are flat or nearly flat, making them subject to intermittent or constant water ponding from rain water that does not drain from the roof surface. As the first line of defense for the seams and penetrations, the sealants must be able to resist the passage of water without being deteriorated by long exposures. Almost by definition, this criterion severely restricts the possibility of using water based sealants as most of these are softened or absorb water when immersed for long periods of time, thus allowing water to pass through them.

Below grade waterproofing, although seldom exposed to hot water, does exist in almost constantly wet conditions. As in roofing and exposed waterproofing, the sealants must not be affected by exposure to standing water and constant humidity.

C. **UV EXPOSURE** This criterion may be ignored for below grade applications but is extremely important in roofing and exposed waterproofing systems. The sealants, when exposed to daylight must withstand ultraviolet radiation without hardening, cracking, or softening to the point that they lose integrity.

D. **HOT/COLD CYCLING** Just as a roof is exposed to very high ambient temperatures, it is exposed to low temperatures, and great temperature swings. On a sunny day in the mid-winter of central Pennsylvania, ambient temperatures may be in the thirties, but a black roofing membrane will reach 150 F. That very evening when ambient temperatures drop into the twenties or lower, the membrane will cool with the atmosphere or re-radiate its absorbed energy, dropping below ambient temperature. This results in a temperature gradient of 130 F or more in a period of several hours. The sealants used with the system must be able to absorb the temperature induced stresses, moving with the membrane, and not become brittle as the temperature drops. Large temperature cycles will occur almost daily for any exposed membrane, anywhere in the USA.

Buried waterproofing membranes will also be affected by temperature swings caused by changes in atmospheric temperature, when the membrane is above the frost line, and by changes in the temperature inside the building. These temperature cycles will not be as great for a buried membrane as for one that is exposed, but must still be considered when evaluating sealants for the waterproofing system.

E. COLOR RETENTION A white sealant must match the color of the surface to which it is applied, and continue to match that surface for the life of the system. This is true of any non black roofing or waterproofing application. Sealants that are not visible do not have to match the membrane, but care must be taken that no ingredients in these sealants will stain or discolor the surfaces around them causing an aesthetically displeasing stripe or spot on the material covering them.

F. DIRT PICKUP Closely tied to color retention, dirt pick up resistance is important to the aesthetics of exposed applications. The sealant should become dirty at the same rate as the membrane, ideally not at all. Likewise it should be as easy to clean as the membrane.

G. APPLICATION IN EXTREME TEMPERATURES Roofing and waterproofing with single-ply membranes go on year round, even in the cold northern climates. Therefore, the application properties of the sealant material at low temperatures must be evaluated. The best sealant in the world is useless if it will not come out of the cartridge at the desired application temperature. At the other extreme, the sealant should be useable in very warm weather. It must not cure solid in the tube or can after a short exposure to typical summer rooftop temperatures. The viscosity and consistency of the sealant should stay fairly constant at any foreseeable application temperature. The adhesion of the sealant to the chosen substrate is also a consideration at both high and low application temperatures. In the United States an application temperature range of -10 F to 110 F can be expected as a minimum. In certain areas temperatures can be colder in the winter, and all over the country the surface temperature will be higher, 160 F to 180 F in the warm months.

H. SLUMP RESISTANCE Of particular importance to sealants that will be applied to a vertical surface is slump or sag resistance. The sealant must be able to resist flowing down the substrate once applied. This property should be checked at all foreseeable application temperatures. Even sealants used strictly in horizontal applications should possess some slump resistance so that they will remain where they have been applied without running.

I. SHRINKAGE RESISTANCE This property is of particular importance to lap sealants whose performance would be diminished by excessive shrinkage. The sealant is often used to form a tiny fillet at the seam edge, creating a smooth transition from one piece of membrane to the next. Resistance to shrinkage is critical if the sealant must hold its shape and create that transition. Likewise in compression seals and pitch pockets, excessive shrinkage cannot be tolerated because the dry sealant could pull away from the substrate creating an open channel for the passage of water into the system. Acceptable shrinkage must be determined on the basis of the expected performance of a sealant in a particular application.

J. RESISTANCE TO EFFLUENTS Effluents can be anything from waste chemicals to cooking oil which is exhausted onto the roofing membrane or is present in ground water which contacts the waterproofing membrane. When resistance to a particular effluent is needed, the entire system, including seams and sealants should be tested. All products that may be exposed to the effluent chemical should be evaluated for their resistance to the material.

The various sealants described above are listed on Table A, together with their most important criteria. For each of these criteria there must be at least one if not several laboratory test methods which can be used to evaluate a particular sealant's performance. Tests and test methods must also be evaluated to determine their usefulness in measuring a particular property. The ideal test method, in this writer's opinion would be easy to perform, extremely repeatable, short in duration, and would yield clear, easily interpreted results which clearly reproduce observed field results.

DESIGN AND EVALUATION OF TEST METHODS

Perhaps the most crucial step in the choosing of a sealant for a particular purpose is the selection of the proper test method to evaluate the important criteria. Ideally the method should be simple to perform, inexpensive, repeatable, and most importantly should measure the intended property, yielding data that are easily interpreted.

Often standard test methods obtained from ASTM, the federal government, ACI (American Concrete Institute), and others can be used as a starting point. The scientist should never be afraid to modify a standard method to suit the needs of a product or service conditions. These standard methods are generally written to suit a broad range of products and the conditions described in them may be too severe, or too lenient for a particular situation. When designing tests for roofing sealants, standard test conditions are often too mild and must be modified to reflect actual conditions on a roof.

In designing tests to evaluate or characterize a roofing or waterproofing sealant, several things to consider are: test temperature, duration of the test, ease of performance of the test, and interpretation of the results.

Test Temperature

When long term performance is being evaluated, a balance between test duration and real world conditions must be struck. For heat aging and hot water immersion tests, 70 C is usually chosen as the test temperature because it is close to the expected upper service temperature of most black roofing systems. By using constant 70 C exposure, aging of the sealant can be accelerated without inducing unusual degradation that would not be experienced in the field. Testing at 50 C is more appropriate for white sealants and membranes because of their lower heat absorption which causes them to remain cooler.

TABLE A -- Most Important Criteria

PRODUCT	CRITERIA
LAP SEALANT	Resistance to heat, humidity, UV, temperature cycling, slump, shrinkage, dirt pickup (white systems), adhesion to membrane, short tack free time
WATER CUT OFF MASTIC	Resistance to heat, humidity, temperature cycling, slump, shrinkage. Must remain permanently tacky and pliable, non-staining
POURABLE SEALER	Resistance to heat, humidity, temperature cycling, shrinkage, and UV, non-staining, adhesion to membrane and penetrations through it
NIGHT SEALANT	Quick cure time, water resistance, must be removable, but have good adhesion to membrane and other roof substrates
IN-SEAM SEALANT	Quick cure time, excellent adhesion to membrane, resistance to heat, temperature cycling, shrinkage, and slump

Test Duration

Long term tests are usually continued for at least 90 days, with evaluations at 30 day intervals during that time. This test protocol provides three data points from which a trend can, hopefully, be determined. Experience has shown that very poor performing materials will show definite signs of deterioration within the 90 day period, and materials that perform well in the field will show little or no change. This depends, of course, on the test being performed. If the technologist is attempting to discriminate between materials which all have fairly good weathering characteristics, longer aging times will be necessary. From six months to one year may be required to produce a change in physical properties of a reliable, weather resistant sealant in heat aging or weatherometry tests.

Because of the long times involved, a sealant should be fully characterized in short term tests like Slump Resistance, Channel Cracking, Thin Film Cracking, etc. before committing test time, manpower and money to extended aging tests.

Ease of Performance

When a test is simple to perform, and the apparatus easy to use by a trained technician, the experimental error introduced into the test results will be greatly diminished. However, when attempting to recreate the outdoor environment in a laboratory, one must consider an enormous array of variables and their interactions. To do this requires tests that evaluate the effect of more than one variable at a time, for example hot/cold cycling, or heat/ humidity/UV exposure. Good apparatus are available to perform these tests automatically, but the machines must be monitored by skilled personnel to keep them functioning properly. Proper maintenance and operation of the test equipment is especially important in tests that require 3 months or longer to perform.

Not all tests need to be complicated or long in duration. For example slump resistance and extrudability can both be measured accurately and repeatably by simple tests that are rapid to perform.

Interpretation of Results

Interpretation of test results is not always simple. It is important to be able to draw clear comparisons between performance in a laboratory test and performance in the field, and to clearly identify differences in performance between candidates evaluated in the same test. For example, if a sealant slumps one-half inch in an ASTM D 2202 test, will that sealant be useable when placed on a membrane seam on a vertical wall? What if the slump result is one inch, or two inches? Or if the skin-over time of a sealant is increased by 30 minutes at standard temperature, will it be unuseable in typical rooftop conditions? The only way to answer these and questions like them is to take the experimental sealant to the field and test it by installing it on a roof or waterproofing site, if there is no in house reference information. When the question is one of long term performance, rather than application properties, it is necessary to rely on data developed when testing similar products if no direct comparison can be made to a control formulation. This is risky business, and when setting pass/fail criteria, it is advisable to be conservative. It is always possible to loosen specification limits when a product proves itself in use, but a field failure can be embarrassing and costly. It is not advisable to introduce a new product to the marketplace without at least a year of field exposure data even when reliable laboratory tests have been performed.

As is always good practice, but especially in long term aging tests, a "control" material having a known performance record of the property being tested should be evaluated using the proposed method. If at all possible, during the test method qualification phase, two controls should be run. One sealant should be expected to pass the test, and one should be expected to fail, based on field performance data.

With care and experience a battery of tests can be developed which will reliably predict the suitability of a sealant for a particular application.

Listed next are several important criteria against which a sealant should be tested, and recommended test methods. This information is summarized on Tables B and C which follow.

HEAT AGING Since the sealant will be exposed at least to intermittent temperature extremes, long term exposure to the average highest temperature exposure is a reasonable starting point. With the exception of night sealant, whose function is strictly temporary, all the sealants used in a particular system should be tested for resistance to heat aging. The length of the test should be 3 months at a minimum, with observations made at monthly intervals. The test sealants should be compared to unexposed controls to gauge changes in color, hardness, flexibility, adhesion, and surface tack. A control sealant whose field performance has been established should be tested together with candidate materials to establish a performance benchmark.

HOT WATER EXPOSURE This property is most easily measured by immersing sealant/substrate composite specimens in water at the upper service temperature. Here again, 70 C is frequently used as the test temperature. All the roofing or waterproofing system's sealant components should be tested for hot water immersion resistance, as they will all be subjected to the effect of standing water at some point in the life of the system. The length of the test is dependent on the intended use of the sealant, but no less than three months exposure should be used for sealants expected to be exposed for the life of a building.

UV EXPOSURE Only those sealants that will be directly exposed to sunlight need to be resistant to ultra violet radiation. This includes the seam sealer, pitch pocket sealant, and any sealant used on counterflashing and terminations. The night sealant, in-seam sealant, and water cut off mastic need not be tested for UV resistance, as they do not require this attribute to perform their function. Ultra violet resistance can be measured in several ways: by exposure to sunlamps for a given period of time, by use of a UV/weathering machine, or by accelerated outdoor exposure. Each method has its advantages and shortcomings. The use of sunlamps is quick, easy and inexpensive to perform. However the radiation from a typical sunlamp is very different from that received from the sun, entailing only a very narrow band of the ultra violet spectrum. Still if a control, whose long term performance in the field is known, is carefully chosen and tested side-by-side with the experimental sealant, this method can be a useful screening tool. More will be said about sunlamps when color retention is discussed.

UV/weathering machines are more sophisticated than sunlamps, offering a broader UV spectrum and the option of cycling temperatures and humidity levels. They are more expensive to purchase and operate than sunlamps, but may provide results which are truer to the "real world".
Recently at Carlisle SynTec Systems a set of EPDM sealant plaques was removed from a UV/weathering machine after 3 years exposure. The sealants showed no appreciable sign of degradation after their stay in the chamber which cycles at 4 hours UV-on at 80 C and 4 hours UV-off at 50 C with condensing humidity. As this particular formulation has a three year track record in the field with no observable degradation, we feel more confident that this material will perform for the required 15 years in normal field conditions.

Accelerated outdoor weathering by use of the EMMAQUA (Equatorial Mounted Mirrors And Water) is probably the most sophisticated UV test, and provides the truest prediction of solar UV stability. The sample is exposed to true sunlight, concentrated by the use of mirrors which accelerates its effects. The test is measured by Langley units, which are a measure of the energy received per unit area on the specimen. For example in a typical EPDM roofing membrane test the sample is exposed until it receives 10,000 Langleys. This unit eliminates the variability of solar radiation from month to month, but introduces some variations in test length from experiment to experiment. EMMAQUA tests typically require much longer than UV/weathering machines to see degradation, over a year in some cases, and are very expensive. An added disadvantage is loss of access to the samples, which require at least a week of travel time to return from the desert test site to the laboratory for actual viewing.

In any accelerated UV or weathering test a control with a known track record in the field must be tested alongside the experimental sealants to establish a benchmark. The test sealants should be compared not only to the control, but to unexposed samples of the same sealants to judge the degree of degradation caused by exposure to the ultraviolet radiation.

HOT/COLD CYCLING Because of the large temperature gradients experienced by an exposed roofing or waterproofing membrane, it is prudent to test a candidate sealant for resistance to thermal shock. In a typical field seam application or water stop detail, the sealant is applied in a thin bead over an angle change or discontinuity. If the membrane or substrate moves as a result of heating and cooling, it is imperative that the sealant be able to bridge the gap and move with the substrate without cracking or peeling. A test developed by Carlisle SynTec Systems specifically for measuring a lap sealant's ability to tolerate thermally induced movement of the membrane is called the Thin Film Cracking Test. This method involves tooling a very thin film, approximately 0.020 in, across a field splice in an elastomeric roofing membrane which has been bonded to a 0.125 in thick aluminum panel. The fresh sealant/ field seam composite is immediately exposed to radiation from an rm sunlamp for 6 to 8 hours. At the end of this period the sealant is examined for visible cracks along the step-off of the field splice, then placed in a freezer at -40 C for 16 hours to induce a thermal shock and load on the splice. At the end of this period the splice is removed from the freezer and examined for cracking. Another cycle of 8 hours sunlamp exposure plus 16 hours in the freezer is then initiated. As with other laboratory tests, it is important to establish a benchmark by using a control sealant with a known performance record in freeze thaw exposure. As this method was developed, two control sealants were tested with every new candidate. One control was expected to fail the test and one was expected to pass based on field experience with both materials.

Freeze-thaw cycling will also provide information about a sealant's low temperature flexibility. If the sealant will be used in an application which is exposed to low winter temperatures, the sealant should be tested to be sure that it will remain flexible at the expected lowest exposure temperature.
This may be a simple bend test where the sealant and intended substrate are bent over a 0.25 inch mandrel (ASTM C 711) or a modification of this test performed in conjunction with the Thin Film Cracking test.

COLOR RETENTION This is one of the properties normally evaluated in UV exposure tests and may be applicable to darkening of white or light colored sealants as well as black materials . In tests using UV lamps one half of the sealant plaque is covered with aluminum foil or other opaque material, then the plaque is exposed to the UV source. At the end of the test the foil is removed and the two sides of the specimen compared to determine the degree of color change. When regular sunlamps are used the sample is typically exposed for 6 hours at a distance of 10 inches from the light source. Weathering machines and outdoor exposure can also be used to determine color retention, but much longer times are required to induce an observable color change.

Any laboratory test must be calibrated by field exposure since unnatural discoloration can be induced by a test machine. Machine testing for color retention can also fail to induce discoloration that actually occurs in the field.

DIRT PICKUP Although a very important consideration for any sealant to be used in a system where aesthetics are a consideration, dirt pickup is a difficult characteristic to quantify. The most reliable test is to install the sealant on the intended membrane in an outdoor test site and observe its dirt pick up resistance for a period of time. Exposure in South Florida is useful because not only atmospheric dirt but also mildew and fungus growth will occur on susceptible materials in the humid Florida atmosphere. It is imperative that a control be used, and advisable to duplicate the test in several locations in different seasons of the year, as a sealant may be more susceptible to dirt pick up in the hot summer months, or in the rainy springtime, than in the fall or winter seasons, or vice-versa. There appear to be no reliable laboratory tests for dirt pick-up resistance and so the scientist must rely on outdoor exposure data which requires months to collect and may be misleading.

LOW TEMPERATURE APPLICATION When a sealant is to be used outdoors all year around, its application properties at cold temperatures must be evaluated. Both the sealant and test membrane should be cooled to the test temperature, and the sealant applied as if it were being installed in the field. For test purposes 0 F is generally used. If the sealant is a gun grade material, its low temperature extrudability should be evaluated both subjectively by manually pumping a quantity from a cold tube, and objectively by use of a press-flow viscosity test as described in ASTM D 2452, but performed at the low application temperature. If the sealant will be tooled after application, then its "toolability" should be evaluated as well. Resistance to wetting the cold membrane and gelling of the cold sealant can cause application problems in the field and should be noted.

SLUMP RESISTANCE Where a sealant is to be used in vertical configurations, as in sealing seams or metal work on a wall, slump or sag resistance becomes an important attribute. The ASTM D 2202 test is an excellent predictor of field performance, provided that the test is performed at a temperature which reflects field conditions. For black membrane roofing that means testing at 70 C and for white membranes at least 50 C. ASTM D 2202 "Test method for Slump of Caulking Compounds and Sealants" entails the use of a special test jig which allows a pre-measured plug of sealant to flow down its vertical face during a preset time period. At the end of the test, the distance the sealant has travelled down the face of the jig is measured and reported. Even when a sealant is intended for use only on horizontal seams, its resistance to slump is important if the sealant is expected to remain in a bead or hold a tooled shape.

When a sealant will be used in below grade waterproofing, it is likely that it will be exposed to high temperatures during application, and therefore its slump resistance should be evaluated.

RESISTANCE TO SHRINKAGE This property is directly related to solids level. The lower the solids content of a sealant the less resistant to shrinkage will be that material. Solids by mass can be measured most easily. This quantity is an indicator of how much a material will shrink, but there is no simple equation to calculate how much a material will shrink at a given solids by mass level. Solids by volume gives a direct indication of shrinkage but generally cannot be accurately measured. If the specific gravity of the sealant, the percent solids by weight of the sealant, and the specific gravity of the solvent(s) are known, the volume solids can be calculated as shown here.

$$\%\ \text{Vol Solids} = \frac{[(100/\text{Gseal}) - (\%\ \text{Solvent}/\text{Gsolv})]}{(100/\text{Gseal})} * 100$$

where % Vol Solids = % solids by volume
Gseal = specific gravity of the sealant
Gsolv = specific gravity of the solvent(s)

% Solvent = 100 - % solids by weight.

Another test used to indicate shrinkage is the Channel Cracking test, a modification of the method described in ASTM C 718. The modified test, using 0.5 by 0.5 by 6.0 in aluminum channel and testing only at 70 c or 50 C (for white) without cold cycling or UV exposure is very useful for evaluating shrinkage of the sealant as it cures, evaluating blistering, and measuring the sealant's resistance to cracking as it dries or cures. The permissible amount of shrinkage must be determined based on the desired aesthetics of a system, as well as the sealant's capability to perform in a thin film. Generally, volume shrinkage of 50% or more is not desirable because the dry sealant bead will be very thin and its long term performance will probably be diminished.

IMMERSION IN EFFLUENT Just as a membrane must be tested for chemical resistance when it is known that a particular substance will contact it, both the seam adhesive and the sealant that will be in contact with the effluent must be tested for resistance to that material. Generally the test consists of immersing a test piece containing the sealant to be used in as close to the intended job site configuration as possible, then removing it after a predetermined time period, and testing it to ascertain whether there has been a change in the physical properties of the test material. In the case of a sealant the test usually consists of comparing the exposed sealant to an unexposed control sample. Loss of adhesion, cracking, softening, hardening, and color change versus the control are the changes to be noted when testing a sealant's resistance to a particular effluent. As with other long term aging tests, a minimum of 90 days exposure to the effluent at the expected field exposure temperature is recommended. If the effluent will contact the membrane and sealants in a standing pond, 70 C would be the recommended test temperature.

TACK FREE TIME Related to dirt pick-up resistance, tack free time is important in characterizing a sealant. ASTM C 679 Test for Tack Free Time, or a modification of this method will provide repeatable data. It is a rather subjective test involving touching a bead of sealant at timed intervals until it does not transfer. The period until no transfer is observed is called the tack free time. In a seam sealant the tack free time is expected to be relatively short, no longer than 30 minutes. Tack free time of pourable sealers is usually a measure of cure rate of the sealer, as most of these materials chemically cure. This is also true of silicone and urethane sealants specified in many systems for sealing the outer edge of metal termination bars and counter flashings. Water cut off mastic should have no measurable tack free time. Permanent tack is crucial to performance of this material, and should be evaluated.

ADHESION/COHESION Adhesion to the substrate is the most basic and critical property of any sealant, and those used in roofing and waterproofing are no exception. A very good quantitative test for adhesion to a particular surface is ASTM C 794 "Test Method for Adhesion in Peel of Elastomeric Joint Sealants" with the modification that the intended surface be used as the test substrate, and aluminum screen be substituted for the airplane cloth. Generally it is preferable to have cohesive failure, even if the actual peel strength is low. This is true for all the sealants used in single ply systems. The theory is that the sealant does not have to act as an adhesive, and is not under stress, so high peel values are not critical. It does however have to resist the infiltration of water into the system. By always failing cohesively, there is no chance that water could be driven under the sealant during freeze-thaw cycling and thereby penetrate into the seam or flow behind the membrane into the roof substrate and building below. Other useful tests for evaluating adhesion and cohesion of a sealant are included as part of immersion, freeze/thaw cycling, heat aging, and UV aging tests.

These often are subjective tests where the sealant is manually peeled from the membrane, or flexed at several intervals during the test period. A comparison of the test sealant to a control is made in terms of per cent cohesive failure, relative peel strength, and number and depth of cracks that appear during flexing.

TABLE B -- Requirements for specific systems

System	Requirements
LAP SEALANT	
EPDM	Non sag, weatherability to match membrane, excellent adhesion to membrane, crack and shrink resistance, applicable over broad temperature range
BUTYL	Same as EPDM; weatherability not as critical since system is not exposed.
NEOPRENE	Same as EPDM
HYPALON	Self levelling, able to apply from squeeze bottle, weatherability to match membrane, excellent adhesion to sheet and reinforcement cord, applicable over a broad temperature range.
PVC	Same as for Hypalon
CPE	Same as for Hypalon
WATER STOP SEALANT	
EPDM	Permanent tack, permanently pliable, excellent adhesion to membrane and a variety of substrates.
BUTYL	SAME
NEOPRENE	SAME
HYPALON	Same, also used for Pourable Sealer
PVC	SAME
CPE	Same, not specified by all manufacturers

System	Requirements (Continued)
NIGHT SEALANT	
EPDM	Little or no effect on membrane, quick through-cure, good adhesion to membrane, but still removable
BUTYL	SAME
NEOPRENE	SAME
HYPALON	Same, no specific product specified generally with the system
PVC	Same as for Hypalon
CPE	Same as for Hypalon
IN-SEAM SEALANT	
EPDM	Rapid through-cure, no swelling of membrane, excellent adhesion to both membrane and splicing cement, high film strength, long term heat resistance
BUTYL	SAME
NEOPRENE	SAME
HYPALON	Not used
PVC	Not used
CPE	Not used
POURABLE SEALER	
ALL SYSTEMS	Adhesion to a variety of substrates, rapid through-cure, broad application temperature range, broad performance temperature range, flow resistance

TABLE C-- Recommended Laboratory Tests To Evaluate Sealant Performance

Property	Test Methods
HEAT RESISTANCE	ASTM C 792 Effects of heat aging on weight loss cracking and chalking of elastomeric joint sealers (@ 70 C), Accelerated weathering, Field testing
WATER RESISTANCE	Hot water immersion at 70 C; Accelerated weathering, Field testing
ULTRA VIOLET RESISTANCE	UV Box exposure, UV lamp exposure, Weatherometer exposure, EMMAQUA, South Florida exposure, Field testing
HOT/COLD CYCLING	Thin film cracking, ASTM C 719 Adhesion and cohesion of elastomeric sealants under cyclic movement, Field testing
COLOR RETENTION	Weatherometer or UV box exposure, UV lamp exposure, EMMAQUA, South Florida exposure, Field testing
DIRT PICK UP RESISTANCE	Outdoor exposure, Field testing
APPLICATION IN EXTREME TEMPERATURES	Low/elevated temperature extrudability (manual), ASTM D 2452 Extrudability of oil and resin base caulking compounds (Press flow viscosity) at application temperatures
SLUMP RESISTANCE	ASTM D 2202 Slump of caulking compounds and sealants, Brookfield viscosity, ASTM D 2452 Extrudability of oil and resin base caulking compounds (Press-flow viscosity)
SHRINKAGE RESISTANCE	ASTM C 718 UV-Cold box exposure of one-part elastomeric solvent release sealants (Modified), Solids, Thin film cracking

Property	Test Methods (Continued)
RESISTANCE TO EFFLUENTS	Immersion in effluents at room temperature and 70 C
TACK FREE TIME	ASTM C 679 Tack free time of elastomeric sealants
ADHESION/COHESION	ASTM C 719 Adhesion and cohesion of elastomeric joint sealants under cyclic movement, ASTM C 794 Adhesion in peel of elastomeric joint sealants, Manual peel after heat aging and/or water immersion

CONCLUSION

Although many different products are specified with the various single-ply roofing and waterproofing systems, the environmental factors acting upon them are identical. Thus similar weatherability tests can be developed to evaluate these products for their specific applications. Good tests evaluate the effect of the forces acting upon the sealants in a measurable and repeatable manner. Even the best and most reliable tests must still be confirmed by field exposure of the candidate sealant for at least a year to build a measure of confidence in the compound under all conditions, including those that are not foreseeable by laboratory testing.

Leon J. Jacob, Kevin J. Payne

POLYURETHANE - A SIMPLE METHOD OF SELECTING POLYURETHANE SEALANTS FOR YOUR PARTICULAR APPLICATION

REFERENCE: Jacob, L. J., Payne, K. J., "POLYURETHANE - A Simple Method of Selecting Polyurethane Sealants for Your Particular Application", Building Sealants: Materials, Properties, and Performance, ASTM STP 1069, Thomas F. O'Connor, editor, American Society for Testing and Materials, Philadelphia, 1990.

ABSTRACT: Numerous polyurethane sealants are available to cover a wide variety of applications. They vary substantially in chemical composition and performance characteristics. This leaves the user with the complex problem of evaluating and identifying the best sealant for a particular application. In this paper, we propose a simple, yet detailed technique to facilitate the selection process. The focus of the test programme is its uniformity and reproducibility. The fundamental feature of this rating chart is the flexibility it offers for a range of applications and types of sealants. Finally, the benefit to the user is the one-value rating which defines the relevant characteristics for a particular application.

KEYWORDS: Polyurethane, polymers, hydrogen, diisocyanates, elastomer, thermoplastic, hydroxyl.

INTRODUCTION

The development of polyurethane from its discovery in 1937 in Germany and the rapid evolution it has attained throughout Europe was stimulated by shortages of natural rubber materials during World War II.

The versatility of these products caught the imagination of development chemists, engineers and designers and resulted in the

Leon Jacob Group Technical Manager, O'Brien Glass Industries
45 Davies Road, PADSTOW NSW 2212 Australia
Kevin Payne Development Engineer, O'Brien Glass Industries
45 Davies Road, PADSTOW NSW 2212 Australia

rapid evolution of the industry which promulgated a wide range of opportunities for polyurethanes to be used as -

- Adhesives
- Sealants
- Insulation materials
- Coating materials, and
- Structural components.

The extent of possible variation within the urethane group of polymers has insured a continuing challenge for the research and development chemist. These products rival other traditional systems and continued investment worldwide is advancing the science.

In this paper we deal with polyurethanes in the first two categories above, i.e. as adhesives and sealants.

MATERIAL DEFINITION

Polyurethanes represent a group of synthetic polymers which are produced by a unique condensation method from the reaction between isocyanate and hydroxyl compounds. These reactions can be designed to produce different grades and types of adhesives and sealant polymers. This will depend on the molecular weight i.e. properties of chain segments, the degree of cross linking, the quantities of filler, and other such characteristics.

Those developed for adhesive and sealant use are predominantly elastomer compounds, that is, products having a high extension and elastic, or pseudo elastic, characteristic similar to vulcanised natural rubber. They are generally thermoplastic. However, thermo setting materials are also available.

COMPOSITION OF SEALANTS

Several one- and two-component urethane sealants are available. Basically they can be categorised into two types.

1. Non-sagging suitable for application in vertical joints.
2. Self levelling particularly designed for horizontal joints.

One-component polyurethane sealants based on butadiene-styrene have also been proposed and other varieties can be formulated for special requirements such as rate of cure, hardness, strength and elasticity.

Two-component urethanes can also be formulated to provide a variety of characteristics.

POLYURETHANE SEALANT TYPES

One-Component System

One-component moisture cured urethane sealants are generally based on isocyanate-terminated pre-polymers of a medium to high molecular weight. The sealant rate of cure is effected by the activity of the functional groups and the presence of catalysts which activate the isocyanate-water curing reaction. The hardness of the cured sealant depends on the density of the cross linking structure and the concentration of aromatic polyurethane or amide groups.

One-component urethane sealants offer the advantages of easier application, that is, no mixing is required, and are either installed with or without the need for substrate priming depending on the formulation and end use. They generally contain a reasonable percentage of solvents to provide fluidity to the sealant and help wet the surface to promote maximum adhesion.

The major shortcoming to these formulations is the slow "cure-through" rate which is due to the very slow diffusion of atmospheric moisture from the surface through the material. Hence, there is a considerable period prior to total cure and heavy applications (that is, large beads) will not cure through entirely. These moisture cured systems normally have a shelf life of 6 to 12 months, depending on the packaging type.

Considerable development activity now centres around the need to produce rapid cure one-component urethane sealants that retain their cured rubber-like consistency without age hardening and without the loss of bond strength.

Two-Component Systems

Two-component polyurethane sealants are generally composed of a blend of liquid isocyanate-terminated pre-polymer and an unpigmented, or pigmented, hydroxyl terminated compound usually known as the curative, or polyol.

The two components (called the base and the catalyst) are either manually mixed or machine dispensed and mixed through a modified Archimedean spiral in the process of transference. With manual mixing it is essential that the two components are totally blended to ensure the polyurethane functions correctly. Continual mixing ensures the intimate blend of polyol and base polymer.

Cure rate of two-component sealants can be readily modified by the inclusion of a variety of catalysts with the hydroxyl component. The overall system provides more versatility to the polyurethane chemist to develop different characteristics of hardness, elasticity, tensile strength, adhesion and cure rate.

Two-component sealants can be produced without the need to contain solvents and can be manufactured to cure through a variety of thicknesses as the reaction is not dependent on the presence of atmospheric moisture.

ADVANTAGES OF POLYURETHANES AS SEALANTS

Polyurethane sealants can be formulated to provide the following characteristics.

1. Strength - both cohesive and adhesive can be varied depending on the requirements of the application.
2. Hardness could be modified for applications requiring soft sealants for setting blocks and even structural components.
3. Surface energy could be adjusted to improve wetting.
4. Excellent chemical resistance and hydrolysis stability.
5. High tear resistance.
6. Ability to vary the cure rates.
7. Exhibit ozone resistance.
8. Provide very low shrinkage even at low temperatures.
9. Suit a variety of joint designs and joint movement.

DISADVANTAGES OF POLYURETHANES AS SEALANTS

1. Not all formulations are colour stable.
2. Will not effectively seal damp joints.
3. Limited package stability for one component sealants.
4. One component sealants generally require longer cure times.
5. May require primers.
6. Poor UV resistance.
7. Isocyanates may be hazardous at the manufacturing and application stages.
8. Because they contain solvents, combustion of by-products may also be hazardous.

APPLICATION

In Australia, polyurethanes are dramatically establishing their credibility and position as insulating glass sealants in the building section of the glass industry.

From very humble beginnings they have now developed into a dominant force in the insulating glass market.

New usage can also be seen in the design and installation of curtain wall structures manufactured using unitized systems incorporating natural stone and other porous substrates.

In the automotive market, polyurethanes are effectively emerging as a dominant sealant. For example, in the fitting of fixed automotive glass, two-component polyurethanes are basically used at the O.E.M. phase while the after market is predominantly catered for by the one-component sealant.

Given the wide range of properties and characteristics and, hence, varieties of urethanes that can be formulated, it becomes extremely difficult for the glass, building or automotive industry to identify the best grade and type of sealant for its particular application.

Imagine the plight of an after-market automotive glass replacement specialist. There are possibly 5 or 6 large international urethane sealant manufacturing companies each offering at least 2 varieties of one-component polyurethanes.

Each company formulates what it believes to be the most appropriate combination of properties and characteristics required for a particular application and naturally highlights the strength and benefits offered by their product, often attempting to clarify this advantage by nominating compliance with a variety of standards.

However, no two urethane sealants are identical.

This forces the user to make a decision as to which product is best without perhaps full knowledge of all the properties of the sealant under consideration. Therefore, some factors that have to be considered in evaluating the products are as follows:

- Strength
- Cure rate
- Skin over time
- Hardness
- Gunability/viscosity
- Need or otherwise for primers, etc.

PHILOSOPHY OF THE RATING CHART

As stated earlier, the diversity of properties and characteristics of any sealant makes the mental evaluation of the figures provided by the sealant manufacture cumbersome and well nigh impossible. As a result of this, Table 1 was developed to facilitate the evaluation of sealants for use with automotive glass. However, it will be shown later that this table can be adapted for the evaluation of urethanes for any application.

The key to the successful use of this rating chart is the initial establishment of the desired characteristics and the working parameters of performance required to obtain the most suitable sealant for the particular application. This is achieved

TABLE 1 -- Urethane Rating Chart

Urethane Name:___________________

		10 GOOD	1 BAD	STANDARD CONDITIONS	WATER IMMERSED	HUMIDITY CABINET	BOIL TEST	1000 HRS QUV	TOTAL
1	Max Hard.(Shore A)	40 to 50	≥60 ≤30	x 1.0=	x 0.5=	x 0.5=	x 1.0=	x 1.0=	/40
2	Av Tensile Stress	≥1 MPa	≤0.1MPa	x 2.0=	x 1.0=		x 1.0=	x 2.0=	/60
3	Av Shear Stress	≥1 MPa	≤0.1MPa	x 2.0=	x 1.0=		x 1.0=	x 2.0=	/60
4	Mode of failure	Cohesive	Adhesive	x 2.0=	x 1.0=		x 0.5=	x 2.0=	/55
5	Skin over time	≤5 mins	≥60 mins	x 1.5=	x 0.5=	x 0.5=			/25
6	Tack Free time	≤30 min	≥90 mins	x 1.5=	x 0.5=	x 0.5=			/25
7	Max Cure Depth	≥5 mm	≤1mm	x 2.0=	x 1.0=				/30
8	Full Cure Shr Str.	≥5 MPa	≤1 MPa	x 2.0=					/20
9	Modulus of Elast.	2.0 MPa	High/Low	x 1.5=					/15
10	Toxicity	None	Extreme	x 1.0=					/10
11	Shelf Life	≥1 year	≤6 mths	x 1.5=					/15
12	Gunability	≤7 kg.	≥25 kg.	x 2.0=					/20
13	Slump	0 mm	≥9 mm	x 2.0=					/20
14	Primer	No	Yes	x 1.5=					/15
15	Special Cleaner	No	Yes	x 1.0=					/10
16	Test Rpts/Tech Inf	Avail.	Not Avl.	x 1.0=					/10
17	Application Instr.	Avail.	Not Avl.	x 0.5=					/5
								TOTAL SCORE	/435

NOTE: Score <50% totally unacceptable
Score >75% highly desirable

by experimentation and practical experience and, most importantly, on the understanding of the function to which the sealant is to be subjected.

Fundamental Criteria

It is important to standardise the test method, criteria and documentation. This will facilitate consistency and reproducibility. It is also necessary to develop a technique for the preparation of standard glass samples which are consistent in both size and quality.

Therefore, every urethane sealant to be evaluated for a particular application will be subjected to all of the tests listed and subsequently rated. The higher the ultimate score, the more acceptable the product.

As an example, Table 1 identifies all the characteristics that are necessary to produce a good sealant for use in the automotive industry, specifically the aftermarket.

Table 1

Lines 1 to 9 deal with the physical properties of the urethane.

Lines 10 to 15 cover user considerations.

Lines 16 and 17 refer to information available to assist the user in the correct application of the sealant.

Test Conditions

Having listed all the factors that are important for a particular application, it is then necessary to identify the variety of test exposure conditions. Where the conditions do not apply, they are left blank.

For example, sealant hardness (line 1). Depending on the application, it could be decided that hardness required is 45 Shore A.

In our application a range of 40-50 Shore A hardness could be acceptable. Consequently any hardness values greater than 60 or less than 30 would be unacceptable. The score given for hardness will vary proportionally from a score of "10" at 45 Shore A, to a score of "0" at 30 and 60 Shore A.

Weighting

In order to be pragmatic about the rating chart, a weighting must be applied to differentiate between the critical and the "like to have" characteristics of the sealant.

Because tensile strength (line 2) is important, we would nominate a higher weighting for standard test conditions. The minimum tensile strength at 1 MPa could be the bench mark. Any test results below this standard would rate a correspondingly reduced score.

On the other hand, application instructions (line 17), though less important, have been considered but with a lower weighting to reflect the reduced importance.

The significance of this chart is its-

1. Simplicity.
2. The consistency of results.
3. The minimisation of bias towards any supplier or product.

This chart is primarily designed as an evaluation tool and not for research and development. Consequently, it is simply a matter of adding or deleting test categories depending on the individual requirements. For example, sealants for which moisture vapour transmission is important could have this category inserted in lieu of, say, tack-free time which may not be necessary.

Alternatively, a totally new chart could be developed, for example, for building sealants in expanding joints.

DISCUSSION

It is appropriate for us to look at each of the items listed in Table 1 in terms of polyurethane for use in the after-market to fit automotive glass. The tests are generally conducted under standard laboratory conditions of 23±3°C and 55±5% relative humidity. In addition, for comparison purposes, some tests are duplicated under high humidity and water immersion conditions.

Weathering/Aging Test

The use of a QUV weathering machine is important to ensure that the sealant can perform satisfactorily when exposed to cyclic UV radiation.

Shore A Hardness (Line 1)

This is a simple test which measures the resistance and resilience to indentation of the cured sealant. It can also be used as a quick measure to identify final cure of the sealant. See ASTM.D.2240.

Tensile/Shear Stress (Lines 2 & 3)

The recommended test method is ASTM D.412. However, it is very important to ensure that the samples are uniform in size and composition and that a standard is maintained during the evaluation.

Modification of this would be to the detriment of both tensile and shear strengths in the short term analysis, i.e. from 0-8 hours, where the strength build up is monitored on an hourly basis and for comparative purposes the average strength is computed.

Mode of Failure (Line 4)

With regard to the Mode of Failure, it is important to know how the sealant will perform under full load conditions. If the test samples fail cohesively, then it can be taken as an indication of good adhesion.

Whereas, if the failure mode is adhesive, there must be concern as to the need for special cleaning and primers with that particular urethane.

Tack Free Time & Depth of Cure (Lines 6 & 7)

These are subjective tests which are designed to give an indication of how long it takes the product to complete the chemical changes required for stability.

For automotive glass sealants, the rate of cure is important because, on occasion (in the after market), it is necessary to use the vehicle prior to complete cure being attained.

This necessitates the determination of polyurethane strength in the short term. That is, 1, 2 or 3 hours after installation.

Because curing takes place from the outside to the centre, the cross section of the area required for the computation of the tensile, or shear, strength could be determined either by a theoretical cross section of the sample or the cross section of the cured material.

This can be determined from the depth of cure test.

Slump Test (Line 13)

The slump test is important to evaluate the thixotropic nature of the sealant and ASTM.C.639 is recommended.

Most of the other tests listed have special significance as far as the automotive glass replacement market is concerned.

CONCLUSION

As stated earlier, polyurethane can be formulated to achieve different parameters and characteristics which leaves the user with a monumental task of trying to evaluate strength, cure rate, hardness, modulus of elasticity, etc. required before selecting the most suitable polyurethane.

In the final analysis, this rating chart is designed primarily to facilitate the selection of the best sealant from a wide range. It is a one-number evaluation of a fairly complex decision process and is a mechanism which can be modified, reduced, or expanded to suit the required characteristics of any urethane sealant for the building, insulating, automotive glass industry, or for any other application requiring sealants.

BIBLIOGRAPHY

Albury, D. W., Jaubury and Beech, J. C., The Influence of Moisture on Building Joint Sealants, Journal of Material Science.

Damusis, A. Sealants, Rheinhold Publishing Corp., New York.

Frisch, Kurt C., Advances in Polyurethane, Science & Technology, Polymer Institute, Uni. of Detroit, Michigan, U.S.A. - February 24-27, 1987.

Oertel, Gunther, Polyurethane Handbook, Hanser Publ., Munich.

Panek, J. R. & Cook, J. P., Construction Sealants & Adhesives, 2nd Edition, Wiley Publ., New York.

Proceedings of the 4th Waterborne & Higher Solids Coatings Symposium, New Orleans, Louisianna.

SAE Australia, Industrial Urethanes, Melbourne - August, 1988.

Tremco Architectural Forum, Sealants & Weatherproofing Systems for Building Construction, New York, May 4, 1978.

John F. Timberlake

SILICON MODIFIED POLYETHERS FOR MOISTURE CURED SEALANTS

REFERENCE: Timberlake, J. F., "Silicon Modified Polyethers for Moisture Cured Sealants," Building Sealants: Materials, Properties and Performance, ASTM STP 1069, Thomas F. O'Connor, editor, American Society for Testing and Materials, Philadelphia, 1990.

ABSTRACT: Silicon modified polyethers are an intermediate for the preparation of one-part and two-part sealants. Silicon modified polyethers consist of a polyoxypropyl-ene backbone with methyldimethoxysilane end groups. Curing with moisture in the presence of catalysts gives flexible polymers with siloxane bonds connecting the poly-ethers. The primary application for these silicon modified polyethers is in construction sealants.

Silicon modified polyethers give soft sealants which have excellent weatherability and durability (joint movement) with little changes in properties after four years under normal use conditions. Weathering data demonstrate that these sealants will last for many years in outdoor exposure. The sealants show good adhesion to most substrates including aluminum, mortar, and stone. Good curing rates are observed in both one-part and two-part sealants over a range of temperatures. The sealants are nonstaining on stone and can be easily painted.

KEYWORDS: SILICON MODIFIED POLYETHER, MOISTURE CURE, SEALANT, METHOXY SILANE, POLYOXYPROPYLENE.

INTRODUCTION

Silicon modified polyethers are a new class of intermediates for sealant preparation. Silicon modified polyethers have a wide range of physical properties which makes them excellent materials for construction sealants. Currently, the construction sealant market in North America [1] utilizes mainly silicone and urethane sealants, with some use of polysulfide. Silicon modified polyethers provide benefits of both urethane and silicone sealants without many of the drawbacks associated with these two technologies.

The silicon modified polyethers have been used very successfully for several years in Japan for the production of construction sealants. More recently, silicon modified polyethers have been introduced into Europe where more stringent health laws have reduced the use of isocyanate containing urethane sealants. The sealants are particularly useful for applications where the good weatherability, non-staining characteristics, and good adhesion of the silicon modified polyethers provide long term usage without

Dr. Timberlake is a Research Scientist at Union Carbide Chemicals and Plastics Company Inc., Technical Center, P. O. Box 8361, South Charleston, WV 25303.

TABLE 1 -- Typical properties of the silicon modified polyethers

Silicon Modified Polyether	SMP-2	SMP-3
Type	very soft	soft
Viscosity at 23°C (poise)[a]	140 ± 40	200 ± 40
Specific gravity at 20°C	1.01	1.01
pH value	~7	~7
Physical state	transparent liquid	
Color	light yellow	
Odor	practically none	

[a]Brookfield viscometer, BM-type, rotor No. 4.

significant changes in properties. This paper details the chemistry of the silicon modified polyethers and an overview of their application to sealant technology.

Two types of silicon modified polyethers are currently available [2]. Silicon modified polyether SMP-2 is a linear polyether with a methyldimethoxysilane group at each end. SMP-2 is used to produce very soft sealants with low modulus and low hardness. Silicon modified polyether SMP-3 is a higher functionality polyether than SMP-2. SMP-3 is used for firmer, but still soft, sealants with somewhat higher modulus and hardness. The properties of SMP-2 and SMP-3 are summarized in Table 1.

CHEMISTRY

The silicon modified polyethers [3] are prepared from high molecular weight polyoxypropylene polyols. The polyol is end-capped with allyl groups followed by hydrosilation to give a polyether end-capped with methyldimethoxysilane groups. The silicon-carbon bond produced by this chemistry is stable to moisture. The polymers also have reduced moisture sensitivity because the polypropylene glycol polyether is hydrophobic.

In the presence of appropriate catalysts, the methoxysilicon groups can be cured by moisture (Diagram 1). The water reacts with the methoxysilicon group to liberate methanol and produce a silanol. Further reaction of the silanol with either another silanol or methoxysilicon group gives the silicon-oxygen-silicon linkage. The final polymer can be thought of as a polyether cured through siloxane groups or as siloxanes "cross-linked" through polyethers. The silicon modified polyethers have the advantage of being non-isocyanate containing and producing very low odor during curing, since the only by-product is methanol.

Diagram 1.

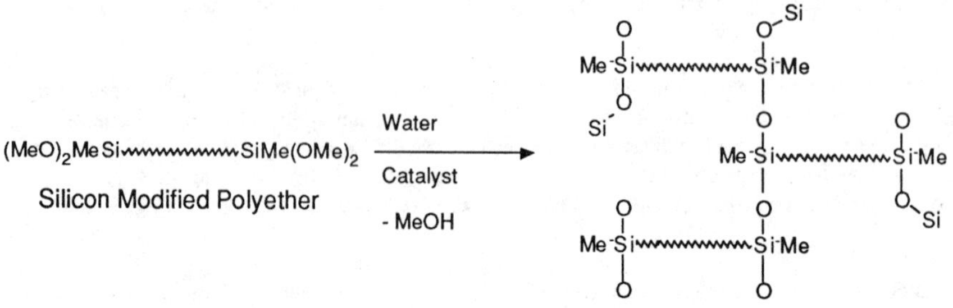

$(MeO)_2MeSi$〰〰$SiMe(OMe)_2$ —Water / Catalyst / - MeOH→ cured network (Me-Si〰〰Si-Me units linked through O-Si)

Silicon Modified Polyether

TABLE 2 -- Physical properties of cured silicon modified polyethers.[a]

Silicon Modified Polyether	SMP-2	SMP-3	60/40 SMP-2/SMP-3
Hardness, Shore A[b]	12	22	17
Elongation, %	233	137	155
Tensile Strength, kg/cm^2 [c]	2.52	4.05	2.81
Modulus at 100%, kg/cm^2 [c]	1.38	3.30	2.09
Tear Strength, kg/cm [d]	0.90	1.09	1.12

[a]Catalyzed by 2 parts dibutyl tin dilaurate and 0.5 parts lauryl amine per 100 parts of silicon modified polyether.
[b]ASTM Method C-661 Indentation Hardness of Elastomeric-type Sealants by Means of a Durometer.
[c]ASTM Method D-412 Rubber Properties in Tension
[d]ASTM Method D-624 Rubber Properties - Tear Resistance

The cured final polymer (unfilled) is a soft, rubbery material. Good surface cure is achieved within several hours, depending on the catalyst (see below), at 23°C and 50% relative humidity. To demonstrate the inherent properties of the polymer, samples of SMP-2, SMP-3, and a mixture of 60% SMP-2/40% SMP-3 were cured without filler. The physical properties of these polymers are given in Table 2. The unfilled polymers have relatively low elongation, tear, and tensile strength. As expected, SMP-3 has a higher modulus and tensile strength than SMP-2 and a lower elongation.

In the following section, it will be seen that the physical properties of the final polymers dramatically improve when properly formulated into sealants. In particular, a tremendous reinforcing effect is observed when a small particle size, surface treated calcium carbonate is used to fill the silicon modified polyethers before curing (see below).

FORMULATIONS

Sealants can easily be formulated from silicon modified polyethers [4,5]. Typical one-part and two-part formulations are given in Table 3. The work for this paper was done with variations on these formulations or on commercial formulations which are similar to these formulations.

Calcium Carbonate Fillers

The use of filler can be key to the cost competitiveness of many sealant formulations. The use of calcium carbonate as a filler also improves the physical properties of the silicon modified polyether sealant. Addition of the calcium carbonate dramatically increases the viscosity of the formulation and shows a thixotropic effect. The viscosity increase is important for application of sealants to vertical surfaces. The final cured sealant exhibits increased modulus, tensile strength, and elongation if proper care is taken in choosing the grade of calcium carbonate to be used.

To examine the effect of various types of calcium carbonate fillers, several fillers were formulated into sealants and tested. The data in Table 4 indicate that the properties of the uncured sealant, such as viscosity and stability, are related to the type of calcium carbonate, but not in a straight forward manner. Calcium carbonates "A" and "B" are both small particle size and surface treated. While they both give good viscosity, the storage stability of "A" is good but "B" gives poor storage stability. Calcium carbonate "C" gives good initial viscosity with a relatively large particle size and no surface treatment, but poor stability. Calcium carbonate "D" gives viscosity which is much too low but good storage stability. The important point is that particular calcium carbonate samples must be tested prior to use to determine if the final properties of the sealant are appropriate for the end use application.

TABLE 3 -- Typical silicon modified polyether sealant formulations (phr)[a].

Type	One Component	Two Component
		Part A
SMP-2	60	100
SMP-3	40	...
Plasticizers	50	55
Filler, Calcium Carbonate	120	120
Filler, Titanium Dioxide	20	8
Surface Improver	...	5
Thixotropic Agent	3	6
Antioxidant	1	1
UV Absorber	1	1
Dehydrating Agent	2	...
Adhesion Promoter	3	...
Hardening Catalyst		Part B
Dibutyltin Dilaurate	2	...
Stannous Octoate	...	3
Co-Catalyst		
Lauryl Amine	0.5	0.5-1
Filler, Titanium Dioxide	...	20
Plasticizer	...	6.5-6.0
TOTAL	302.5	326

[a]Parts per hundred parts of resin by weight.

Similar effects are seen on the properties of the cured sealant (Table 4). Calcium carbonate "A" and "D" give good modulus, tensile strength, and elongation. Calcium carbonate "B" and "C" give reduced elongation and high modulus even though "B" is similar to "A" and "C" is similar to "D" in measured physical properties. For the purposes of this study, calcium carbonate "A" was used.

Catalysts

Some of the tin catalysts which are used in Japan are currently unavailable in the United States. Of those tin catalysts available in the U.S., dibutyl tin dilaurate (DBTDL) has been used successfully in commercial one-component sealants, but requires that a primary amine, e.g. lauryl amine, be used as a co-catalyst. In an effort to identify a tin catalyst which does not require an amine co-catalyst, a number of alternative catalysts have been examined for curing these sealants.

TABLE 4 -- Effect of calcium carbonate on properties of two-component sealants[a].

Calcium Carbonate	A	B	C	D
Particle Size, 10^{-6} m	0.08	0.03	1.0	4.0
Surface Treatment	Fatty Acid	Resin Acid	none	none
Sealant Viscosity, poise[b]	18,600	11,400	16,800	240
Storage Stability	good	poor	poor	good
Modulus @ 150%, kg/cm^2 [c]	2.8	7.1	4.8	2.6
Tensile at Break, kg/cm^2 [c]	9.2	10.2	9.1	6.8
Elongation at Break, % [c]	650	410	430	730

[a]Two-part sealant formulation - Part A: SMP-2 (100), dioctyl phthalate (15), dehydrating agent (1.5), calcium carbonate (100). Part B: dibutyltin phthalate (0.5), zinc oxide (1.8), dioctyl phthalate (0.6).

[b]Brookfield viscometer, BS-type, Rotor No. 7.

[c]Japanese Industrial Standard K 6301.

To test these new catalysts, SMP-2 (100 parts by weight) was mixed with various tin catalysts (2 parts by weight) and the mixture was moisture cured under standard conditions [6]. The tack free (surface cure) time was determined. The following reactivity order was observed:

UL-11A > T-1 ~ UL-8 >> { SUL-3, SUL-4, UL-22, UL-24, UL-28, UL-29, UL-38
T-5, T-9, T-12, T-120, T-125, T-131 }

The last group of catalysts all took longer than 24 hours to become tack free.

The same catalysts were also tested using lauryl amine (0.5 parts by weight) added to the above mixture. The following reactivity order was observed:

UL-11A ~ UL-28 > T-9 ~ UL-38 > T-1 ~ T-125 > UL-8 ~ T-12 >> { SUL-3, SUL-4
UL-22, UL-24, UL-29
T-5, T-120, T-131 }

Again the last group was not surface cured after 24 hours. These results are not totally consistent with the structures of the catalysts. For example, T-12/SUL-4 are both dibutyl tin dilaurate and T-1/SUL-3 are both dibutyl tin diacetate. The pairs should have given equal cure rates. Since T-12 was faster than SUL-4 and T-1 was faster than SUL-3, further work will be carried out to determine the reasons for the differences between them and whether the differences carry through into the formulated sealants.

The catalysts UL-11A, T-1, and UL-8 appear to be likely candidates for catalysts which will be effective without an amine. These catalysts, along with UL-28, UL-38, and T-125, may be effective at lower levels than the DBTDL/lauryl amine combination. Further work is in progress to check these cure rates in the formulated sealants.

The speed of cure is not the only criterion for choosing a catalyst. For example, the Fomrez® UL-11A cured the fastest, but gave a polymer that was softer than most of the other active catalysts even though the polymer appeared to be fully cured. It must be noted that a fully formulated sealant may have different cure characteristics than these test samples. Dibutyl tin dilaurate (DABCO® T-12) gives a tack free surface in about 7 hours in this test, but cures in 3-4 hours in a properly formulated sealant. Additionally, switching to a different calcium carbonate gave cures greater than 24 hours.

A special case is DABCO® T-9 (stannous octoate) which gave quick initial cure, but did not cure all the way through the polymer in this study. The catalyst was deactivated by the moisture or the oxygen in the air. This phenomenon is well known for stannous octoate catalysts in urethane systems [7], but is important only for the one-part silicon modified sealants. In two-part systems, T-9 can be used and provides good curing (see below) because the cure rate is so much faster that the cure is complete before the catalyst is deactivated by absorbed oxygen.

Long term stability studies have been done on sealants prepared using dibutyl tin dilaurate for one-part and stannous octoate for two-part sealants, but not on any of the other catalysts available in the United States. Care must be taken in choosing which catalyst to use for a formulated sealant and consideration given to the interactions of the formulation ingredients.

CURING RATES

Silicon modified polyethers produce sealants which cure well over a wide range of conditions. It is important to have the wide range of curing conditions to insure that the sealants can be used in cold and hot climates. These good curing characteristics will impact on the ability to use sealants in all areas of the country and during most weather conditions. Two-part sealants can be formulated which cure to the ultimate properties within seven days over the temperature range from 5°C to 70°C with little effect from changes in humidity. Table 5 summarizes the data for the properties for a two-part

TABLE 5 -- Curing rates for two-part sealants[a] under varying conditions

Curing Conditions		Modulus at 150% Elongation (kg/cm^2)[b]	Tensile at Break (kg/cm^2)[b]	Elongation at Break (%)[b]
Temp. (°C)	Time (days)			
-10	7	0.7	3.0	740
	21	1.7	5.9	610
5	3	1.3	5.7	710
	7	1.8	5.7	600
23	3	1.9	6.7	610
	7	1.8	5.6	600
50	3	1.9	6.9	630
	7	1.7	5.7	560
70	3	1.5	5.3	620
	7	1.6	5.6	620

[a]Two-part formulation: SMP-2 (100), dioctyl phthalate (50), calcium carbonate (110), titanium dioxide (28), thixotrope (6), carbon black (0.5), antioxidant (1), UV absorber (1), stannous octoate (3), lauryl amine (1).
[b]Japanese Industrial Standard K 6301.

TABLE 6 -- Modulus at 150% elongation of various two-part sealants after seven days curing[a]

	Sealant Type			
Temp. °C)	Silicon Modified Polyether[b]	Silicone A[c]	Silicone B[d]	Polysulfide[e]
-10	0.7	0	0	0
5	1.8	0.1	0.2	0.6
23	1.8	1.1	0.5	1.3
50	1.7	0.4	0.2	1.9
70	1.6	0	0	1.4

[a]Japanese Industrial Standard K 6301. Units are kg/cm^2.
[b] Formulation as in Table 3.
[c] Medium modulus commercial silicone sealant.
[d] Low modulus commercial silicone sealant.
[e] Commercial polysulfide sealant.

TABLE 7 -- Curing depth of one-part silicon modified polyether sealant[a] at different temperatures

Temperature = 5°C[b]						
Curing Time (days)	1	2	4	8	11	14
Curing Depth (mm)	2.9	3.1	3.8	7.5	9.2	10.3
Temperature = 23°C[b]						
Curing Time (days)	1	2	4	8	11	14
Curing Depth (mm)	2.8	5.0	6.1	10.5	11.0	12.7

[a]Same as One-part sealant in Table 3 except that the dibutyl tin dilaurate and lauryl amine were replace with 2 phr of dibutyl tin diacetylacetonate.
[b]Relative humidity 50-60%.

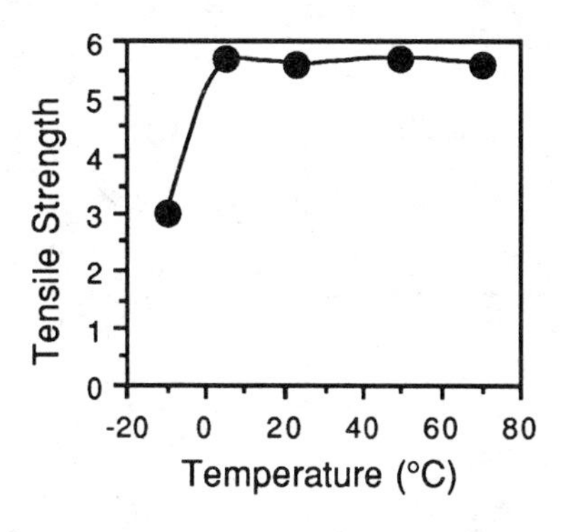

FIG. 1 -- Tensile strength of two-part sealants after 7 days as a function of curing temperature

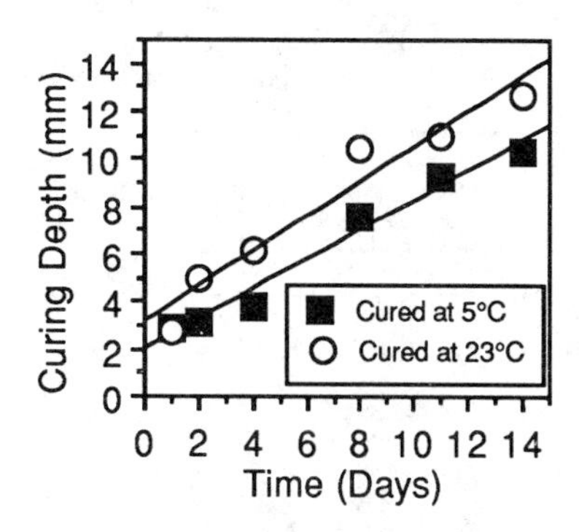

FIG. 2 -- Curing depth versus time for one-part sealants at two temperatures

sealant using SMP-2. Even at -10°C the sealant cures in seven days and reaches its ultimate properties in 21 days, implying that silicon modified polyether sealants can be used in all but the most severe winter conditions. Figure 1 shows that good cure is obtained within 7 days if the temperature is at or above freezing.

The data in Table 6 compares the curing rates of several types of commercial sealants. The two-part silicone sealants do not cure well at low and high temperatures after seven days. The polysulfide sealants gives a slow cure at low temperatures. By comparison, the silicon modified polyether sealant cures well at high temperature and has cured significantly at low temperatures in seven days.

The one-part sealant curing data is summarized in Table 7. At normal humidity, the one-part sealant cures on the surface within hours and to a depth of 10 mm or more within two weeks. Good cure and relatively quick tack free surface are obtained as low as 5°C. As you can see from Figure 2, the curing rate of the one-part sealant is good at both temperatures. Lower relative humidity will slow the cure, but the ultimate properties should be the same when cured.

WEATHERING

Excellent weathering is one of the features which distinguishes silicon modified sealants as good construction sealants. Extended use in Japan as construction sealants has given good long term use with little change in physical properties. The data in Table 8 show that there is little change in the sealant properties under accelerated weathering studies. Even after three to four months of testing, the one-part and two-part sealants give little change as indicated by the modulus, elongation, and tensile strength. Figure 3 shows that the change in tensile strength with time is small. If anything, the one-part sealant may be getting stronger and the two-part sealant demonstrates only a slight decrease with time.

Similar results were obtained for the two-part sealant under normal exterior use conditions shown in Table 9. After four years, the sealant properties are well within the acceptable ranges for construction sealants with only a slight decrease in elongation and increase in modulus. The surface of the sealant showed no cracking, splitting, or discoloration. Silicone sealants gave similar results. The polyurethane sealant had severe surface cracking and yellowing. The polysulfide sealant had lost much of its elasticity and split during normal joint movement.

TABLE 8 -- Accelerated weathering tests[a] for silicon modified polyether sealants

Time (hours)	Modulus at 150% Elongation (kg/cm^2)[b]	Tensile at Break (kg/cm^2)[b]	Elongation at Break (%)[b]
One-Part Sealant[c]			
0	7.8	11.7	326
240	7.5	13.6	290
480	8.6	13.1	257
720	7.7	13.4	273
960	8.6	13.6	257
1200	8.3	13.2	247
1440	8.2	15.5	308
1980	8.1	13.4	273
2520	7.8	12.4	257
Two-Part Sealant[d]			
0	2.8	9.9	765
480	3.3	10.4	732
960	3.4	9.6	690
1200	3.7	10.6	663
1980	3.6	10.0	657
2520	3.6	9.8	617
3000	3.7	8.7	607

[a] Sunshine WOM, black panel temperature of 63°C, water sprinkle for 12 minutes in every 120 minutes.
[b] Japanese Industrial Standard K 6301.
[c] One-part commercial sealant - exact formulation unknown.
[d] Two-part sealant as in Table 3.

TABLE 9 -- Normal use weathering data[a] for silicon modified polyether sealants

Time (years)	Modulus at 150% Elongation (kg/cm^2)[b]	Tensile at Break (kg/cm^2)[b]	Elongation at Break (%)[b]
Two-Part Sealant[c]			
0	2.8	9.9	765
1	4.4	13.4	742
4	4.5	9.7	542

[a] Outdoor weathering in Kobe, Japan, south facing, 45° angle.
[b] Japanese Industrial Standard K 6301.
[c] Two-part sealant as in Table 3.

Another set of tests for durability of silicon modified polyether sealants is summarized in Table 10. Both one- and two-part sealants show only minor changes in the physical properties after being subjected to 90°C in an oven for up to 29 weeks. These sealants were both commercial formulations similar to Table 3 marketed in Japan for construction sealants. These tests all show that sealants prepared from silicon modified polyethers will provide good service in a variety of applications.

TABLE 10 -- Heat resistance at 90°C[a] of silicon modified polyether sealants

Time (weeks)	Modulus at 150% Elongation (kg/cm^2)[a]	Tensile at Break (kg/cm^2)[a]	Elongation at Break (%)[a]
One-Part Sealant[b]			
0	7.9	12.0	328
2	7.1	11.0	373
7	6.6	10.2	328
13	6.1	9.2	332
17	6.7	9.4	318
23	6.0	8.0	273
26	5.2	7.6	320
29	5.1	11.5	250
Two-Part Sealant[c]			
0	2.8	9.4	745
1	2.8	12.8	880
7	3.1	11.3	805
13	2.8	10.6	800
23	2.8	9.4	785
29	2.7	8.9	748

[a]Samples tested after being subjected to 90°C by test method JIS K 6301 type III.
[b]One-part commercial sealant - exact formulation unknown.
[c]Two-part sealant as in Table 3.

Another aspect of weathering is the appearance of the sealant and the substrate around the sealant after years on the building. When used for sealing between granite, stone, or other substrates, these silicon modified polyether sealants do not cause staining of the surface around the joint. The staining observed in silicone sealants is caused by silicone polymers bleeding from the sealant, spreading on the surface and on the substrate around the sealant, and accumulating dirt. After seven years of outdoor exposure, the surface near the sealant formulated as in Table 3 looks as good as the general surface of the granite.

ADHESION AND PAINTABILITY

Silicon modified polyether sealants have excellent adhesion to a wide variety of substrates. The data in Table 11 shows that these sealants adhere well to mortar, stone, metals, and plastics. The good adhesion is maintained after weathering so that the sealant does not separate from the substrate after years of use. Accelerated weathering sample (Figure 3) still give cohesive failure with only slight changes in the overall properties of the sealant. The adhesion is important to prevent a crack from forming between the sealant and the building, reducing moisture and air movement.

Unlike many silicone sealants, paints will adhere well to the surface of silicon modified polyether sealants. Using water based latex paints, the cross hatch peel test shows no loss of adhesion between the sealant and the paint on either one-part or two-part sealants. Other paints adhere well to the sealants, but tests should be performed before large areas of sealant are covered with paint. In applications where the joint movement is large, the paint must be able to stretch with the sealant to prevent paint cracking.

The curing of silicon modified polyether sealants produces siloxane bonds similar to a neutral cure silicone sealant, but the main portion of the polymer is still polyether. As a result the surface properties

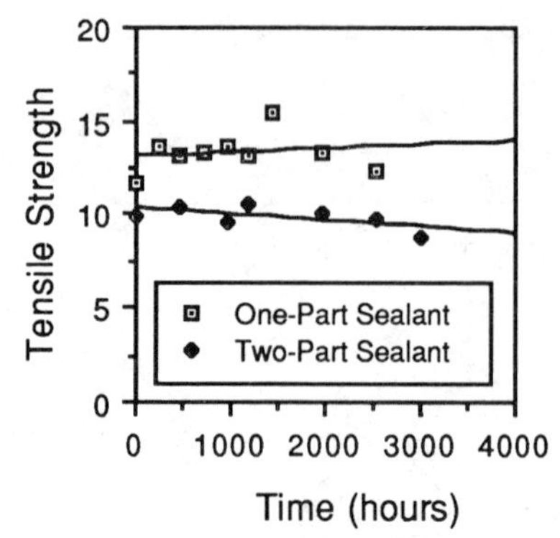

FIG.3 -- Accelerated weathering data for one- and two-part sealants.

TABLE 11 -- Adhesion of one-part silicon modified polyether sealants[a] to several substrates[b]

Substrate	Adhesive Strength (kg/cm)
Aluminum	4.0
Stainless Steel	10.0
Granite	5.0
Mortar	10.3
Plywood	10.3
PVC Plate	10.0

[a]One-part sealant with formulation in Table 3 except that the catalyst was dibutyl tin diacetylacetonate.
[b]Test method: 180° peel, 10 mm wide.

of the silicon modified polyether sealants are similar to urethane sealants, which are also mostly polyether.

SUMMARY

Silicon modified polyethers are ideally suited to formulation into construction sealants. Both one-part and two-part sealants have excellent weathering characteristics, adhesion to a variety of substrates, and are non-staining for mortar, stone, and metal. For painted structures, the silicon modified polyether sealants show excellent paintability. The silicon modified polyethers are easy to formulate with excellent long term storage stability, workability, and curing characteristics, even at low temperatures.

Silicon modified polyethers offer architects and contractors another alternative for sealant application which provides a unique combination of properties. Their performance provides a broad range of essential physical and esthetic properties which are maintained for many years in actual use.

ACKNOWLEDGEMENTS

The author wishes to thank K. Isayama and co-workers for their assistance in preparing this paper.

REFERENCES

1. Meyer, R. E., "Sealants," in Encyclopedia of Polymer Science and Technology, Vol.15, John Wiley and Sons, New York, 1989, pp. 131-145.
2. Silicon modified polyethers SMP-2 and SMP-3 were obtained as Silmod™ 20A and Silmod™ 300 polyethers, respectively, from Union Carbide Chemicals and Plastics Company Inc., Danbury, CT.
3. Isayama, K., and Hatano, I., *US Patent 3,971,751*, "Vulcanizable Silylether Terminated Polymer," July 27, 1976.
4. Takase, J., Hirose, T., and Isayama, K., *US Patent 4,444,974*, "Room Temperature Curing Composition," April 24, 1984.

5. Mita, T., Nakaishi, H., Takase, J., Isayama, K., and Tani, N., *US Patent 4,507,469*, "Curable Composition Comprising Polyether Having Silyl Group," March 26, 1985.
6. Tin catalysts Fomrez® SUL-3, SUL-4, UL-8, UL-11A, UL-22, UL-24, UL-28, UL-29, and UL-38 were furnished by Witco Corporation, New York, NY. Tin catalysts DABCO® T-1, T-5, T-9, T-12, T-120, T-125, and T-131 were furnished by Air Products and Chemicals, Inc., Allentown, PA
7. Saunders, J. H., and Frisch, K. C., "V. Formation of Urethane Foams," in Polyurethanes Chemistry and Technology, Part I, Chemistry, Interscience Publishers, New York, 1962, pp. 219-260.

Michael J. Schroeder, Edward E. Hovis

SEALANT BACK-UP MATERIAL AND NEW DEVELOPMENTS

REFERENCE: Schroeder, M.J. and Hovis, E.E., "Sealant Back-up Material and New Developments", Building Sealants: Material, Properties and Performance, ASTM STP 1069, Thomas F. O'Connor, editor, American Society for Testing and Materials, Philadelphia 1990.

ABSTRACT: During polyolefin extrusion trials in the first quarter of 1988, we were experimenting with polyolefin blends trying to develop a high resilient backing material which was suitable for use in a joint with slow curing, low modulus, self-leveling sealants. One of the requirements for this new backing material was better compression recovery than traditional closed cell polyethylene foams.

While working with polyolefin blends to increase recovery, it was discovered that a dimensionally stable, open cell/closed cell structure could be extruded with the added benefit of non-outgassing when ruptured.

KEYWORDS: closed cell, open cell, open cell/closed cell structure, nongassing

[1]Years ago, in the days before the curtain wall, large buildings were made of thick masonry walls which did not exhibit much temperature change and consequently, little or no movement of the joints. These joints were first packed with oakum or other fibrous materials and then sealed with a non-flexible mortar. When buildings had massive masonry walls, the fact that they leaked was often hidden by their ability to absorb rain without letting it into the interior space.

Mr. Hovis and Mr. Schroeder are employed by Applied Extrusion Technologies, Inc., Middletown Industrial Park, Middletown, DE 19709.

With the advent of modern architecture and the invention of curtain wall construction techniques, joints became an important design condition in the building construction industry.

In the middle 1960's, architects discovered that joints should be designed around the movement capabilities of elastomeric sealants and backer rods needed to be considered as an integral part of the joint design. See Figure 1

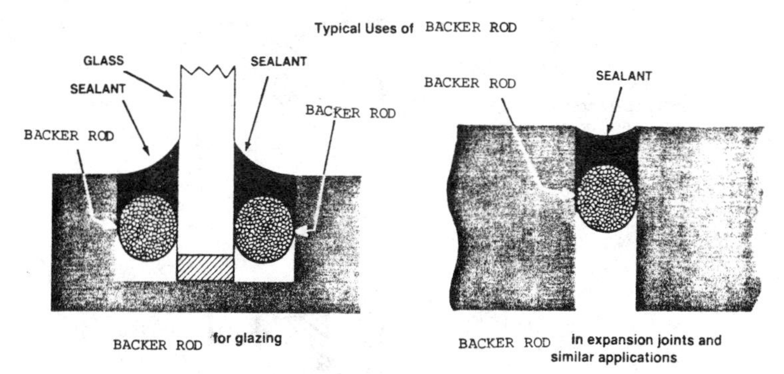

FIGURE 1

The purpose of using backer rod is to limit the depth of applied sealant, to provide a proper shape to the sealant and to act as a bondbreaker preventing back side adhesion.

GOOD DESIGN

Joint width wide enough to accommodate movement.

Joint sawed deep enough to allow backer rod sealant placement.

Proper backer rod used and placed.

Sealant tooled to ¼ inch below the pavement surface.

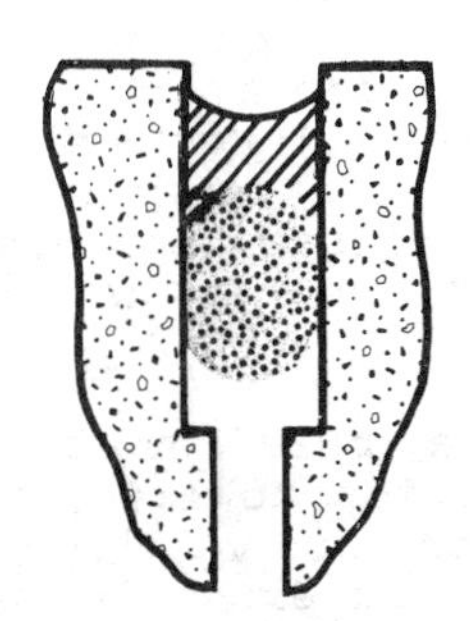

FIGURE 2

Figure 2 is an example of a good design and illustrates the importance and need for backer rod. The joint width is wide enough to accommodate movement. The joint is sawed deep enough to allow backer rod sealant placement, the proper backer rod is used and placed and the sealant is tooled below the surface per sealant manufacturers recommendations.

POOR DESIGNS

Sealant not tooled to proper depth.

Sealant bead too thick; wrong width-to-depth ratio.

A

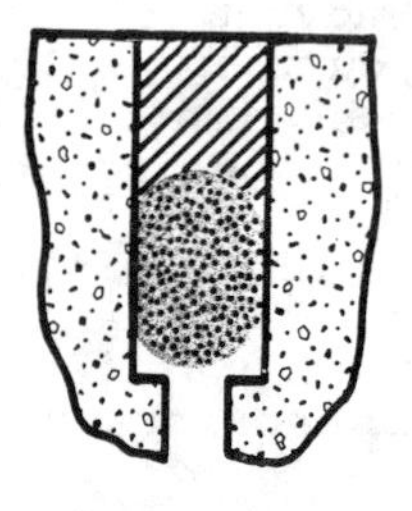

No backer rod; bead too thick; no support to tool against.

B

No backer rod; sealant bonding to back of joint; joint cut too shallow.

C

FIGURE 3

Figure 3 shows three examples of poor design. In (A) the sealant is not tooled to the proper depth, the sealant bead is too thick and has the wrong width-to-depth ratio. In (B) no backer rod is used, the bead is too thick and there is no support to tool against. In (C) no backer rod has been used, the sealant is bonding to the bottom side of the joint and the joint is cut too shallow.

This is a brief history of the development of backer rod. Prior to 1960, newspapers, twisted paper, stitched cotton or welt cord were used as sealant backing to prevent excessive use of sealants.

In 1960, closed cell polyethylene foam was developed. It was developed because architects and industry leaders recognized the need for backer rod to limit the amount of applied sealant; to assist the sealant in forming the proper shape factor; and to act as a bondbreaker to prevent third side adhesion of the sealant to the bottom of the substrate. Closed cell polyethylene foam backer rod answered the needs of the architects in the 1960's.

Closed cell backer rod became the standard of the industry because unlike alternate backer rods, it did not absorb water, it was clean, uniform, resilient and it was easy to use.

During this time, there were certain advancements made by the sealant formulators with the development of low modulus, slow curing sealants which created problems that were associated with closed cell backer rod. During the installation of the backer rod, often times it was punctured or ruptured by the applicator.

The low modulus sealant took up to 21 days to cure, and during that time the substrates would cycle, pumping the air from within the ruptured cells of the backer rod into the soft, uncured sealant, producing bubbles in the sealant. This is an undesirable phenomena known as outgassing of backer rod. It created a need for a backer rod that did not outgas when ruptured. Because of this need, open cell backer rod was developed in the 1970's.

At that time, it was believed to be impossible to make open cell foam on a continuous direct extrusion line.

Open cell urethane foam is a nongassing backer rod, however the drawback was that this sponge-like material would absorb or wick water and/or retain moisture. This is undesirable because moisture on the substrate severely affects the adhesion properties of sealants, and moisture in a back-up might cause water sensitive sealants to soften and fail.

For almost 20 years, a need existed for a backer rod that would combine the best of both open and closed cell backer rods. A need existed for a new product that was non-absorbing, nongassing when ruptured, easily conformable to irregular joints, and affordable.

During extrusion trials in the first quarter of 1988, experiments were run with polyolefin blends in order to develop a higher resiliency backing material which was suitable for use in a joint with slow curing, low modulus, self-leveling sealants. One of the requirements for this new backing material was better compression recovery than traditional closed cell polyethylene foams.

While working with polyolefin blends to increase conformability and compression recovery, it was discovered that a dimensionally stable, open cell/closed cell foam structure backer rod could be extruded with the added benefit of non-outgassing when ruptured.

Upon further investigation and after numerous, production and end user trials, it was confirmed that this newly discovered foam fulfilled many of the needs for a new generation of backer materials. The compression deflection of the new rod is less than 50% of traditional closed cell backer rods, and compression recovery about 12% higher allowing one size to be used over a wide joint width range. In fact, another desirable property is that it easily conforms to uneven and irregular joint openings or varying joint widths.

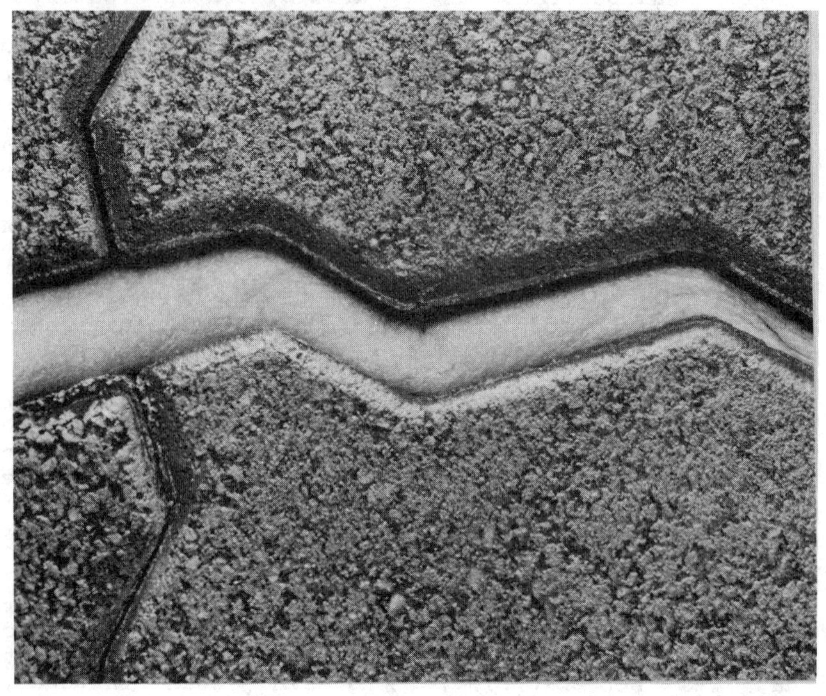

FIGURE 4

In Figure 4 we see one size of open/closed cell backer rod that is accommodating more than 100% variance in the joint width. Because one size easily conforms to these extremes, only half the number of sizes of open/closed cell backer rod are required for the industry. This translates into substantial reduction of inventories; a benefit to both distributor and applicator.

The higher compression recoveries than closed cell backer rod allows the rod to provide better conformability to joint movement than closed cell rod. See Figure 4

Compression Recovery Testing

The scope of this test is to measure the original outside diameter and then compress the rod 50% holding it in compression for 16 hours. After that time, the rod is released from compression and the diameter of the compressed dimension is measured at several intervals. These diameters are recorded at the end of 1 minute, 15 minutes, 30 minutes, 1 hour, 2 hours, and 24 hours. These data points give us a curve on the percent of recovery. See Figure 5

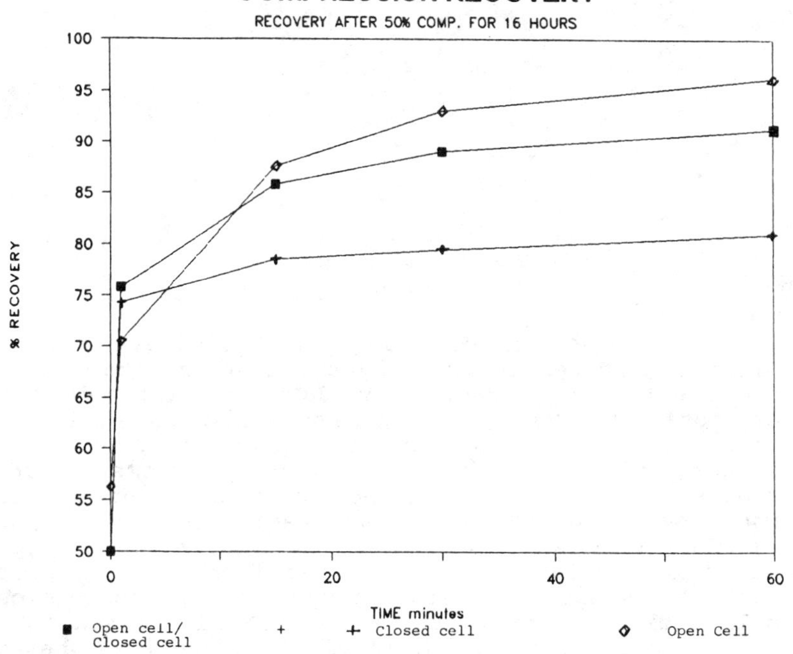

FIGURE 5

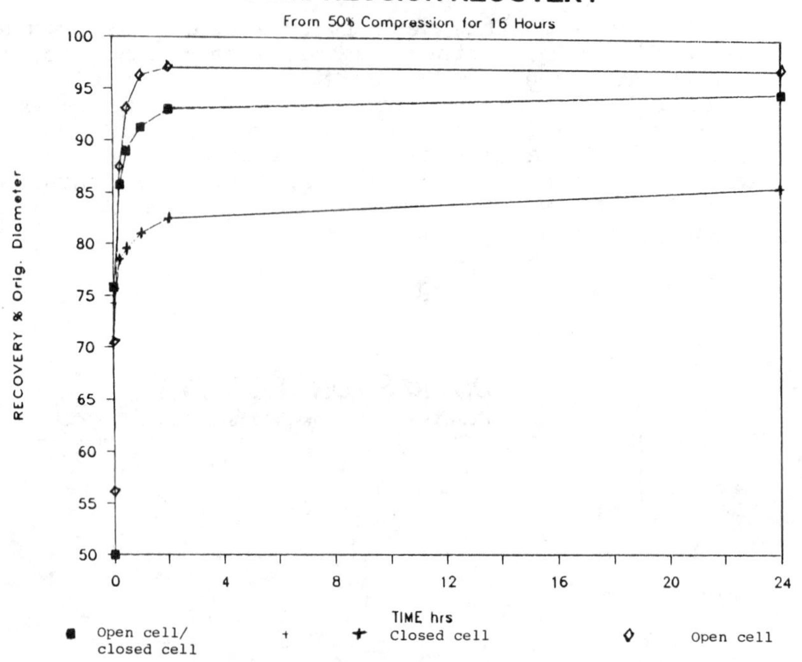

FIGURE 6

The curves in Figures 5 & 6 indicate that the first 30 minutes is probably the most important to the end use of the product. The data show that about 90% of total recovery is attained after the first 30 minutes. The significance of high resilience in a backer rod is applicable to joint movement during sealant cure. For example, on a horizontal joint, where self-leveling or pourable sealants are used, the backing material has to dam the bottom of the joint reservoir, preventing sink holes from forming in the sealant. Typically, slow recovery of compressed closed cell backing material can allow the liquid sealant to flow around it. This may result in back-side adhesion, improper shape factor, and shorter service life of the seal.

After numerous outgassing tests, it has been shown that when tested by the proposed ASTM method, open/closed cell rod does not outgas when punctured.

Some time ago, ASTM proposed an Outgassing Test Method for determining outgassing for joint filler materials used as back-up for sealants in building joints. The scope of this method is a procedure for determining outgassing of joint fillers if they are ruptured during or after their installation and the rupture occurred before the sealant attains a cure. The summary of the method is to take a 12-inch sample of joint filler material and place it between non-porous substrates in a compressed condition. This simulates the placing of backing material on a job site situation. The backing material is then purposely ruptured at evenly spaced intervals along the top of the backing to a depth of 1/2 of its diameter. A sealant is applied immediately after rupturing and then tooled to make a typical joint as outlined in ASTM C-962. After tooling, the sample is placed in an air circulating oven at 50°C for an hour. It is then removed and compressed 12-1/2% and in the compressed state, returned to the oven. After another hour in the oven, the sample is again removed and compressed another 12-1/2% and again returned to the oven where it remains for another two hours at 50°C. The sample is then removed from the oven and the sealant is allowed to finish curing at room temperature still in the compressed state. The sealant is then removed from the joint and the backside, the side against the backer rod, examined and then slit along its entire length, cutting the sealant in half from top to bottom. The size of the voids are then estimated. The significance of this test method is that a typical joint is formed with the punctured backer rod. The joint is compressed and heated to simulate the conditions of the sunny side of the building. After this, the sealant is examined for sizes of voids present, if any. This data test method should serve as an indicator of potential sealing problems that could occur if a backing material is flawed in getting to the job site or abused in the installation process. The data could also serve to warn the user that with some backing material, caution or special precautions should be used in the installation. It is important to note that this is just one test method which is an attempt to create a bubble which can be associated with rupturing of the backing material. It is not the only cause of bubbles in sealants and should not be construed that way. When performing this test method, we could repeatedly and routinely, on a predetermined basis, create bubbles in soft, uncured sealant when closed cell backer rods were ruptured. We could not, however, produce bubbles when the same type method was used to rupture open cell urethane backing materials. We could not produce bubbles in open cell/closed cell backer rod either.

The test apparatus consists of a 7" x 12" base plate (A) (See Figure 7). On this is set a 1½"x12" metal angle bolted to the base (B). Next to this is set a 1½"x12" metal angle (C). A similar set-up is done on the other side of the base plate so that the result is a 12" simulated joint (D) formed between the 2 metal channels held in place by angle irons. It's in this joint that the backing material (E) and the sealant (F) will be placed. On one side the bolt from the angle iron to the base is in a slotted hole (G) so the channel can be moved to accommodate different size materials. The vertical part of one length of angle iron has 2 threaded holes about 3" in from each end and 7/8" up from the base. 3/8" stove bolts are put in threaded holes (H). The turning of these bolts will provide the compression of the joint required to do the test. See Figure 7 for end view.

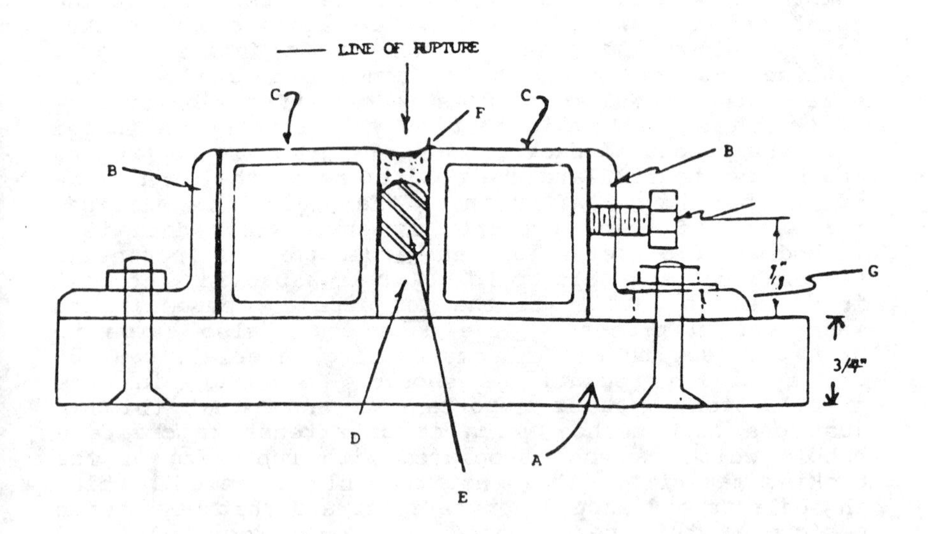

FIGURE 7
END VIEW OF OUTGASSING TEST FIXTURE
A MOVABLE VICE TO SIMULATE JOINT MOVEMENT

Open/closed cell backer rod answers the needs that the industry has been looking for in the past 20 years which are non-absorbing, nongassing when ruptured, easily compressible, fits joint irregularities, and above all, economical.

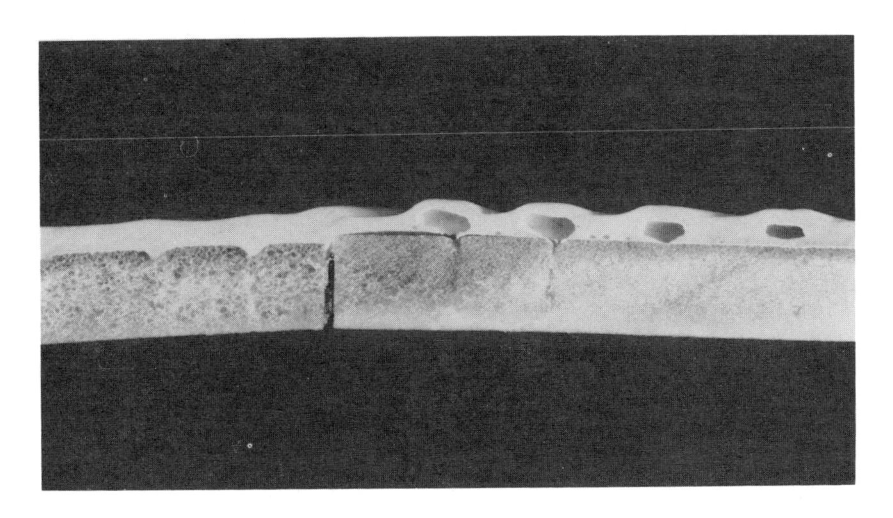

FIGURE 8

On the right in Figure 8 is closed cell backer rod. On the left is open/closed cell backer rod. Backer rod was placed in the joint reservoir and purposely punctured at one inch intervals to simulate field experiences. A low modulus, slow curing sealant was immediately gunned over the ruptured rod, tooled, and placed in an oven per the proposed ASTM test method for outgassing. After the sealant was fully cured, the sample was dissected to observe for outgassing bubbles.

Figure 8 clearly demonstrates that open/closed cell backer rod on the left does not outgas when ruptured while closed cell backer rod on the right has caused undesirable bubbles in the sealant.

REFERENCES

[1] Building Seals & Sealants, ASTM STP606, Back-up Materials, page 325

Sealant Joint Performance

Per G. Burström

EUROPEAN EXPERIENCES OF SEALANTS IN SERVICE, CORRELATION TO RESULTS FROM LABORATORY TESTS

REFERENCE: Burstrom, P. G., "**European Experiences of Sealants in Service, Correlation to Results from Laboratory Tests,**" Building Sealants: Materials, Properties, and Performance, ASTM STP 1069, Thomas F. O'Connor, editor, American Society for Testing and Materials, Philadephia, 1990.

ABSTRACT: Various examinations of sealant joints show that there are many problems with leakage, even in fairly new buildings. Two main reasons have been that the sealant properties are overrated and that the joint design or workmanship is poor due to lack of knowledge about the basic requirements and the material properties.

The failure mechanism of an elastic sealant is described. This serves as a basis for a suggested engineering approach for designing sealant joints. A methodology is described by which aging and durability properties of a sealant can be evaluated. Examples of test results for various sealants are given.

Today there are sealants on the European market with excellent durability and high movement capability. The conditions are therefore good for a service life of at least 25 years for many selants. The problem is to get these sealants designed and used in the proper way.

KEYWORDS: Joint, sealant, elastic, movement capability, durability, aging, tensile stress.

The building technology for high-rise buildings changed considerably in Europe during the fifties. Non-bearing concrete panels began to be used in the facades. Curtain walls of aluminium and glass became more and more common in buildings with steel frame systems. The combination of various materials and the use of large building elements concentrated the problems on the joints.

From the beginning various kinds of oil based materials were

Dr. Burström is senior lecturer at the Division of Building Materials, Lund Technological University, P.O.Box 118, 221 00 Lund, Sweden.

used in the joints. In 1955 the first polysulfide based sealant was used in Europe. Since then butyl-, acrylic-, polyurethane- and siliconebased materials have entered the market.

RESULTS FROM EXAMINATIONS OF JOINTS

A couple of examinations have been made e.g. in Germany and Sweden concerning the status of the joints in facades, se also (1).

In a German investigation published in 1976 more than 2.6 million meters of facade and window joints were examined. The frequency of damage in the facade joints was 31 % during 1958-1965 and 11 % during 1970-1975. The sealants used were polysulfides, butyls, silicones, polyurethanes and acrylics.

The main reasons for the damage to buildings from 1958-1965 were divided almost equally between design failures, material failures and application failures. In buildings from 1970-1975 it was estimated to be about 20 % design failure, 16 % material failure and 64 % application failure.

From inspection of joints in Sweden it was estimated in 1981 that there was an acute need of restoration in about 2.1 million of meters of failed joints in houses built mainly during the sixties.

Another Swedish investigation published in 1986 and comprising 608 000 flats showed facade joint problems in about 10 % of all flats. 25 % of these problems were in houses built before 1959, 48 % in houses built during 1960-1973 and 27 % in houses built after 1974. All these problems were not due to sealant failures, but also included gasket problems and problems with open drained joints.

According to the referred examinations it is reasonable to believe that there are a lot of problems with sealed joints not only in Germany and Sweden but also in other European countries. There are evidently also many problems in fairly new buildings.

WHY DO SEALANT JOINTS FAIL?

Introduction

There can be a lot of reasons for sealant failure. However, there are two main reasons: 1. The properties of the sealant used are overrated or not fully known. 2. The joint design or the workmanship is poor due to lack of knowledge about the basic requirements and the material properties.

The test methods and specifications for sealants have changed a lot since the first attempts were made late in the fifties. A detailed treatment of sealant specifications is given in (2). It is evident that the movement capability of many sealants was overrated earlier. Practical experiences and results from laboratory testing have therefore gradually lowered the allowable movement capability for the best sealants from ±25 % to totally 25 % in many standards in Europe today.

However, some of these older sealants functioned surprisingly well over a long period of time. One reason for this could be that the joint movements did not actually reach these high levels. In many joints between concrete panels the temperature dependent movements were counteracted by the moisture dependent movements, thus reducing the actual movements.

The knowledge about the aging and durability properties of sealants has increased greatly during the years. But still, no test methods exist which can give us all the answers concerning the long-term performance of a sealant. Therefore, many sealants have failed due to unexpected poor aging properties.

The lack of knowledge as a reason for sealant failure, was earlier and still is a very common cause of damages in Europe. The background for this is very significant and described in (3): "Sealing is a minor item when the cost of a building is considered. It has none of the glamor or impact of structural concrete, structural steel, ... Thus, the importance of sealing is often overlooked, and the subject rarely appears in the curricula of architectural and engineering schools".

Failure mechanism of an elastic sealant

Previously it was very common to present the elongation at break as one of the most important properties of a sealant. A high value, of say 500 %, was a guarantee of good function in the joint. Still such advertising figures are published. However, today we know that even such a sealant can fail in a joint where the movements are only 25 %. How is this possible? An example, taken from (1), could give an explanation.

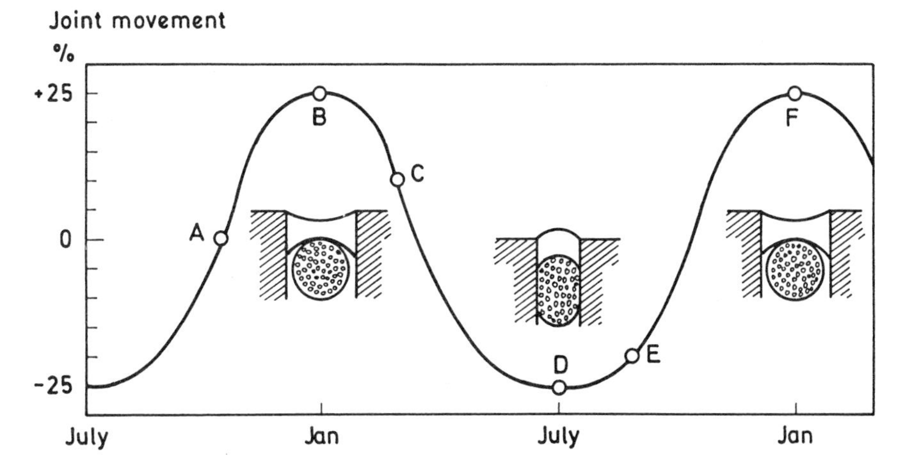

FIG. 1 Example of yearly dependent joint movement. Cf. Fig. 2a and b.

Look at Fig. 1. A sealant is applied in the autumn, point A. To simplify matters, this point is supposed to be where equal movements occur (25 %). During a decrease in temperature the joint opens to point B where the sealant is elongated 25 % (calculated according to the original sealant width = 20 mm). The joint width at this point is consequently 25 mm.

The "stress-strain-relaxation-curve" that has been formed during the joint movement A-B is concerned can be seen in Fig. 2a. The vertical broken line at point B shows the amount of stress relaxation that takes place during the total time when the sealant is elongated.

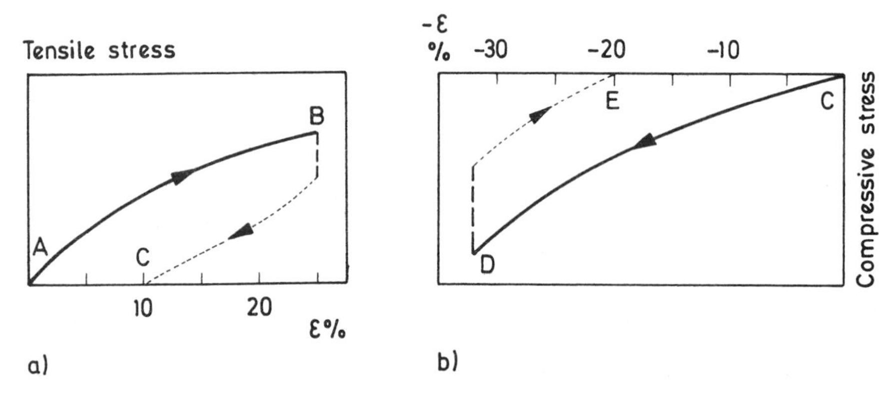

FIG. 2 a) Tensile stresses induced in a sealant and remaining deformations according to the example in Fig. 1.
b) Compressive stresses induced in a sealant and remaining deformations according to the example in Fig. 1.

During an increase in temperature the joint starts to close and the sealant starts to become compressed. At a certain level, in this example when the deformation is 10 %, zero stress in the sealant is reached (the dotted line in Fig. 2a). Thus, a new sealant width is formed (= 20 + 2 = 22 mm). This happens at point C in Figs. 1 and 2a.

The joint continues to close and reaches its narrowest value at point D where the original joint width is compressed 25 %. However, the new sealant width will now be compressed 2 + 5 = 7 mm and the percentage compression is consequently 7/22 · 100 % = 32 %. The corresponding "stress-strain-relaxation-curve" can be seen in Fig. 2b.

During the following opening of the joint the sealant is allowed to elongate, but zero stress is reached at 20 % compression in our example (point E in Figs.1 and 2b). Thus, the new sealant width is now 22 - 20·22/100 = 17.6 mm. Therefore, the elongation up to point F in Fig. 1 is = (20 - 17.6) + 5 = 7.4 mm. Now, the sealant is forced to accommodate an elongation of 7.4/17.6·100 % = 42 %.

This example shows that, in spite of the nominal percentage of deformation being ±25 % calculated from the original width, the real deformations are always increasing. This induces both higher cohesive and adhesive stresses and therefore the joint will fail.

The amount of stress relaxation causes problems. This amount is time-, temperature- and stress-dependent. By reducing the allowable joint movements to e.g. ±12.5 % the stresses induced in the sealant would be reduced and consequently the amount of stress relaxation in the sealant. This might have prevented a failure.

AN ENGINEERING APPROACH FOR DESIGNING SEALANT JOINTS

A structural engineer using e.g. steel or concrete must design the structure in order to ensure that the allowed stresses in the materials are not exceeded.

A similar approach should be used when designing a joint for sealants. The loads in the structural design, correspond in the joint to the performance requirements, including the ability to allow movements. The allowed stresses during the structural calculations correspond in the joint design to the allowed deformations in the sealants but also to the stresses arising in the sealants at a certain deformation. In order to make this possible we must

* produce relevant data for sealants, e.g. on movement capability, stress levels at certain deformations and aging properties

* present these data in a way and with the use of a vocabulary which the architect and the engineer is familiar with.

DETERMINATION OF AGING AND DURABILITY PROPERTIES OF SEALANTS

Introduction

New sealants are constantly being introduced on the market. Are they better or worse, for a given application, than the materials already available? Two major questions must be answered: 1. How is the durability of the sealant against weathering, i.e. heat, water, UV-light and so on? 2. What is the movement capability of the sealant?

In the laboratory it is possible to examine how certain isolated factors causing aging affects e.g. the mechanical properties. These factors can be combined in various ways in order to accelerate the aging of the material. In practice, however, these factors cooperate in a very complex way at the same time as varying movements affect the sealant. It is therefore recommended that outdoor-tests should be carried out parallell to the ordinary laboratory tests.

In (1) the equipment is described where sealant specimens can be installed outdoors and at the same time be exposed to the weather and to the movements induced by the variations in temperature and moisture content. The movements are constantly measured and the specimens are examined regularly. Research on similar equipment has been reported from e.g. Canada, New Zealand and West-Germany.

The results from the laboratory tests and the outdoor tests should then be compared to the experiences which are obtained from examining the sealants in real buildings. For future evaluations of new sealants it is possible to obtain very useful and reliable design values within one year with the described principle.

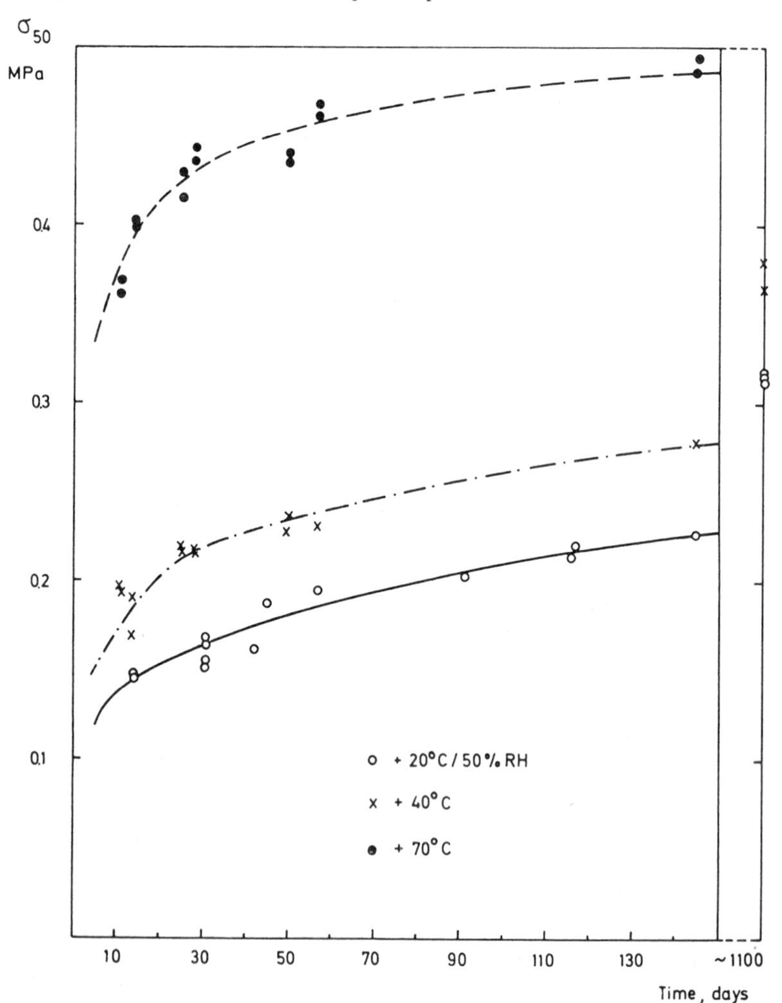

FIG. 3 Tensile stress at 50 % elongation for a 1-component polyurethane sealant as a function of storing time at different temperatures. Initial curing 7 days at +20°C/50 % RH. (1 MPa = 145 psi).

Examples of test results according to the described methodology

Several sealants of various types have been tested by the author according to the above mentioned principles. Three examples are given.

Some laboratory results are shown for a 1-component polyurethane sealant in Fig. 3. The curing rate is rather slow and the curing still continues after about 3 years in a climate of +20°C/50 % RH. If the temperature is increased there is a remarkable increase in the hardness in spite of the relative humidity being very low at the higher temperatures.

During outdoor exposure tests the increasing hardness successively led to growing adhesive failures. In joints in real buildings this sealant showed both adhesive and cohesive failures after some years.

Similar tests for a 2-component polyurethane sealant showed tensile stress values at 50 % elongation of about 0.26 MPa (about 38 psi) after 3 weeks and still after 3 years. This sealant has functioned very well during both outdoor tests and in actual buildings.

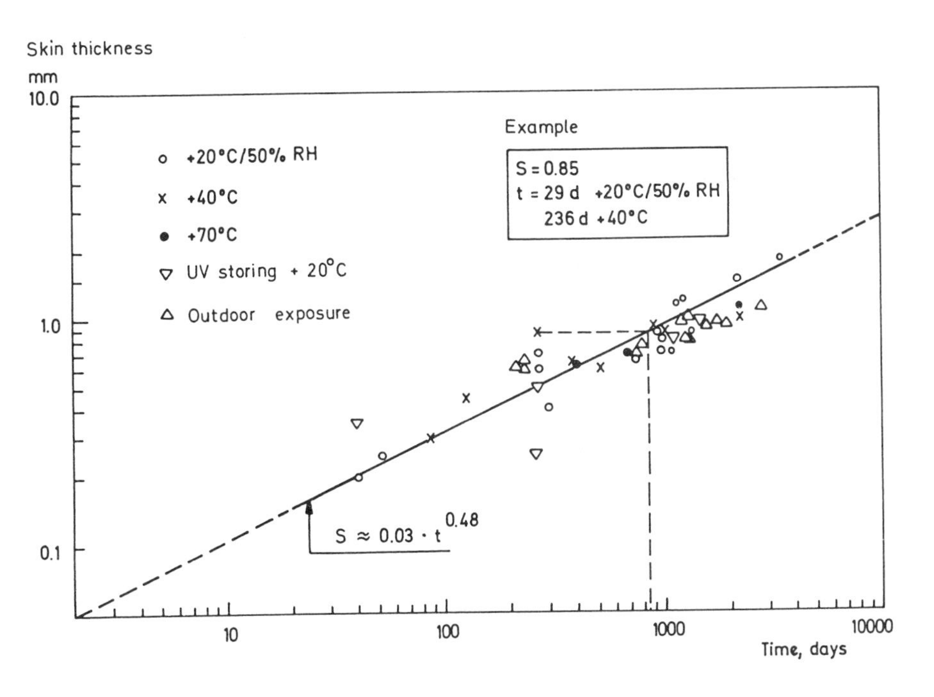

FIG. 4 Surface skin thickness as a function of storing time at +20°C/50 % RH for an oil based sealant. Example of calculating the equivalent time at +20°C for specimens stored at +40°C. The values of material stored at +40 and +70°C are calculated values to equivalent times at +20°C according to the example in the Fig.

Many oil based sealants have been used over the years. The ability of an oil based sealant to function is, apart from the ductility of the skin, also greatly dependent on the fact that the thickness of the surface skin does not increase too quickly. Many of the oil based materials used in practice have shown poor quality.

However, one sealant has functioned surprisingly well during both outdoor tests and in joints in actual buildings. This sealant is characterized by a skin with a high ductility and a low rate of skin formation. The last property is shown in Fig. 4.

Tests have also been going on for several years with a silicone and a solvent acrylic sealant used as glazing materials in wooden windows with very good results. In particular the acrylic sealant has therefore found very wide use for this purpose in northern Europe.

REFERENCES

(1) Burström, P. G., "Aging and Deformation Properties of Building Joint Sealants," Report TVBM-1002, Division of Building Materials, University of Lund, Sweden, 1979.

(2) Panek, J. R., "A Review of Sealant Specifications Throughout the World," in Building Seals and Sealants, ASTM STP 606, Philadelphia, 1976, pp. 134-150.

(3) Klosowski, J. M., Sealants in Construction, Marcel Dekker Inc., New York, 1989.

Mark F. Williams and Barbara Lamp Williams

SEALANT USAGE FOR EXTERIOR INSULATION & FINISH SYSTEMS

REFERENCE: Williams, M.F. and Williams, B.L., "Sealant Usage for Exterior Insulation & Finish Systems", Building Sealants: Materials, Properties, and Performance, ASTM STP 1069, Thomas F. O' Connor, editor, American Society for Testing and Materials, Philadelphia, 1990.

ABSTRACT: Finish coat delamination is a prime reason for sealant joint failure in certain Exterior Insulation & Finish Systems (EIFS). Several factors contribute to the delamination process and subsequent failure at sealant joints. General design, specification and application procedures for sealant joints on EIFS are discussed with specific recommendations for avoiding this common cause of failure.

KEYWORDS: Exterior Insulation & Finish Systems (EIFS), synthetic stucco, delamination, sealant joint failure.

INTRODUCTION

Proper sealant joint design and application are critical for optimal performance of Exterior Insulation & Finish Systems (EIFS). Sealant joint failure on EIFS buildings is common, with both cohesive and adhesive failures observed. One type of failure appears to be uniquely related to the material properties of certain EIFS product formulations as well as application guidelines. Based on numerous investigations of EIFS failure, the authors discuss how sealant joints are adversely affected by finish coat delamination which causes a cohesive failure of the sealant substrate.

Mark F. Williams and Barbara Lamp Williams are president and vice-president, respectively, of Kenney | Williams | Williams, Inc., located at 945 Tennis Avenue, Maple Glen, PA 19002.

OVERVIEW OF EIFS

Components of EIFS

EIFS refers to exterior cladding composed of rigid insulation and wet applied finishes. Generically, EIFS consists of a lamina and insulation board. For purposes of general orientation, a brief discussion of EIFS and its related component, the building substrate, is provided.

Lamina: The lamina is a composite layer consisting of three parts: finish coat, base coat and reinforcement. The finish coat—the system's outermost surface—provides color and texture to the building. The base coat—the foundation for the finish coat—holds the system reinforcement and, most importantly, serves as the primary barrier against water penetration of the wall assembly. Finally, EIFS reinforcement is typically external--fiberglass mesh or metal lath--either of which may be combined with internal fiber reinforcement mixed in the base compound.

Rigid insulation board: Rigid insulation board provides thermal resistance. Because the insulation isolates the lamina from the substrate, the potential for system cracking is reduced. Typical kinds of insulation used in EIFS include molded expanded polystyrene (MEPS), extruded expanded polystyrene (XEPS), semi-rigid fiberglass, polyisocyanurate board and mineral/rock wool. The actual type and density of insulation used vary by individual manufacturer.

Substrate: Although not a component of EIFS as such, the substrate is nonetheless an integral part of EIFS claddings. It represents the innermost point of contact between the system and building. Gypsum board substrate with steel studs is used most often in U.S. installations. Masonry, concrete and cementitious board represent other substrates. EIFS is attached to the substrate by adhesives, mechanical fasteners, or a combination of both. For adhesively attached systems, base coat compound is typically used to adhere the insulation board to the substrate.

Types of EIFS

The Exterior Insulation Manufacturers Association (EIMA) currently recognizes two classes of EIFS: polymer-based (PB) and polymer-modified (PM). According to the EIMA classification, PB systems utilize a 100 percent polymer-based finish coat. By contrast, PM systems utilize a polymer-modified cementitious finish coat. The EIMA classification does not account for the "hybrid" product lines offered by some manufacturers; however, it does provide a useful framework for understanding the basic types of EIFS available.

PB systems: PB systems are the most frequently installed type of EIFS in the U.S. today. PB systems are sometimes referred to as "thin coat," "soft coat" or "flexible" systems. Although proprietary products vary, the most common PB system configuration includes a polymer-based finish coat, a base coat of portland cement and polymer, fiberglass mesh reinforcement and MEPS insulation board. Aside from their use of a polymer-based finish coat, PB systems typically specify a "thin" base coat, roughly 1.6 to 2.4 mm (1/16 to 3/32 in.); include one or two layers of external fiberglass mesh; and use adhesive attachment.

PM systems: The less frequently specified PM systems incorporate a "cementitious" finish coat. In general, PM coatings have a higher cement content which results in a thicker system; hence, PM systems are sometimes termed "thick coat," "hard coat" or "rigid" systems. Commonly, PM systems rely on mechanically attached fiberglass mesh or metal lath reinforcement. Some systems use internal chopped fiber reinforcement in combination with external reinforcement. XEPS board is the typical insulation used. Finally, PM systems utilize accessories at sealant joints, items seldom specified in PB systems. Typically, accessories are composed of plastic or metal, and include corner reinforcements and control/expansion joints.

In sum, PB systems rely to a greater extent on polymers, incorporate thinner laminas and typically use adhesive attachment. These combined factors heighten concern about the integrity of PB systems, especially if sealant joints fail. In contrast, PM systems have thicker laminas of higher cementitious content, are mechanically attached and use accessories. For these reasons, finish coat delamination and subsequent sealant joint failure appear to be less prevalent in these systems. Therefore, this paper focuses on the system of greatest concern with respect to finish coat delamination and sealant joint failure—PB systems.

SEALANT JOINT FAILURE ON EIFS PB SYSTEMS

Materials and Properties

EIFS-clad structures, like all buildings, are subject to the vicissitudes of weather and thermal movement. Sealant joints are needed to accommodate and control the effects of these factors on the building. Moisture tightness is absolutely essential for EIFS claddings. Moisture within EIFS layers causes system components to deteriorate, which can jeopardize wall integrity and necessitate costly repairs. Sealant joints play a critical preventive role in this regard.

Currently, the general industry practice for PB systems is to seal to the finish coat; thus, the PB finish coat serves as the sealant substrate. Although individual manufacturers have developed their own proprietary finish coat formulas, and product variations do exist, PB finish coats share similar characteristics because of their polymer base.

One characteristic of most finish coats is their tendency to soften when exposed to moisture for extended periods of time. Finish coat softening leads to delamination such that the finish coat pulls away from the lamina composite, specifically at sealant joints. It appears that the bond between the finish coat and the sealant is stronger than the bond between the finish and base coats. Cohesive failure of EIFS and subsequent sealant joint failure are the result.

Performance: Field and Laboratory Findings

We have repeatedly conducted 30-minute water tests in the field whereby 15.14 to 18.93 liters/min (4 to 5 gal/min) of water is sprayed from a 16 mm (5/8 in.) garden hose approximately 1.83 to 2.44 m (6 to 8 ft) away from PB wall surfaces. After a 30-minute exposure, the PB finish coat has typically softened to a point where it is easily scraped from the base coat with a blunt object or fingernail.

Microscopic examination of PB finish coats lends further insight into the softening process observed in field testing. Taken from a sample installed to manufacturer specification, Fig. 1 shows the cross-section of a PB specimen magnified X20. The porosity of the finish coat contrasts with the relatively non-porous appearance of the underlying base coat. Fig. 2 shows a face view of the same finish coat specimen at X30. The voids seen on the surface function as "conduits," hollow formations capable of carrying unwanted moisture into the EIFS finish coat, which brings about softening and delamination. Microscopic inspection helps explain why the primary barrier against moisture in most EIFS systems is the base coat.

Field observations and microscopic inspections are supported by laboratory findings. A laboratory investigation was undertaken in order to observe the effect of moisture on PB sealant joints. A PB system sample containing a properly installed sealant joint with recommended seal-to-finish-coat details was obtained from a leading EIFS manufacturer. The sealant joint incorporated a multicomponent polyurethane sealant and a closed-cell jacketed back-up rod. Sample specimens measured (L x W x D) approximately 368.3 mm x 50.8 mm x 57.15 mm (14 1/2 in. x 2 in. x 2 1/4 in.). The procedure utilized a test and a control specimen as follows:

1. Using digital calipers, the sealant joint width of each specimen was measured. The initial width measurement for the test and control speci-

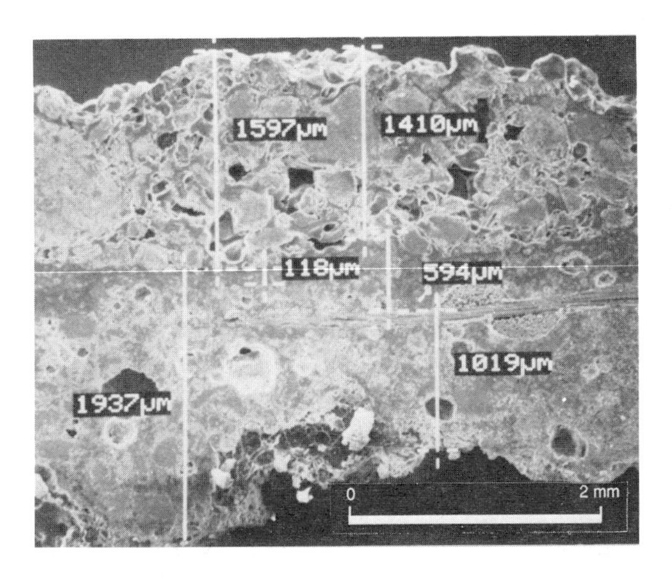

FIG. 1 — Cross-section of PB lamina at X20.

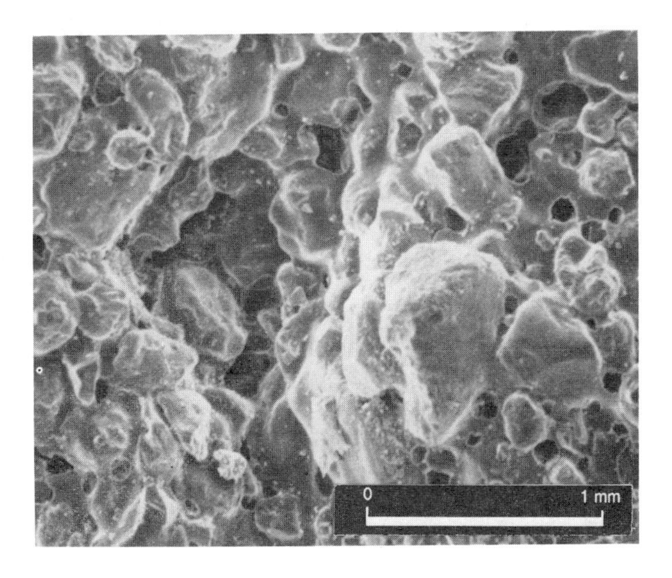

FIG. 2 — Face view of PB finish coat at X30.

mens was 22.22 mm (7/8 in.). The width-to-depth ratio of each joint was approximately 2 to 1.

2. Each specimen was subjected to tensile force such that both sides of the joint detail were grasped and pulled outward until the manufacturer's maximum extension of 25 percent was attained. Specimen joints were then permitted to return to their original width.

3. The face of the test specimen was then exposed to moist conditions for approximately 3 hours; specifically, a moist paper towel was placed on one side of the joint at the juncture of lamina and sealant. The control specimen was maintained in dry conditions.

4. Step 2 (application of tensile force) was repeated, followed by a repetition of Step 1 (joint measurement). Joint failure on the test specimen obviated measurement.

The control specimen remained fully intact with the sealant joint adhered. By contrast, the sealant joint of the test specimen evidenced failure. It is important to emphasize that only minimal tensile force was applied to the test specimen before a loss of attachment due to finish coat delamination occurred (Fig. 3). A decisive break was visible between the PB finish coat and base coat at the moistened side of the joint. Close visual examination of the test specimen revealed that the finish coat aggregate, while remaining bonded to the sealant, had delaminated from the EIFS base coat (Fig. 4).

It is conceivable that the initial application of tensile force (Step 2) may have weakened joint bonding to a degree not detected by visual inspection. To address this possibility, a second test specimen was subjected to the procedure with Step 2 omitted. The results were the same: unmistakable failure at the moistened juncture of sealant and EIFS due to finish coat delamination.

Moisture applied to the finish coat of certain PB systems which use seal-to-finish-coat details causes cohesive failure of EIFS at sealant joints. Failure occurs regardless of tensile stress conditions. On the basis of field, microscopic and laboratory findings, several conclusions can be made:

1. The polymers of certain PB finish coats will soften in extended conditions of moisture.

2. The voids of PB finish coats are capable of conducting/holding moisture.

3. The finish coat softening and moisture retention characteristics of certain PB systems cause sealant joints to be particularly vulnerable to failure.

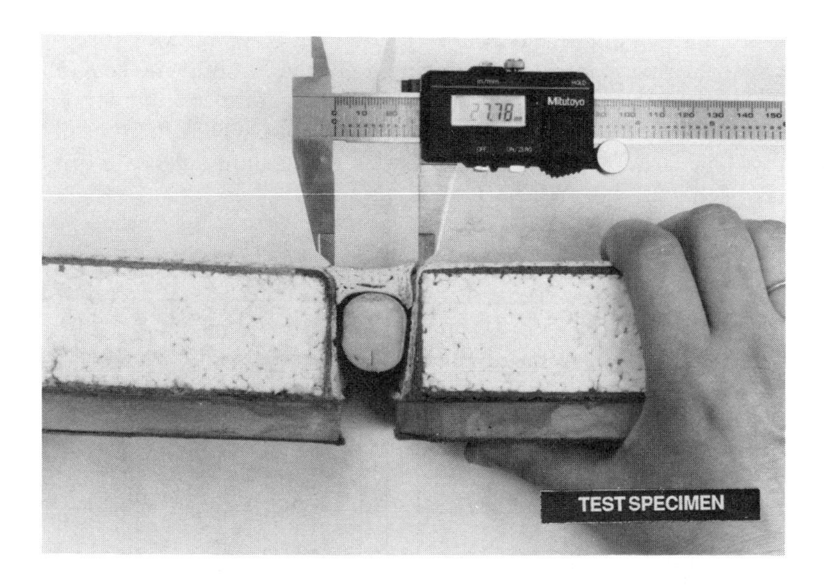

FIG. 3 — Loss of attachment due to finish coat delamination.

FIG. 4 — Cohesive EIFS failure: Finish coat delaminates from the base coat while maintaining bond with sealant.

It is important to understand how the foregoing material properties cause the failure of PB sealant joints in the field. When the finish coat is carried from the EIFS face into the joint interface, moisture is able to enter the finish coat at an oblique angle and penetrate the joint (Fig. 5). Moisture from the lamina surface percolates through finish coat voids into the sealant cavity and causes finish coat softening, delamination and sealant joint failure. Moisture is less able to penetrate the base coat which is typically more dense. Hence, the use of the base coat as the sealant substrate diminishes the risk of this kind of failure (Fig. 6).

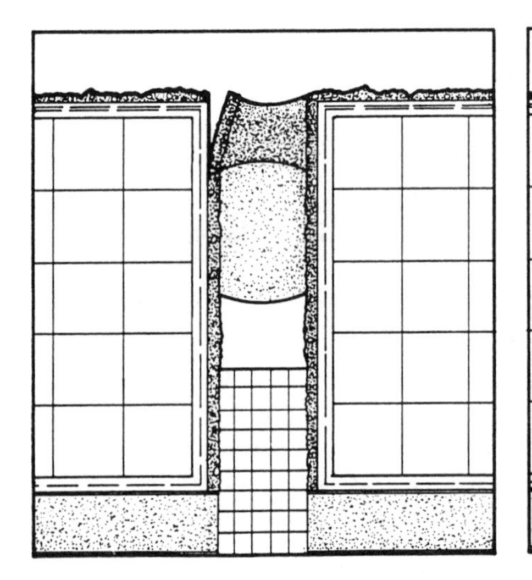

FIG. 5 — PB sealant joint with seal-to-finish-coat detail.

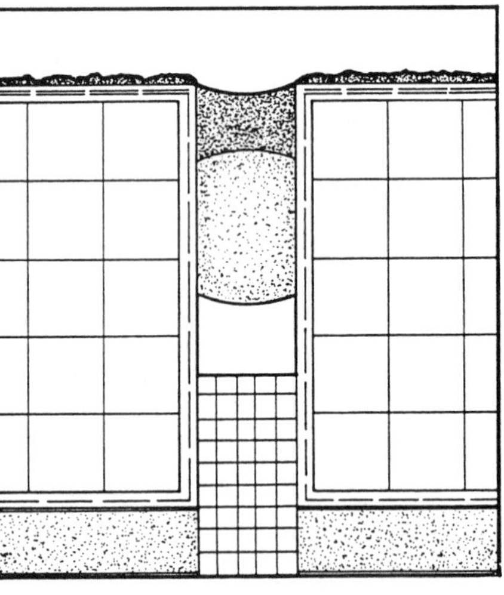

FIG. 6 — PB sealant joint with seal-to-base-coat detail.

PREVENTIVE MEASURES

General Considerations

The process of finish coat delamination and sealant joint failure is aggravated by conditions related to the design, specification and application of EIFS and sealant joints. Although these conditions are many, a few are especially important and merit brief discussion:

1. coefficient of thermal linear expansion
2. finish coat aggregate and texture
3. back-up rods
4. backwrapping
5. quality workmanship

Coefficient of thermal linear expansion: Thermal movement is commonly recognized as a major cause of joint distress. Standard sealant guidelines for joint design are intended to accommodate thermal movement as well as other forces affecting joint integrity. For example, it is recommended that the minimum size of the joint be four times the anticipated movement.

Based on numerous investigations of sealant joint failure, the authors believe that joint movement on EIFS structures is often greater than anticipated. One means of more accurately assessing the potential for thermal movement is the coefficient of thermal linear expansion. Unfortunately, coefficients are seldom available for EIFS claddings—and for good reason. Given the variational differences in EIFS materials, related building components and construction dimensions, no single coefficient can adequately represent EIFS as a cladding type. The provision of coefficients by the manufacturers of EIFS wall assemblies would help assess actual performance requirements and lead to better joint design.

One reason joint movement often exceeds design expectation is color selection. Color has a direct effect on thermal joint movement. Light and heat are reflected by pale colors and absorbed by dark ones. Dark color absorption increases thermally induced movement which, in turn, strains improperly designed sealant joints. Sealant joint design should take into consideration the reflection value of the selected color. The higher the reflectivity, the less a particular color is affected by thermal factors. If dark colors are used, it may be necessary to limit panel size, increase the number of joints or enlarge joint width.

Finish coat aggregate and texture: The aggregate composition and texture of the finish coat can accelerate delamination and sealant joint failure. Aggregate composition has a direct bearing on finish coat porosity and compaction, both of which affect moisture intrusion and retention. Texture behaves in a similar manner: it creates voids capable of retaining moisture.

Aggregate composition is a factor of particle size and gradation as reflected in the granulometric curve (proportion and distribution of different-sized aggregate particles). Some EIFS finish coats are "gap graded"—predominantly large and small aggregate particles; others are continuously graded with a complete range of large-medium-small particles. "Gap graded" materials do not compact well, resulting in an open finish coat with voids. In addition, the tooling of sealant against a "gap graded" material is more difficult.

To minimize the effects of aggregate gradation and texture on sealant joints, a few rules of thumb apply. First, continuously graded materials are preferable to "gap graded" materials. Second, the more removed a texture is from a smooth, sand-like finish, the greater the likelihood of finish coat voids. Finally, coarser aggregates and textures are more difficult to tool against.

Back-up rods: Back-up rods, correctly specified and installed, are critical to good sealant joints. Joint depth on EIFS should not exceed joint width, with a maximum depth of 12.8 mm (1/2 in.) stipulated. A width-to-depth ratio of 2 to 1 is generally recommended by sealant manufacturers. To control the depth of joints, back-up rods are used. Back-up rods vary in material composition. Opened-cell rods retain moisture, whereas closed-cell rods are less likely to hold moisture. Water vapor migrating from the interior to the exterior of the wall may condense and be retained in opened-cell rods. In addition, moisture may enter the system from exterior sources due to a failure of system details, failure of sealant, or impact damage to the wall itself. In order to minimize the effects of unwanted moisture, closed-cell rods should be used. Care should be taken during installation to avoid puncturing the jacket of closed-cell rods since this leads to "outgassing" and results in air bubbles in the sealant material.

Backwrapping: PB systems are typically "backwrapped" at all exposed edges of the insulation board; that is, the lamina is returned from the system face, over the edge and to the back of the insulation. An exception to this backwrapping directive is EIFS installations which utilize accessories. In this case, the lamina terminates at the edge of the accessory.

Backwrapping helps prevent moisture from entering EIFS layers by creating a continuous lamina. Mesh must be fully embedded; neither mesh nor insulation should be exposed at the joint interface. "Hooping" or buckling of mesh at this interface is especially problematic. Buckled mesh conducts water and potentiates finish coat softening by keeping finish coat surfaces moist. It also diminishes the integrity of the system because a partially embedded mesh represents a weakened lamina.

Quality workmanship: All aspects of EIFS installation demand skilled and careful workmanship. Field applicators must inspect and prepare surfaces to insure clean, dry and sound sealant substrates, free of loose particles and contaminants. Although workmanship impacts virtually all aspects of EIFS installation, two aspects are particularly relevant to sealant joint failure: mixing of sealant and tooling.

Both PB and PM systems use some form of polyurethane sealant, an elastomeric material capable of maintaining a good seal when completely cured. A multicomponent sealant consisting of a base compound, a curing agent and a coloring compound is generally specified. Multicomponent sealants are mixed at the site just prior to application. If recommended mixing proportions are not followed, the sealant may be chemically incompatible with the lamina or insulation board and cause deterioration. Careful attention must also be given to the complete dispersion of curing and coloring agents in the base compound; however, overmixing must be avoided.

Joints should be neatly tooled to compact the sealant and eliminate air pockets or voids. Tooling also provides a smooth finished appearance to sealant joints. Numerous EIFS failures, poorly tooled or lacking tooling altogether, have been investigated by the authors. These failures underscore the importance of tooling as well as quality workmanship in achieving proper adhesion between sealant and substrate.

Specific Recommendations

In addition to the foregoing general considerations, specific recommendations can be given with the intention of preventing sealant joint failure as a result of finish coat delamination. Based on consistent field and materials laboratory findings, it is our opinion that certain PB finish coats are not suitable substrates to receive sealant, even when a recommended surface primer is used. Specifically, finish coats with "gap-graded" aggregates, as opposed to continuously graded aggregates, may prove unsuitable. Accordingly, we suggest that substrates other than the finish coat be used and specifically recommend the following:

1. seal to base coat
2. seal to accessories
3. install double seals

Seal to base coat: One option is to use the base coat as the sealant substrate (Fig. 6). With this option, the finish coat is carried to the edge of the sealant joint on the building surface and stopped. In contrast with standard practice, the finish coat should not be applied to the joint interface. Standard application and tooling of the sealant material to the base coat then follows. By omitting the finish coat from the joint interface, moisture conducted through surface voids in the finish coat will be less likely to enter joints.

Seal to accessories: Use of accessories represents a second detailing option. Again, the sealant material is applied to a substrate other than the finish coat. In some cases, the use of accessories may preclude the need for sealant altogether. Currently, few U.S. PB manufacturers incorporate accessories into their systems. By contrast, European EIFS systems have long relied on accessories for system detailing. We have studied the European use of accessories and, given the success of this approach, anticipate increased U.S. interest in their use.

Install double seals: Regardless of which option is used—seal to base coat or seal to accessories—a double seal with vent is recommended. The double seal is "added" insurance against moisture intrusion, and the vent discharges moisture.

Several manufacturers of PB systems already recommend the use of double seals on sealant joints. The installation specifications of these manufacturers vary, but typically a secondary seal is placed in the wall construction, separate from the EIFS cladding; the primary seal is detailed to the EIFS finish coat. Since the finish coat still serves as the substrate for one of the seals, the procedure as currently recommended by certain manufacturers does not preclude finish coat delamination and primary sealant failure. At best, this particular approach to double sealing recognizes that should the primary seal fail, the secondary seal will keep moisture from the substrate.

In contrast, we recommend a double seal approach wherein a primary seal to the base coat or accessory is backed up by a secondary seal. If possible, the secondary seal should be placed within the EIFS component layers. However, the actual detailing of any double seal requires a thorough assessment of specific project conditions and construction allowances.

CONCLUSION

Careful attention must be given to sealant design and application on EIFS claddings. There is a need for the EIFS industry, sealant manufacturers and design professionals to work together for a better understanding of how materials, properties and performance contribute to the success and failure of sealant joints on EIFS.

Karen L. Warseck

WHY CONSTRUCTION SEALANTS FAIL -- AN OVERVIEW

REFERENCE: Warseck, K. L., **"Why Construction Sealants Fail--An Overview,"** Building Sealants: Materials, Properties, and Performance, ASTM STP 1069, Thomas F. O'Connor, editor, American Society for Testing and Materials, Philadephia, 1990. [1]

Abstract: This article is intended as an overview of why construction sealants fail in the "real world". These reasons can be categorized by design errors, application errors and materials failures. Design failures are those which occur from the work of the design professional, such as improper joint sizing. Application failures are those that occur due to problems during installation, such as lack of surface preparation. Materials failures result not only from problems with the sealant itself, but also because of the substrates to which they are applied.

Keywords: Sealants, joint design, sealant selection, tooling, substrate failures

When faced with the complexity of the construction industry, one sometimes forgets the basic tenets of keeping building joints watertight. The failure of construction sealants is generally one or more of the following:

1. Failure of the sealant to adhere to the substrate.
2. Tearing apart of the sealant itself.
3. Discoloration of the sealant or substrate.
4. Hardening of the sealant.
5. Craze cracking.

Why these happen can be summarized in a few words -- lack of attention to detail. Too often, since the sealants are a small percentage of the work, they are perfunctorily specified, easily substituted and haphazardly applied. To create a successful joint requires meticulous joint design, selection of appropriate sealant materials and painstaking

Karen L. Warseck is President of Building Diagnostics Associates, 1816 Sherman Street, Hollywood, Florida 33020.

application. In the "real world", however, even the best design and the best attempts at proper installation can be undone by job site conditions and uncontrollable events.

Joint Design and Building Movement

The building is going to move. The movement may be caused by expansion and contraction of the building materials due to heat absorption or loss, moisture absorption or loss, by dynamic forces such as wind, or even vibration of mechanical equipment or movement of live loads. Too often, this movement may be forgotten by design professionals in the attempt to provide the owner with an aesthetically pleasing building. Since movement or control joints are not especially handsome, they tend to be placed as far apart as possible. These joints are detailed excessively wide to make up for the lack of sufficient number. Another common mistake is to detail them in sufficient number but require the size to be as small as possible. Both are invitations to failure.

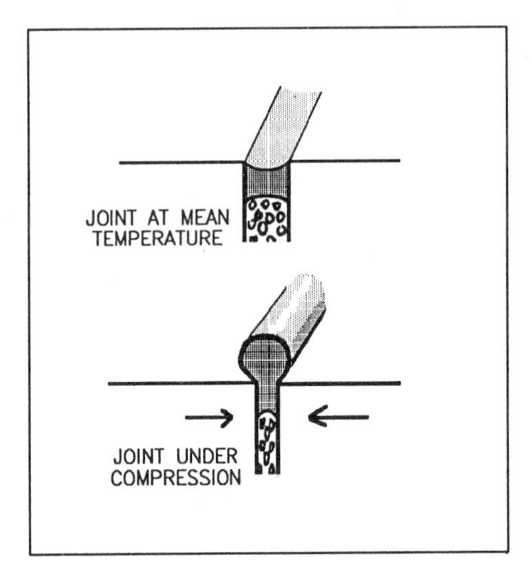

Figure 1 - Extruded sealant.

If the joints that are detailed are spaced too far from each other or are too small or too few, the building will simply create its own. Most often this will manifest itself as cracking in the exterior walls, but the fact that the joints are incorrectly sized or located may also be revealed by bending or bowing out of the walls, crushing at the joints, or shearing of curtain wall fasteners or masonry ties. If the joints are too narrow (less than 1/4 inch (.6 cm)), the expansion of the substrate can cause the joints to close too much, extruding the sealant. (Figure 1) When the building contracts, the extruded sealant is no longer in the joint and leaks result. If a joint is too wide (generally, greater than about one inch (2.5 cm), but it depends on the type of sealant -- some can go as much as two inches (5.0 cm)), the sealant may sag out of it. In addition, a wide joint requires a deep sealant bead to avoid cohesive failure. The problem is, the deeper the bead, the less able it is to stretch.

Like a thick rubber band, a bead that is too deep for the width of the joint limits the sealant's ability to stretch and retract with the movement of the substrate. Because of this, the forces attempting to stretch the thick bead will cause undue stress at the bond line, and cause adhesive failure. To keep this from happening, a general rule in joint design is that the depth of the sealant should be no more than 1/2 the joint width, and never more than 1:1 at mean temperature, generally to a maximum depth of 1/2 inch (1.3 cm). (Figure 2)

Sealant Selection

The most common design failure in sealant selection is that the chosen sealant lacks sufficient movement capacity for its intended use. Part of the problem lies in the manufacturers' imprecise descriptions of their products. One of the most common sealant product descriptions is the word "performance". Most manufacturers literature uses "performance" without definition, to the confusion of anyone attempting to evaluate the differences and similarities of the multitude of sealant products and formulations available.

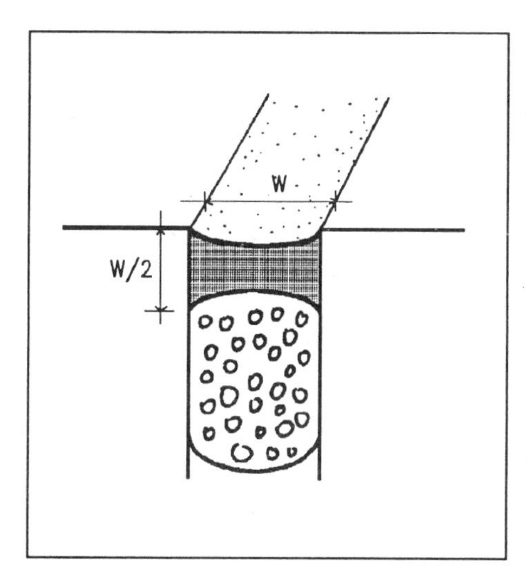

Figure 2 - Proper Depth to Width Ratio.

"Performance", as used in the literature, relates to the amount of expected joint movement and the capacity of the sealant to elongate and recover and is divided into high, medium and low. The range of movement capability fitting the description "high", "medium" or "low" performance varies widely from one manufacturer to another. The manufacturers' literature has given "low" performance sealants top joint movement capability as high as plus or minus ten, or as low as plus or minus five percent of the joint width. Medium performance sealants have been stated as being able to reach a high of anywhere between 5% to 12.5% movement, depending on whose sales literature it is. High performance will take movement in joints greater than 12.5%, but some manufacturers claim a sealant must move 25% of the joint width to be termed "high performance".

Where this becomes extremely important is, that unless the elastic characteristics of a sealant are defined in a comparable manner from manufacturer to manufacturer, the prospects of good joint design are remote, indeed. The design professional must read the literature very carefully in order to be able to make intelligent comparisons. A sealant that will not take the anticipated movement will either pull away from the substrate or cohesively fail.

During the design of the building, incompatible sealants may be placed in close proximity to each other without realization by the designer that there will be a problem. As an example, a silicone sealant may be properly specified as the weather seal around aluminum windows. However, if the aluminum windows are to be installed in a building that incorporates an exterior insulation and finish system (EIFS), the silicone sealant may come in contact with the polyurethane sealant used to seal the joints of the EIFS where the window sill or head crosses a vertical EIFS joint. This juxtaposition may not be noticeable during the design stage, but can lead to problems later on.

Another problem that is almost always ignored during the design phase is access to the joint once the building has been erected. This is especially true in double sealed joints. In some cases, there isn't any. In other cases, double sealed joints are only accessible from the exterior and the interior joint is impossible to reach from the outside.

Recovery and Incompatibility

Recovery characteristics are also an important consideration in sealant selection. Some sealants can stretch but once subjected to tensile stresses, do not easily return to their original shape (stress relaxation). (Figure 3) Others will remain bulged out after being compressed and do not easily stretch out again (compression set). (Figure 4) In each case, the sealant has developed a memory which will eventually cause failure. A stress relaxed sealant will assume a distorted shape when the joint closes. When the joint reopens, the sealant will not stretch as it should from its distorted shape and fail. A sealant with a compression set will adhesively fail due to the tensile stress developed as the joint opens.

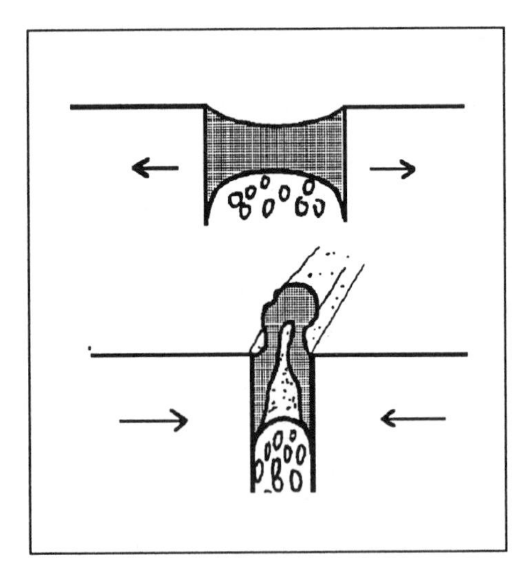

Figure 3 - Stress Relaxed Sealant.

Incompatibility between the sealant and substrate or sealant and the rest of the assembly of which it is a part can also be the direct cause of sealant failure. This is especially true in rehabilitation projects. Too often, the specifications for the original installation are no longer available, and there is no way to determine the composition of the existing sealant. Thus, a specified sealant may not be compatible with the existing one. Total removal is then required. However, it is almost impossible to remove every trace of an existing sealant, and no matter how good the contractor, there will almost always be conditions where the inaccessibility of the joint precludes proper cleaning of the surface.

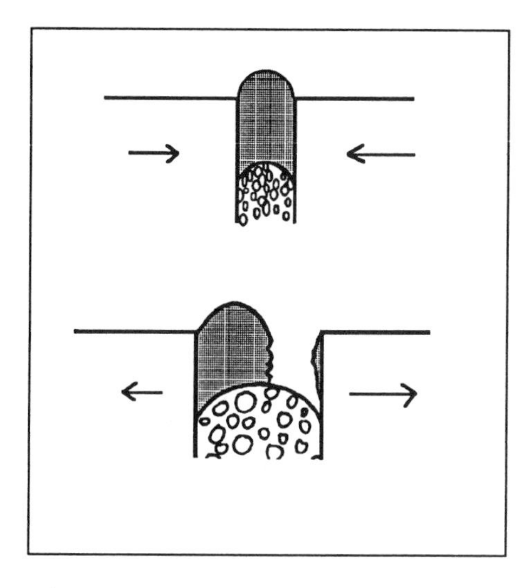

Figure 4 - Sealant with a compression set.

Even sealants of the same generic type can be incompatible with each other. For example, it is widely known that acetoxy silicone sealants are not compatible with some multi-component silicone sealants used in the manufacture of insulating glass units. Installing these two sealants in close proximity has been shown to cause failure of the structural seal of the unit. Other symptoms of incompatibility problems include staining or etching of the substrate and disintegration, discoloration or hardening of the sealant.

Primers must also be chosen for their compatibility with both the sealant and the substrate. Adhesion problems will occur if the primer is not specified and applied when required.

Problems due to improper sealant selection are also caused by failure of the specifier to consider the environment in which the sealant is expected to perform. Some of the considerations are:

1) Resistance to puncture and vandalism
2) Resistance to ultraviolet rays
3) Chemical resistance
4) Abrasion resistance
5) Exposure to extreme weathering
6) Difficulty of access for application and repairs
7) Possibility of continued submersion
8) Dirt pickup due to slow cure or dust attraction

A sealant installed where people can reach it will almost always be poked, prodded and picked apart. A sealant that will do well in the northeast may have severe problems in south Florida where sunlight is bright and ultraviolet is strong. In fact, polyurethane sealants in south Florida have failed due to ultraviolet degradation in as little as five years. These types of environmental factors are often overlooked. Failure to take these factors into consideration can result in adhesive failure, cohesive failure, craze cracking, hardening, disintegration, color changes or other types of premature failure.

Sealant Backings and Bondbreaker Tape

Sealant backings are used to form the depth to width ratio, act as a bondbreaker and also provide a firm surface against which tooling can be accomplished. Failure to specify an appropriate bondbreaker, whether sealant backing or tape, will result in adhesion of the sealant on more than two sides of the joint. A joint with three sided adhesion will fail cohesively, adhesively or both. (Figure 5)

Sealant backings, popularly known as backer rods, are of two types, open cell polyurethane and closed cell polyethylene. Each type has specific areas where it should or should not be specified. For example, open cell sealant backings should not be used where moisture absorption into the sealant backing can be a problem, including horizontal joints and submerged joints. Bondbreaker tape should be used only where there is a firm bottom surface and when the joint is so shallow that a sealant backing will not fit. Since sealant backings are held in place by compression, it is important that the sealant backing selected is about 30% greater than the maximum expected joint opening. If the sealant backing is too small, it will move farther back into the joint when the sealant is pressed on during tooling. As a result, the depth to width ratio of the joint will be compromised and the sealant bead will be too thick to freely move.

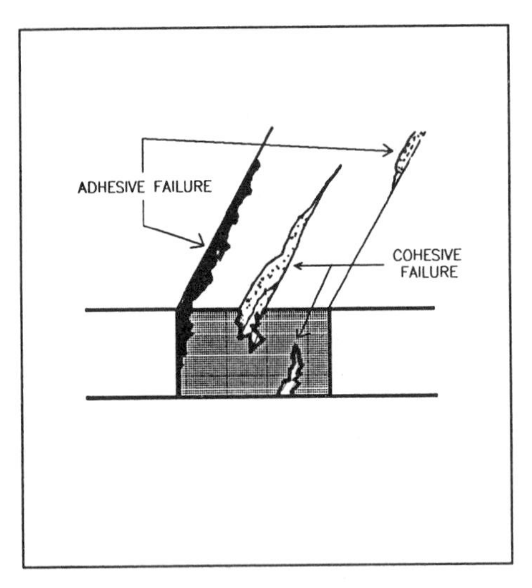

Figure 5 - Three sided adhesion.

Paints

In order to provide a monolithic looking surface, often designers will specify painting, allowing the paint to be applied directly over the joints. Caulks, meant merely to be used in joints with little or no movement, can usually be painted with no ill effects. Elastomeric sealants, however, are used for the specific reason that they are able to and supposed to stretch and relax. Most paints are not formulated for nor intended to take the kind and amount of movement to which sealants are subjected. When joints are painted over, the paint may adhere to the sealant. When the joint moves, the adhered paint will crack, causing stresses or even tearing in the sealant.

Installation Problems

Even when the joints have been correctly sized and placed by the design professional and the sealant selection has been the proper one for the anticipated movement, substrate and climatic conditions, there is no guarantee that the sealant and related construction can or will be installed correctly.

All manufacturers state in their recommended application procedures that the substrate must be clean and dry for the sealant to properly adhere. The most common of all adhesion problems begin when this basic rule is violated and the substrate is improperly or insufficiently cleaned. Some of these cleaning deficiencies include not cleaning at all, using contaminated or dirty solvent, using the wrong solvent for that particular sealant and/or substrate, allowing contaminated solvent to dry on the substrate, using contaminated rags or brushes or using a rag containing lint.

Unfortunately, all too often, the cleaning of the joint in preparation for sealing cannot be complete due to the location of the joint. This is often the case in resealing conditions. If the joint is accessible only from the interior, and needs to be sealed at the exterior edge, cleaning, and indeed sealing, will be next to impossible. Then, too, solvent wiping prior to sealing may also be complicated by weather conditions. If the ambient temperature is warm, it may be well within range of application for the sealant. However, due to the warmth, the cleaning solvent may dry before it can be wiped off. This can result in contaminants being left or redeposited on the substrate. In either case, the result can be inferior adhesion.

The second most prevalent adhesion problem is caused by the improper use of primers. Some of the most common problems are not using the primer at all, using too much primer, using the wrong primer for the specified sealant, using the wrong primer for the substrate, or not allowing the primer to dry completely before applying the sealant.

The weather during sealant application is an extremely important factor rarely considered. Applying organic sealants at too low a temperature will cause an increase in viscosity. Thus the sealant may be difficult to apply without gaps and voids and may be too thick to properly tool. In addition, cold air is usually less humid, which can retard the cure of certain sealants. Also, when the substrate has contracted due to cold temperatures, the joint is wide open. If the sealant is applied then to a narrow joint, as the substrate expands and the joint closes, the sealant may be squeezed out of the joint entirely. Too high a temperature can cause the sealant to sag or even flow out of the joint. If the sealant sags, the bead formed is of an uneven thickness, causing differential stresses in the bead and eventual failure. High temperatures can also cause a premature skinning over of the sealant bead causing craze cracking.

Premature skinning over of the sealant will also slow the release of solvents in solvent releasing sealants. This can cause blistering in the bead.

When the sealant is applied at high temperatures, it is applied to a joint at its narrowest dimension. When the temperature drops and the substrate contracts, the joint will open to its largest dimension. Because the sealant was applied at a high temperature, the bead of sealant used in the joint will be too small to stretch across the open joint. Thus, it may fail cohesively or be pulled away from the substrate. If there is a large variation in temperature before the sealant has cured, it may cause adhesive failure and/or craze cracking in the partially cured sealant.

Tooling

Tooling is needed to compress the sealant, push it against the sealant backing, assure contact with both sides of the joint, and provide the preferred hourglass shape. By tooling the joint, the air pockets that may have been created during gunning are eliminated. If air pockets remain in the sealant, they may expand during hot weather and rupture the sealant. Lack of proper tooling will result in a deformed bead that will not have the ability to easily stretch and may rupture. (Figure 6) Another result may be that the sealant bead does not develop sufficient bond area and, when stretched, pulls away from the substrate. Unless the sealant is to be installed against bondbreaker tape, with a firm surface behind it, tooling requires proper placement of the sealant backing in order to be effective. This is not always possible. Even in the best constructed buildings, the joint width may vary between the top and bottom of the building. As a result, the sealant backing will be tight in some areas and loose in others. Tooling will cause the sealant backing to move, changing the depth of the sealant bead from one place to another. If the sealant backing moves, the sealant cannot be forced against it and the bead does not form enough bond area to avoid adhesion failure.

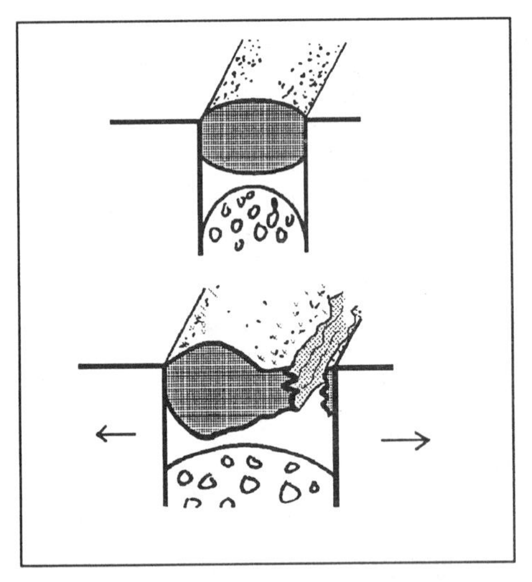

Figure 6 - Improper tooling of sealant bead.

Variations in the size of the joint will also affect the sizing of bondbreaker tapes. As the joint varies in width, the bondbreaker tape may be too small in some places and too large in others. If the bondbreaker tape is too small, it will allow the sealant to adhere to the back of the joint as well as the sides. The resulting three sided adhesion will cause the sealant to fail cohesively if the adhesion is strong or adhesively if the adhesion is weak, or both. Too large a bondbreaker tape will cause the tape to wrap around the sides of the joint resulting in loss of bond area and adhesive failure. In Figure 7, the bond area depth is shown as "D".

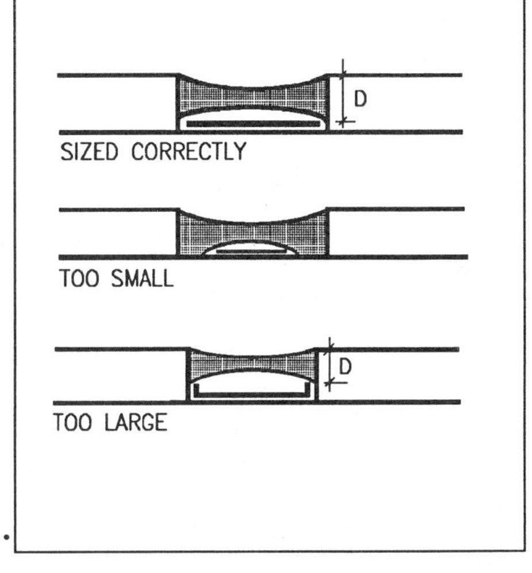

Figure 7 - Bondbreaker sizing problems.

Finally, the edges of joints are not necessarily smooth and without jagged edges. Irregular edges in the joint can result in air pockets being formed during installation due to the sealant missing the indented surfaces. This results in voids between the sealant and substrate if the sealant is not tooled to push the sealant in tightly. Cohesive failure can result if foreign materials such as mortar remains in the joint. The mortar, firmly adhered to the masonry, will cause the sealant bead to be different depths at different locations in the joint. When movement of the substrate causes the sealant to stretch, the uneven size of the sealant bead will cause differential stresses within the bead itself. This may cause the bead to fail at the areas where the greatest stress occurs.

Materials Problems

The most frequent materials problem during application is the unauthorized substitution of the specified sealant. Many sealants look alike, but they do not work alike.

In the field, the most common materials problems result from the improper preparation of two and three part sealants. One trade association involved with exterior finishes recommends the use of multipart sealants in the field to assure quality control. This is not in touch with reality. Although manufacturers attempt to minimize mix proportion failures by preparing containers to be mixed one to one, too many times inexperienced mechanics or laborers are used to mix product for application. Since these people are not experienced with the use of multipart sealants, spillage of one component is ignored and not compensated for, mixtures are whipped together, or too much is mixed at one time. Spillage can result in incorrect mix ratios, with the result being that the sealant does not cure properly. Mixing the sealant at too high of a rate of speed can introduce air into the sealant mixture. Air bubbles can expand during hot weather causing ruptures in the bead. If too much sealant is mixed at one time, the material may begin to cure before it is installed.

Storage at site can also contribute to sealant failure. Storage facilities at job sites is often primitive at best, and material can easily be left out to freeze or overheat.

Substrate Failures

Sometimes the material to which the sealant will eventually be applied can be the source of the problems, if not anticipated and planned for.

Masonry sealers, anti-graffiti compounds and "waterproofing" compounds and coatings vary widely in their composition and formulation. However, their main intent is to provide a surface that will repel dirt, paint, and/or water. They can also have the same effect with the sealant on surfaces treated with those materials. Some sealant manufacturers, recognizing the widespread use of such compounds have pretested the materials to determine compatibility, and the materials have proven themselves. However, these materials sometimes undergo frequent and drastic formulation changes, and can vary widely from batch to batch, so the actual material being applied may differ from that which was tested and approved. Some of the materials and compounds even have a base compound with low solvent resistance that dissolves into a clear, gummy residue upon contact with the sealant or primer, making adhesion impossible.

Site conditions during the installation of the waterproofing coating usually do not compare with those tested in the lab, and make extrapolation from the lab adhesion tests to the job site difficult at best, useless at worst. Even the application thickness can affect the sealant adhesion. A sealer that is applied at a thin rate may not affect the sealant while a thicker coat may not allow the sealant to adhere at all.

Since these materials are clear and difficult to detect, an unsuspecting sealant applicator may not realize that they are there. This will result in failure of the joint if testing has not been done on the actual material in situ. Ideally, the sealant application should be done prior to installation of the coating or waterproofing compound. However, the contractor may not realize that the two materials are incompatible or be aware that scheduling the sealant application first is required. Many times job condi
tions will be such that the sealant contractor is not able to get to the work at the proper time in scheduling and, in an attempt to finish the building on time, the coating applicator does his work first. No matter how careful the applicator, it is extremely difficult to apply such materials

onto the surface without contaminating the joints. The result is that the sealant does not adhere to the substrate and failure results. Form release agents, if applied according to the manufacturer's recommendations, many times do not affect sealant adhesion. However, too thick of an application may leave a brittle film that can flake off. Sealants applied to those areas will come off with the form release agent. Petroleum based release agents may be incompatible with the sealant. Thin, sharp edges of a concrete joint will spall if the stress on the sealant is stronger than the tensile strength of the concrete.

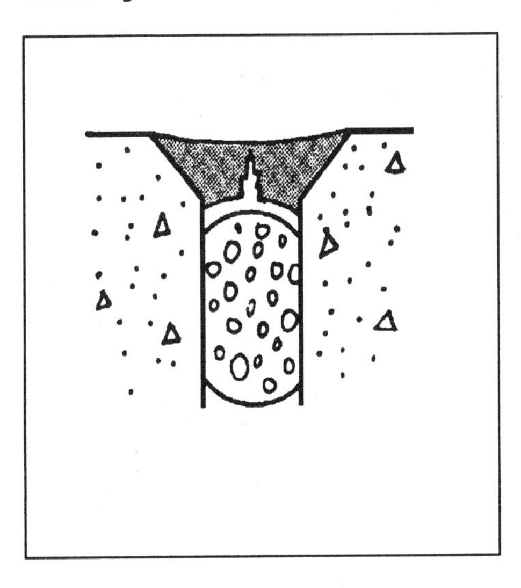

Figure 8 - Sealant bead at neck of chamfer.

One way to minimize this problem is to chamfer the edges of the joint. However, the top of the sealant bead must be kept below the bottom of the chamfer or tearing may occur, since the neck of the bead will attempt to move as much as the face. (Figure 8) In addition, the width of the sealant joint below the chamfer must be sufficient to allow for expected movement.

Concrete should be fully cured before attempting to install sealants to be sure that it is sufficiently cured to avoid adhesion problems on a damp substrate. However, in the real world, the time between pouring the concrete and installation of the sealant is minimized to keep construction projects on schedule. The result is that the sealant is applied to surface dry concrete. Further hydration of the concrete will bring moisture up to the surface, creating a damp situation below the sealant. Since almost no sealants will stick to damp substrates, the adhesion fails. Sealant selection should be done with care for all masonry and concrete substrates. Some sealants will bleed into those porous substrates causing substrate discoloration. Acid cure sealants may chemically react with and etch limestone or marble.

Aluminum coatings can contain mill contaminants, oils, graphite and carbon residues and oxides that act as release agents for elastomeric sealants. Some baked on coatings will also make adhesion difficult. Variations in fabrication and coating makes each batch different, further complicating the process, since even pretesting the aluminum may not be representative of the whole installation.. Galvanized steel has a history of poor sealant adhesion. Due to the sacrificial nature of the galvanizing, the surface will erode, making long term performance of an elastomeric sealant difficult.

Wood, in addition to the joint movement due to swelling and shrinkage, has other special requirements in joint design and sealant selection. Woods that are naturally resistant to water, such as redwood and teak, contain natural oils that may affect the adhesion. Unpainted wood is not a good substrate for any sealant. Since the wood will absorb moisture, adhesion problems will eventually result. If the wood is painted, the paint that is chosen must be compatible with the sealant, and firmly adhered to the wood. If the paint is not fully adhered, the sealant will pull the paint completely off the wood, causing leakage.

Conclusion

All members of the building team have a part to play in minimizing sealant failures. The design professional must understand the dynamics of building movement and the characteristics of sealants that make them suitable for particular applications. Care should be taken to be sure that the specified sealant will be compatible with the substrate and all other materials with which it will come in contact and that it will be an appropriate choice for adverse environmental conditions. The design professional should also specify and insist on sealant adhesion testing in the field under real life conditions.

The contractor must understand and enforce proper mixing, storage and application procedures, especially those relating to cleaning and priming the substrate. He/she must also be aware of scheduling conflicts and avoid them as best as possible.

The sealant manufacturer must represent his product in such a way as to make it understandable to those who must specify and install it. This includes not only precise definition of terms, but also making known to the specifier when to use as well as when not to use the product. Furthermore, the manufacturer must not rely on laboratory results to prove the product's suitability in the field. Once these important lessons are learned, sealant failures, other than those due to natural weathering and aging, should be minimized.

[1] Warseck, K. L., "Why Sealant Joints Fail," Architecture, Dec. 1986, pp. 91-95. The author has adopted and updated her previous publication for ASTM presentation and publication.

Present and Future Trends

Julian R. Panek

STATUS OF HIGH-RISE BUILDING SEALANTS IN THE U.S.

REFERENCE: Panek, J. R., **"Status of High-Rise Building Sealants in the U.S.,"** Building Sealants: Materials, Properties, and Performance, ASTM STP 1069, Thomas F. O'Connor, editor, American Society for Testing and Materials, Philadephia, 1990.

ABSTRACT: Polysulfide sealants were the only elastomeric sealants available for the modern curtainwall beginning in the early 1950's. They had rapid growth until the early 60's with sales of 3 million kgs., when solvent acrylic, urethane, and silicone sealants were introduced and broke the monopoly of polysulfides. By 1975 the above four sealants had sales in the range of 3 to 4.5 million kgs. By 1988 silicone sales were 10 million kgs followed by urethane with 6 million kgs. Polysulfide and solvent acrylic sealants totaled 3 million kgs. Butyl sealants and acrylic latex sealants are not really competitive with the above.

KEYWORDS: curtainwall, movement capability, recovery, stopless glazing, construction adhesive sealant

INTRODUCTION

This paper is only concerned with elastomeric type sealants [1 and 2] for use on high-rise buildings. These materials have greater movement capabilities and have the properties to meet existing or developing specifications of ASTM Committee C-24 and the Federal specifications. Sealants for the over-the-counter market are not included but sales are given for this group to compare the size of the market. These materials are generally manufactured to lower quality standards which may be satisfactory for home use but may or may not meet existing standards. These materials are distributed through hardware stores, building supply houses, and our modern "super markets".

Julian R. Panek, Specialized Consulting Services, 1261 Madison Drive, Yardley, Pa. 19067

EARLY HISTORY

Polysulfide sealants were the only elastomeric sealants available for the new modern curtainwall in the early 1950's. Their growth was very rapid until it peaked at about 1975 at which time sales were in the range of 4.6 million kgs. Their movement capability was ± 25%. In the early 60's Rohm and Haas and Union Carbide introduced solvent acrylic polymer solutions at high solids which could be used to formulate solvent acrylic sealants having a more limited movement capability of ± 12.5%. However these sealants had lower price and could be used in areas having limited movement around door and window frames and heavy masonry construction and found a ready market. They became competitive with polysulfide in some areas. In the middle 60's urethane technology based on the availability of diisocyanates in combination with polyols, polyamines, and other chemicals established an entire generic class of polymers and sealants. These were now competitive with polysulfides in performance and broke their monopoly of the building sealant market. From here on it was all downhill for polysulfides. The urethanes had better movement capabilities, were cheaper, had better weather resistance, and could be readily manufactured. They became a good competitor for this market. The silicones in the early 60's had high modulus, were more expensive, and had poor adhesion to masonry and therefore not very popular. In the early 70's the silicones were modified to give medium modulus, and a considerably improved movement capability of ± 50% and were also more competitively priced. This in combination with better weather resistance and adhesion to all surfaces with or without primers made these materials highly desirable. By 1975 polysulfides, urethanes, solvent acrylics and silicones all had sales in the range of 4.5 million kgs. The only deficiency with silicones is that they can stain masonry due to the formation of a silicone oil which occurs during polymerization of the base polymers.

STATISTICS

Table 1 compares the volume sales of the various elastomeric sealants over the years. A comparison of the values shows the rise and fall of polysulfide sealants over the years. The decrease in volume from 1975 was partly due to competition among the polysulfide sealant manufacturers and competition from other generic sealants. The urethanes had excellent growth up to 1975 with a more gradual growth over the years partly at the expense of polysulfide, but also because of their non-staining performance on masonry buildings. Once the medium modulus silicone sealants were introduced in the middle 70's they enjoyed a rapidly growing market which has been expanded to include their use as a construction adhesive for structural glazing applications. It is the only sealant which has essentially the 100% recovery needed for this application which has also seen fantastic growth over the last 15 years. The solvent acrylic sealants even though they have lower movement capability still find use around door and window frame assemblies and some masonry construction because of their lower cost. Their sales over the last 10 years has leveled off at about 2.7 million kgs. and will probably retain this volume in the future. Acrylic latex sealants in the past have been primarily used indoors because of water sensitivity and limited movement. In the last several years with improvement in formulations they have been recommended for outdoor applications. Nevertheless, they do not appear to have all the necessary properties for dynamic joint movement.

Table 1

High-rise building construction sealants

	1965	1975	1980	1988
Polysulfide	3.0	4.5	2.3	0.9
Urethane	1.4	4.0	4.5	5.8
Silicone	0.9	3.1	6.3	9.8
Solvent acrylic	2.7	3.6	2.7	2.2
Acrylic latex	-	0.2	1.1	4.5
Butyl	1.8	4.5	4.5	4.5
Hypalon	-	0.5	0.5	0.5
total	9.8	20.4	21.9	28.2
Over-the-counter	36	54	71	93

Note: all volumes in million kilograms of sealant

MOVEMENT CAPABILITIES

The polysulfide sealants were limited to ± 25% movement capability. The urethane sealants had the same movement capabilities, but one manufacturer claimed +40% and -25%. In 1987 several manufacturers were claiming ± 50% movement in order to be competitive with the medium modulus silicone sealants. The first silicones only had ± 25% capability, but in the middle 70's the newer medium modulus modifications had a ± 50% capability. In 1980 low modulus silicones with +100% and-50% capability were introduced. The solvent acrylic sealants only claimed ± 12.5% movement. The acrylic latex sealants originally claimed ± 10% but in 1987 are claiming + 25% and - 25% capability. Cured butyl sealants have a ± 5% to ± 10% movement.

Hypalon which is a chlorosulfonated polyethylene based sealant also claims ± 12.5%.

SEALANT SPECIFICATIONS

The early sealant specifications are referenced in a separate paper [3]. The first polysulfide sealant specification ASA A116.1 1960 was developed by the Building Trades Specification Committee an independent group of sealant manufacturers who requested ASA now ANSI to sponsor this document. The specification was adopted in 1959 at a General Conference at which time ASTM was elected the Society responsible for all future revisions. This resulted in the formation of ASTM Committee C-24 on Building Seals and Sealants. This specification was adopted by several organizations throughout the world including the Sealants Manufacturers Council in England. This specification served as a model for specifications being developed by the National Bureau of Standards. Arthur Hockman of NBS began developing specifications for sealants to be used on Federal and State buildings and his first specification was issued in 1961. In the following years Hockman issued additional specifications including many revisions and amendments until by 1968 the final specifications were as follows:

TT-S-00227E for multi-part sealants
TT-S-00230C for 1 part sealants
TT-S-001543A for 1 part silicone sealants
TT-S-001657 for butyl sealants

The 1543A specification was very similar to 230C except that among others the test specimen for tensile adhesion was thinner to enable the high modulus silicone sealants at that time to meet this portion of the specification. The above specifications became models for Standards all over the world, and many of their features, particularly the famous Hockman test Cycle for extension-compression were accepted or modified to suit the various climates throughout the world [4]. When C-24 issued its building sealant specification C-920 in 1979, NBS withdrew from further development of sealant specifications and recommended that where applicable C-920 be accepted in place of the Federal specifications. The three Federal Specifications in addition to C-920 covered two classes of movement namely ± 12.5%, and ± 25%. C-24 is working on a modification of the limits in C-920. There is a Canadian Specification for silicones with a ± 40% movement capability. There is no specification for solvent acrylic sealants, but C-24 expects to issue a specification in 1990. In the last 15 years several manufacturers have been referencing 230a which is not apropos for solvent acrylic sealants. ASTM Specification C-834 is written for latex sealants but in the last several years several manufacturers have been claiming to meet part of the performance requirements of C-920. They want some of the limits to be broadened in order to fall under the umbrella of the C-920 specification. C-24 is developing a specification for silicone structural sealants which will be issued in 1990.

PREDICTIONS

In my opinion. the future position sees silicones sales growing. Urethanes will hold in a relatively static position mainly because they do not stain masonry. Polysulfide sealant has little future in the building market. but will be used in marine and in aircraft where it will probably never be replaced. Solvent acrylic sealant sales are slipping but there may still be areas with limited movement where they will be used because of lower price. The acrylic latex sealants are attempting to enter a more competitive market area and much depends on sealant formulations. Butyl sealant is still used to repair neoprene gaskets, and is needed as a mastic for sound adsorption, and as a joint filler between metal interlocking panels and overlapping panels on metal buildings and hidden joints.

REFERENCES

[1] Building Seals and Sealants **ASTM STP 606** Julian R. Panek, editor,American Society for Testing and Materials, Philadelphia, 1976.

[2] Panek, J.R. and Cook J.P., Construction Sealants and Adhesives 2nd edition 1984 John Wiley and Sons Inc.

[3] Panek, J.R. "History of the First Building Sealant Specifications", Symposium on Building Sealants: Materials Properties and Performance, ASTM STP 1069, Thomas F. O'Connor editor, American Society for testing and Materials, Philadelphia, 1990

[4] Beech, J.C. "A Review of test Methods for the Assessment of the Movement Capability of Building Sealants" Publication SP-94 American concrete Institute 1986

Andrew Charles Yanoviak, AIA, CSI

ARCHITECTURAL DESIGN CHALLENGES for SEALANT TECHNOLOGY and DESIGN STANDARDS

REFERENCE: Yanoviak, A. C., **"Architectural Design Challenges for Sealant Technology and Design Standards,"** Building Sealants: Materials, Properties, and Performance, ASTM STP 1069, Thomas F. O'Connor, editor, American Society for Testing and Materials, Philadephia, 1990.

ABSTRACT: A historical overview with specific examples of challenges presented by architects and architectural designers for building owners, contractors, suppliers, manufacturers, and formulators of design standards for seals and sealants and glazing systems. A historical trace of technological responses and product solutions to field challenges and problems. Specific examples of developments in applications technology as it relates to various materials used to seal the building exterior. Reflective hindsights, insights, and foresights on the needs for quality performance and design standards in the construction industry. Brief references to the applicable building codes and standards and their development in the area of building seals and sealants.

An illustrated slide lecture presentation of completed successful works of architecture, and product useage misapplication problems. Creative and innovative design challenges presented by architects making an "expressive statement". Inventive technological responses to these challenges by manufacturers of seals and building sealants and glazing systems. Examples of innovative products produced by research and design teams within industry which have advanced architectural design technology. Certain examples of water leakage problems on highrise construction in Hawaii regarding glazing systems for window walls and curtain walls.

KEYWORDS: highrise, towers, architecture, design, historical, construction, glazing systems, sealants

Andrew Charles Yanoviak, AIA, CSI is an architect-planner consultant with Andrew Charles Yanoviak, AIA; 1188 Bishop Street/Suite 3011, Honolulu, Hawaii 96813

Architects and engineers associated with highrise building design and construction have creative and innovative minds. In addition to challenging themselves, they challenge building owners, developers, designers, technicians, suppliers, manufacturers, testing facilities, and code development and enforcement agencies. They also challenge those of us involved with design and construction industry standards development.

It has been my personal experience over some period of time, that it is next to impossible to keep up or even attempt to get ahead of these avant garde designers: including, not only the architectural and engineering acrobatic "stunt" artists, but also the inventive and innovative industrial designers who keep on coming up with new and advanced sealant products and glazing technology systems.

Even before owners and developers of highrise office towers were looking for unique and distinctive corporate symbols of power, stability and progressiveness with their "signature" buildings, architectural design gymnasts were hard at work contorting and distorting the extruded "box" or "vertical slab" associated with the "Cigarette Carton" vernacular of typical highrise building forms.

Reflecting back on the historical development of the origins and advances in the generation of highrise building forms, it is interesting to keep in mind the simultaneous technological developments which made many of these rather spectacular and marvelous achievements possible.

Once upon a time, there were no rolled structural steel shapes for girders, beams, joist, purlins, and columns. Architects and engineers were attempting to create the very first highrise structures in a manner somewhat similar to the Gothic cathedrals. Like their Gothic counterparts,pre-Modern architects and engineers piled stone upon stone and used metal anchor ties to achieve lateral stability as they designed and constructed multi-story warehouses, industrial buildings, and administrative office structures. A cross-section cut through one of the Gothic cathedral buttresses and the wall-section of these pre-Modern structures were almost identical; in that, they achieved their greater than normal heights at the expense of increasing the thickness of their supporting walls at every lower story. This technique in structural engineering resulted in wall thicknesses of from three to four feet at the base of the structure to five to six feet in massive stone space-consuming "foundations".

Depending on how far back we want to go into the seminal origins and historical developments of form generation in architecture, it would probably suffice to say that we can clearly trace the beginnings of Modern skeletal framework design expressions with infilled facade panels to the ancient Grecian Erecthion temple neighboring the magnificent Doric ordered Parthenon on the Acropolis in Athens. Of course, admittedly this is strictly artistic expression rather than being truly representative of inherent structural principles with designated flows of forces through the skeletal framing members. This is because the vertical columns and the horizontal beams or lintels are made up of small pieces of marble stonework that are quite literally glued together. Furthermore, the infilled panels are also made up of small pieces of trimmed and dressed marble stonework that is load-bearing and part of the overall structural support system along with the pilaster type columns and architraval beams. It is interesting to note however, that the Erecthion also had window openings which are very rare in ancient Greek temples. Whether or not they were glazed is still a mystery.

In direct contrast with the Erecthion, modern pioneer architects around the turn of this century abandoned the Beaux Artes artistic approach to the creation of architecture, an also rejected the insincere and ecclectic Neoclassicism of their contemporaries. Consequently, they adopted the brick masonry and steel structural expression of architect Peter Behrens of Germany where three of the five masters of Modern architecture once apprenticed; i.e., Le Corbusier, Mies van der Rohe and Walter Gropius. From the standpoint of Modern glazing systems and the expression of rather large glazed openings in between vividly expressed and pronounced structural supports, Walter Gropius who founded the Bauhaus school of design brought the school of architecture at Harvard University to world reknowned prominence, is generally credited with this milestone achievement in the chapters of architectural design styles.

As a consequence, the stage was set for Modern architects such as Ludwig Mies van der Rohe to create structural steel skeletal frameworks fully exposed to the natural weather elements, and with non-structural infill panels of brick masonry or glass. A lot of model building codes throughout the United States responded to this new architectural design idiom, and the initiation of new building design standards also followed their development. The Barcelona Pavilion by Mies van der Rohe in 1929 vividly demonstrated the "open space" house plan capabilities along with other works of Le Modular by Le Corbusier, Walter Gropius, Alvar Aalto, and Frank Lloyd Wright. These seminal works of architecture imparted a new sense of freedom and liberal means of expression to both architects and owners as well as manufacturers of building products. At the same time, they brought about whole new sets of challenges and problems especially with regard to water infiltration. There was a time in certain parts of the United States where it was considered somewhat fashionable and actually a status symbol to have a leaking building designed by a famous American architect -- that was prior to the culmination of our current litigious society where law suits for construction defects and deficiencies abound.

To make a long story short, a quick trace through some of the outstanding architectural design works by Mies van der Rohe after World War II, will pave the way for latter day glazing system challenges and solutions. Of notable interest is his Farnsworth House in Illinois which was designed in 1946, and which served as a major source of inspiration for the "Glass House" of his disciple Philip Johnson in New Canaan, Connecticut in 1949. Mies van der Rohe had a unique opportunity to create several of the campus buildings at the Illinois Institute of Technology which ultimately led to the twin Lake Shore Drive apartment towers of 26 stories each. Subsequently, in the mid-50's Johnson and Mies collaborated on the Seagram Building tower in New York City. Many architectural historians view this work as a climatic event similar to the Parthenon with excellent design proportions and the end result of a stylistic trend. However, it was followed by what seemed to be an endless series of highrise boxes giving rise to "curtain wall" design and construction where prefabricated metal and glass replaced the predecessor infilled masonry panels, and "stick systems" of metal window mullions prevailed as the predominant design expression of the times.

Frank Lloyd Wright had a major influence on the "destruction of the box" and his principles and design philosophies have prevailed in penetrating the Mies van der Rohe "straightjacket". Wright was heavily influenced by the structural design principles inherent in the works of the ancient Roman architects and engineers with their great concrete arches and domes. The Roman spans far exceeded the limitations of the "post and lintel" expressions of the ancient Grecian and Egyptian architects. We have a more recent example of the influence of his work and philosophy in Honolulu with the design of the Bank of Hawaii building in the Financial Plaza complex designed by architect Leo S. Wou, FAIA with its concommitant glazing challenges. However his designs in the mid-30's and late 40's for the Johnson Wax administrative offices and laboratory tower in Racine, Wisconsin have also had a tremendous influence on even our Post-Modern architects. Wright's Johnson tower complex incorporated vaulted columns reminiscent of both Gothic and Romanesque spatial expressions and cantilevered structural systems that were quite daring. However, contemporary architects have not yet quite caught up to Wright's soaring cantilevers and vaulted constructions or his inventive use of tubular "bamboo" glass for unique glazing systems.

Frank Lloyd Wright's design for the Price Tower office building in Bartlesville, Oklahoma which is based on hexagonal geometry, has also served as a source of inspiration for our own Executive Center residential and commercial office highrise complex in downtown Honolulu, which was designed by Jo Paul Rognstad, AIA, and on which I had an opportunity to serve as a design consultant. The top of this highrise tower has a pyramidal roof, which is actually the upright corner of an inverted cube, which forms a geometrical shape called a tetrahedron. The original three-dimensional space frame model was designed and constructed by myself with a glazing system similar to the "Glass Cathedral" designed by Philip Johnson, FAIA; however, it was not approved for construction by our local Building Department because a helicopter could not land on its roof to evacuate occupants during a fire.

Before we get into the American Telephone and Telegraph highrise tower by Philip Johnson in New York City, whose "Chippendale China-closet" scrollwork is perceived as the turning point into the Post Modernism expressionistic movement in highrise tower design in the United States, we will very quickly review some of the earlier works which challenged glaziers and glazing manufacturers as well as writers of glazing and sealant design and construction standards.

As can be seen in the Seagram Tower in New York City and in other similar highrise towers, many architects during Modern times have classically designed their towers as a symbolically extruded classical architectural column, complete with "base" and "capital" configurations and styling elements. These expressions of artistic taste have often challenged not only structural engineers, but also glazing designers and technicians as with the World Trade Center in New York City by the late Minoru Yamasaki, FAIA. Hugh Stubbins, FAIA also challenged the constructors of his Citibank tower in the heart of Manhattan with its assymmetrical pyramidal rooftop cap or "capital". With his Transamerica tower in San Francisco, William Pereirra, FAIA stunned the architectural community with his gently sloping obelisk or extruded truncated mastaba form taken from the ancient Egyptians; however, because of the design configuration, the sloping walls performed in a fashion similar to a sloping roof, but the sloping glazing system immediately failed due to improper design integration. Remedial repairs were required to render the overall exterior skin system successful.

The John Hancock tower in Chicago designed in the late 60's by Skidmore, Owings and Merrill is again a sloping highrise of 100 stories in height comprised of both office, residential and commercial use floors. Its volumetric shape very nearly reflects to form of the interior elevator core with more shafts at the base than at the top of the tower, which is also true of the Transamerica tower in San Francisco. Not only its vertical and horizontal structural steel columns and beams are vividly expressed on the exterior facades, but also the diagonal steel wind bracing for lateral stability of the tower itself is boldly and dramatically expressed on the building's exterior. Again, the window openings were uniquely shaped and many glazing system challenges were offered to the construction industry.

It was Philip Johnson who with his partner John Burgee, AIA in the early 70's began the total "destruction of the box" with the multi-cubed I.D.S. Center in Minneapolis. This was not only a lesson in prismatic geometry, but with the multi-faceted mirrored glass reflections, it definitely presented challenges for quality control with regard to both glazing systems and sealants. In the mid-70's, Johnson and Burgee created the striking Pennzoil twin towers in Houston, Texas that were only separated by five feet, and were sculpturally capped with "pitched" sloping roof/wall skylights. In between the two towers at ground level, there is a multi-story commercial lobby space with dramatically designed sloping skylights as a part of a diagonal roof structure with trussed tubular steel supports, that certainly offered many challenges to the designers as well as the construction industry.

Also in Minneapolis in the early 70's, architect Gunnar Birkerts, FAIA created a gently sloping cantenary facade incorporating an upside-down arch for a highrise office building with multiple challenges for the glazing and construction industry. As artists, architects demand a certain graphical treatise and printout in their facades expressing clear design concepts which are often a major technological challenge, especially when movement systems are involved. In certain buildings on the same facade, the design architect will call for the combination of "curtain wall" as well as "window wall" glazing systems, which leads to many technological design challenges. The mixture of suspended, bolted, and fixed or operative vents and lights on the same building facade also offers major technological design challenges. This is especially true in geographical areas where seismic design criteria are required by the building codes and standards, as well as hurricane wind forces with their associated hydrostatic pressures.

As with the Sears tower in Chicago, since most highrise towers are designed as cantilevered beams or slabs, the structural engineering concept involving the "bundle of tubes" theory has a dynamic affect on the movement systems which need to be accommodated by the glazing systems and their anchorage design details and specifications. We are now also in an era where we find examples of highrise reinforced concrete structures sitting perched on top of structural steel highrise bases, with entirely different movement criteria for the glazing systems attributable to both wind and seismic forces which again need to be accommodated in the design details.

In the early 80's when Philip Johnson, FAIA revealed

his "Chippendale" highrise design motif and AT&T as a corporate client relished their signature building as a profound masterpiece of good taste and avant garde design, the tidal flood gates were opened. Therefore, during this past decade we have seen a remarkable assemblage of even some undignified, inelegant and rather garish signature type Post-Modern highrise towers in our major metropolitan urban centers. Perhaps the most striking in this particular category would be the Portland, Oregon office building by Post-Modern pioneer architect Michael Graves. Another might be the Proctor & Gamble General Offices in Cincinnati, Ohio by architects Kohn, Pederson, Fox and Associates. Still another might be the Pittsburgh Plate Glass "PPG Place" in Pennsylvania where Johnson and Burgee have created a multi-faceted undulating mirrored glass edifice with a whole series of pyramidal spires on the exterior facades.

One of the most profound and daring architectural design "stunt" acts of the late 80's is the creation of the Texas Commerce Tower in Dallas by Skidmore, Owings, and Merrill. This 55-story signature edifice has a gaping six story high hole labeled "sky window" by the designers, which begins about a dozen floors down from the top of the office tower. A 9 July 1987 cover story article in the Engineering News Record mentions this Dallas structure as having a "Crazy curtain wall" in that, "The entire 600,000 square feet of curtain wall hasrequired 15 different types of glass in more than 600 different sizes ... in this lively architectural shape." The ENR article further mentions that, "... the curtain wall subcontractor, is busy putting the many varied sizes and shapes of sloped and curved insulated glass into place around the sky window and sky lobby, there are 54 different sizes ... none of these have 90-degree corners ..." One Liberty Center in the city of Philadelphia designed by architect Helmut Jahn, who also designed the new United Airlines skylighted terminal buildings at the O'Hare airport in Chicago, is yet another example of avant garde multi-faceted highrise office tower shapes which continue to challenge the sealant technology and glazing industries. And it is their on-going capacity to fulfill this continuing marketplace need caused by both avant garde architects and engineers and their corporate signature clients, that in reality permits these unique architectural forms to be created and realized.

Ieoh Ming Pei, FAIA, whom many architectural educators consider to be the leading and most eminent world class architect practicing today, has created a very original "Oc-tet" vertical space frame tower composed of a classical geometrical assemblage of octahedrons and tetrahedrons. This 1200 feet tall rotating helical spiral tower for the Bank of China in Hong Kong was designed in conjunction with Leslie E. Robertson, P.E., structural engineer and the construction was completed during 1989. The structural design of the highrise building tower is of composite steel and concrete, and somewhat like the John Hancock building in Chicago, it expresses its diagonal bracing members on the exterior facade. Also, somewhat like the Pennzoil sloping lobby skylight glazing systems between the towers, the new Bank of China building also employs similar design and construction challenges for sealants and the accommodation of movement systems.

Having had firsthand experiences myself with angled and splayed vertical and horizontal and diagonal glazing systems as a design consultant to Warner Gayle Boone, AIA who used to work for Ming Pei, I can very well appreciate the need for three-dimensional models as well as full-size mock-up samples of intersecting joinery of glazing and sealant systems. These are definitely needed to augment the typical design and construction drawings as well as shop drawings when manufacturers and contractors in the field are presented with the atypical challenges in fabricating and erecting these complex and innovative geometrical shapes. In addition to my design consultation experiences on the Waikiki Trade Center, I also had the opportunity to work with Warner Boone on a mixed use commercial and residential project in Singapore where I was able to employ triangular geometries which I had thoroughly researched on a previous project.

Fumihiko Maki, architect and originator of the Metabolist Design Group in Japan, has recently designed and constructed a unique urban helix spiral structure of 12 stories in downtown Tokyo. The exterior front facade along the main street thoroughfare is multi-varied and multi-faceted and again a challenge to design detailers and constructors of glazing and exterior wall systems. The Metabolist Group in Japan some 20 years ago, was a close association of progressive architectural design research architects who proposed avant garde forms for future megastructures with a special emphasis on organic and biomorphological forms related to systems in nature.

Richard Rogers, RIBA, architect was awarded the Royal Gold Medal for Architecture for his 12 story high Lloyd's of London collossal masterpiece of glass and steel, which is a "high technology" design expression that has been forecast to be very much emulated and copied as a prototype of office building design in the near future. Rogers uses trussed tubular steel on the exterior facade in lieu of mullions for the multi-storied atrium space. Some critics consider the Rogers design to be flambouyant but tasteful. Perhaps his creative design work has already inspired other architects who have a strong desire to replicate the Crystal Palace design of Sir Joseph Paxton, London architect in the early 1850's. For example, the August 1989 issue of Architectural Record features an avant garde penthouse suite of offices atop an existing structure by Wolf D. Prix and Helmut Swiczinsky of Coop Himmeblau, architects in Vienna, Austria. Some observers feel that their work borders on

"Deconstructivist Architecture" which is the latest stylistic movement to follow "Post-Modernism". Again, from the standpoint of glazing systems and sealant technology, the generation and construction of these forms to produce quality workmanship and to prevent water infiltration, is very challenging indeed.

Some architectural historians and observers of design trends and stylistic movements in architecture, attribute this latest "Deconstructivist Architecture" movement to Frank Gehry, a Los Angeles based architect. As a form and space generator, he is a very creative and innovative architect. He has generated several new and innovative forms and spaces from the wild happenstance and "collision" of several forms or rooms or buildings intersecting with one another in a "pile-up" accidental arrangement. Even though he has designed several residences and lowrise institutional buildings in this particular idiom, he has not yet published any of his highrise designs expressing this unique philosophy of form and space generation. However, on the basis of results published and exhibited in recent design competitions, his influence is rather pervasive, especially among young impressionable design architects and architectural students.

Norman Foster, RIBA, architect of London who designed the recently constructed Hong Kong and Shanghai Bank headquarters in Hong Knog has received many acclamations and awards for his professional creative work. The structural system of the building is unique in that the office floors are suspended in tiers of 6 stories from an overall 43 story structural steel framework. Typhoon wind loading on the building tower causing it to sway out of vertical plumb alignment was a major consideration in the detailed design of the exterior cladding systems.

Gymnastic type suspension systems defying gravity are also making their way into innovative glazing systems. For example, the USAA Financial Center in San Antonio, Texas features a very impressive entrance lobby with glass suspension systems over four stories in height. Also, the recently completed Delta Air Lines Terminal at the Los Angeles International Airport by Gensler and Associates, architects features a suspended glass installation at the main entry from the skybridge connection.

In addition to the challenges presented by design architects, owners, and developers involved with the signature type innovative highrise office buildings just discussed, there are also other challenges presented to manufacturers, suppliers, contractors and creators of design and construction industry standards. For example, in Honolulu we have more reinforced concrete highrise office and residential condominium and hotel buildings than structural steel. Therefore, our design details are more suitable to warehouse or industrial shop or garage construction than to residential or commercial construction.

This is because we often use structural columns and spandrle beams or edges of reinforced concrete floor slabs to provide our rough window openings without the benefit of reglets, reveals, or stops, or stepped window sill configurations. Consequently, we have many situations where remedial repairs are required within a very few years after occupancy, due to severe water infiltration as well as life-safety problems and code violations with window frames that are not properly anchored and are therefore on the verge of falling out of the building.

In Honolulu, invariably our highrise commercial office and residential condominium and hotel structures are air conditioned; therefore, we can create rather severe pressure differentials between the exterior and interior spaces. In many instances, we have found that both our selected manufactured window designs and their internal metal to metal sealant systems are inadequate, and consequently, rainwater during severe storm conditions including high winds has a tendency to be sucked or wicked into the building interior. Of course, water infiltration damages can result not only to interior architectural finishes, but also to interior furnishings and appliances. Consequently, construction litigation generally results due to the failure of initial expectations and the inability of the contractors, architects, and engineers to provide satisfactory remedial repair design solutions.

These climatological conditions as well as inadequate design details are further aggravated in many instances by the lack of good quality concrete construction. Therefore, the rough window openings end up outside of the American Concrete Institute's (ACI) acceptable realm of tolerances for both vertical plumbness and horizontal levelness. Remedial repairs are often difficult and expensive and sometimes just a "quick fix" without reliable warranties or guarantees. The problems are further aggravated by the fact that most manufacturers ship their anodized aluminum window frames to Hawaii "K.D."; i.e., in a "Knocked-Down" disassembled state. Depending on the subcontractor, approved shop drawings, specifications, and manufacturer's assembly drawings and written instructions, sometimes in the worst cases, these K.D. anodized aluminum window frames are assembled on the concrete floor slab of the room where they are to be installed, rather than shop fabricated. Sometimes, the intersecting metal component parts are cut and drilled and shaped to fit with portable hand tools and without jigs or templates; in which cases, daylight is generally visible at these metal to metal joints, and the application of sealants is virtually non existent.

Another major problem is that in several instances, the remedial repair solutions proposed by manufacturer's troubleshooters and other "experts" have also failed, and in certain cases have actually caused more severe water damage and infiltration problems. In one particular case, an out-of-town consultant specified a caulking sealant for interior use which caused a condominium apartment resident to be hospitalized for inhalation of toxic fumes, when the label on the cartridge very clearly stated "For Exterior Use ONLY".

In "window wall" design systems, the problems originate with the selection of a particular window from the window manufacturer's technical catalog literature. Generally, this is because the manufactured product being advertised and promoted is not what is eventually delivered to the project site, even though the specification selection process designates same by the manufacturer's catalog identification number. This is in addition to the "K.D." shipping problem mentioned earlier. Because of tool and die problems on the part of the manufacturer, the metal extrusions and profiles may vary in both dimension and strength. Fabrication, assembly, and approved shop drawings may vary widely in design details and joinery and sealant conditions leading to further problems in field assembly and installation.

Further problems exist in the Architectural Aluminum Manufacturer's Association (AAMA) and American National Standards Institute (ANSI) standards and instructional literature and certification programs. Even though these standards are adopted almost verbatim by most model building code organizations in the United States, they in fact only represent the very minimum requirements for water infiltration resistance in high storm conditions. However, a very problematic representation on the part of the manufacturers and AAMA is with regard to their testing certification programs. This is because the approved testing procedures often fail to address actual field installation conditions from the standpoint of both design detailing and also construction quality control. In these particular instances, chances are that these glazing systems will not lend themselves to be remedied by the addition of metal angles or plates or caulking sealants.

In the first place, most aluminum sliding or single or double hung windows that are tested by the manufacturer are too small for residential condominium or commercial use. And the corrolary to this, is that unless a window larger than or equal to the rough window opening on the job site is tested by the manufacturer, all other sizes tested are inapplicable to that project. Furthermore, both AAMA and ANSI will permit windows to be tested and certified on the basis of using wooden frames in lieu of concrete, masonry, or structural steel rough openings. In addition, the AAMA certified testing program does not include any provision whatsoever for the use of sub-frames; i.e., sub-heads, sub-jambs, and sub-sills. Therefore, the same identical window tested with or without sub-frames will produce different results; i.e., complete failure or marginal success regarding the potential for eventual water infiltration problems.

Another problem with the AAMA certified testing program is that if the manufacturer substitutes operating hardware, rollers, weatherstripping, etc. which he generally procures from a variety of suppliers, the assembled window will produce different test results. Furthermore, the AAMA certification label does not apply to window wall systems; i.e., the combination of similar or different windows within the same opening or in conjunction with customized "stick systems". In other words, if two AAMA label certified sliding windows are installed side by side within the same window opening separated by a vertical metal mullion, the AAMA labels of certification are nullified. Also, if an AAMA certified picture sliding window that has been labeled is combined with one or more AAMA certified double hung windows in the same rough window opening, again the certification labels for all of the individual windows within that window wall system are nullified.

For these and other reasons, it is very important that the specifications of the architect contain a provision for sample field tests at different elevations and orientations of windows installed in mock-up apartments or office bays, in conjunction with the results provided by the wind tunnel tests. In most projects with performance versus prescriptive specifications, and with generic versus proprietary specifications, the architect is not really aware of what he is getting as a final window product. His contract document drawings may only provide a generic guide to the type of window desired; however, the contractor may be entitled and permitted to make a few product line substitutions. Therefore, when the architect and structural engineer have an opportunity to review the substituted window shop drawings, this may in fact be their very first introduction to the product line. Most architects would do well to request sample profile extrusions and fittings of the window type and manufacturer being approved. This is because AAMA will allow the certified and tested metal window to be sawn into smaller component parts as a part of the file on the testing program.

With regard to "curtain wall" design, one of the most important lessons learned in Honolulu is the need for spandrel beams in a reinforced concrete structure to compensate for deflection. Also, concrete is subjected to the phenomena of "creep" shrinkage over time, and in a highrise tower this can be appreciable, and should be calculated by the structural engineer-of-record to prevent excessive loadings on the glass and mullions. While it is acceptable to use the "curtain wall" glazing system with poured-in-place reinforced concrete construction, it is preferable not to attempt to use same with precast or prestressed concrete elements due to the excessive movement systems experienced and the inability to sucessfully manage same.

Field experience has demonstrated that caulking sealants applied without backer rods and without proper cleaning and surface preparation for new construction or remedial repairs, can be a worthless exercise.

In closing, I would like to share with you an exterior view of a "window wall" glazing system that appears to be a "curtain wall" system from all outward appearances. This facade expression is part of a twin tower mirrored glass highrise office complex in downtown Honolulu. Finally, we have the "Blending of Art & Science" as an advertisement by Rhodorsil Architectural Silicone Sealants featuring the spectacular pyramidal transparently glazed entrance to the Grand Louvre in Paris, France by architect I.M. Pei and Partners. For some architects, this brilliant achievement was the most significant architectural event of 1989 as noted in the December issue of the Hawaii Architect journal. It certainly presented numerous architectural design challenges for sealant technology and design standards on an international scale of development.

* * * * *

Author Index

Subject Index

I

J

L

M

N

O

P

Q

R

S